AN INTRODUCTION
TO MECHANICAL
VIBRATIONS

AN INTRODUCTION TO MECHANICAL VIBRATIONS

THIRD EDITION

Robert F. Steidel, Jr.

WILEY

JOHN WILEY & SONS

New York   **Chichester**   **Brisbane**   **Toronto**   **Singapore**

Library of Congress Cataloging in Publications Data:

Steidel, Robert F., 1926–
   An introduction to mechanical vibrations.

   Bibliography: p.
   Includes index.
   1. Vibration.   I. Title.
TA355.S74   1989   620.3   88-20508
ISBN 0-471-84545-0

Printed in the United States of America

10 9 8 7 6 5 4 3 2 1

PREFACE

In the nine years since the second edition was published, the field of mechanical vibration has not changed, but the way we handle vibration problems and systems has changed. Specifically, the use of computers has made it much easier to handle vibration systems with many degrees of freedom. For the beginning student, this means that generalized coordinates, principal coordinates, normal modes and mode shapes are topics that cannot be left for a later course that he or she may never take. The study of several degrees of freedom must now be a part of an introductory course in mechanical vibration.

In this newest edition, as in the previous ones, I have stressed problems and examples. I believe that the measure of a textbook is its problems, and it is by the problem collection that I would like to have this book judged. There are 556 problems and worked examples in the third edition, nearly one-fourth more than the second edition. Many problems from the second edition reappear in the third. This is by intent. Each problem is selected and placed strategically. If a problem was a good teaching vehicle in the earlier editions, it has been retained. Some, which were not so good, have been removed. I usually assign about 60 problems in a 15-week semester course. I do not recommend more. At this assignment rate, there should be an ample problem supply for course use over several years.

It is absolutely necessary for an undergraduate course in mechanical vibration to proceed further than one degree of freedom. The concepts of generalized coordinates, principal coordinates, normal modes, coupling, eigenvalues, and the like, cannot be taught, if you are limited to one degree of freedom. They can be demonstrated with two degrees of freedom, and quite adequately. In fact, there is a good argument that there is no need to include more than two. Every concept that needs to be demonstrated in many degrees of freedom can be demonstrated with two degrees of freedom. I share this view. The only qualification to this statement would be the introduction of matrix equations. Matrix equations and matrix methods are necessary to handle many degrees of freedom, whereas they can be used, but are not necessary, for two. My practice has been to introduce matrix equations for two degrees of freedom and to use these simple equations with numerical techniques. The fact that students could determine natural frequencies and mode shapes without resorting to numerical techniques does not concern me. I require a numerical solution for a specific problem. If a student also finds an analytic solution to an assigned problem, he or she will be doing a problem twice. That is repetition, and repetition is a form of learning.

The book has been reorganized, somewhat, to make it easier to reach multiple degrees of freedom in one semester, without skimping on fundamental concepts and principles. In particular, Rayleigh's principle now is placed within the chapter on two degrees of freedom. Random vibration has been removed from the chapter on forced damped vibration and appears as a chapter by itself. The small section on nonlinear vibration, which had been included at the tag end of forced undamped vibration, to which it added very little, has been expanded to another separate chapter. This makes it easier to include or exclude either, as a matter of instructor choice. Finally, the very large chapter on discrete systems with multiple degrees of freedom has been divided into two chapters: the first is on matrix methods and the second is on finite elements. This leaves the new Chapter Eleven, on two degrees of freedom, as the largest in the book. It also now has the most problems. In my own teaching of mechanical vibration, I do spend a lot of time on two degrees of freedom.

I would like to acknowledge, with special thanks, Professor Jerald Henderson, of The University of California at Davis, who read and criticized the final manuscript. He has a remarkable eye for detail and rigor, which I do not have, and this talent was put to a strong test over these last few weeks. His effort to make the finished text better has certainly raised its final quality.

I, again, urge you to read the prefaces to the first and second editions. These contain my thoughts on the teaching of mechanical vibration, which have not changed.

Last, but by no means least, I would like to thank my wife, Jean, for her patience and support through three editions.

ROBERT F. STEIDEL, JR.

Berkeley, California
October, 1988

PREFACE TO SECOND EDITION

In the preface to the first edition I established my reasons for writing still another vibration text, together with my thoughts on teaching mechanical vibration. Teaching the subject certainly did affect the somewhat unusual arrangement of the topics in the book, and my thoughts on teaching mechanical vibration have not changed since the first edition of this book was published. For example, I still feel strongly that it is a pedagogical mistake to incorporate damping any earlier than Chapter 6.

As a field, mechanical vibration has not changed much since the first edition. There is, perhaps, more interest in systems with multiple degrees of freedom, but this makes a thorough understanding of basic principles even more necessary. Random vibration has receded in importance, as could have been predicted. The only real source of random vibration is the rocket motor, so applications of random vibration are inevitably tied to the space program. When the space program declined, our interest in random vibration declined.

There are no big changes to announce in the present edition, but there are several small changes. For example, my explanation of hysteretic damping has been expanded and the problem collection has been expanded. I am still convinced that learning mechanical vibration is best accomplished by the working of problems. If you make a comparison between the two editions, you will notice that I have added about 50 new problems and omitted some that are no longer useful. In particular, there are more problems on nonlinear vibration.

This edition uses the System Internationale (S.I.) throughout. There has been considerable debate over what units should be used in mechanical vibration, much more debate than the issue merits. Since damping ratios, frequency ratios, and amplitude ratios are dimensionless, the units used do not really matter. Most vibration measurements are measured in millimeters and recorded on strip charts running at a rate calibrated in millimeters per second. Oscillograph readings are calibrated in millimeters. S. I. units seemed logical to me, and for this edition, I have abandoned conventional units in favor of them.

The subject of torsional vibration (now Chapter 9) has been separated from the subject of discrete systems (Chapter 10). I included both topics in one large chapter in the first edition because I felt that the early work of Holzer and W. Ker Wilson was a valuable background for multiple degrees of freedom systems. These men accomplished a great deal without modern computers. I felt that the Holzer method, although archaic, was a prelude to our modern numerical methods. In learning how to handle multiple mass systems, some historical background is helpful. The point is now established; there is not need to belabor it. Chapter 9 can be included or excluded as the reader or instructor sees fit.

For the present edition, I am indebted to Professor Jerald Henderson of the University of California, Davis, and Professor C. Daniel Mote, Jr., of Berkeley, who as close personal friends, have given me all the thoughts and opinions of this book that only close personal friends can give. Their criticism has been invaluable. I would also like to express my thanks to Professor Diaz-Jimenez of the Universidad de America in Bogata, Columbia. He and I have had detailed correspondence on specific problems, and I have learned from him. It is largely because of his convincing letters that a few of my more ambiguous problems have been removed. I hope that the replacements fare better.

I urge you to read the first preface, and I hope that you will enjoy your experiences with mechanical vibration.

ROBERT F. STEIDEL, JR.

Berkeley, California 1978

PREFACE TO FIRST EDITION

In reading a preface to an introductory text in mechanical vibrations, I suppose that the foremost question in mind is why should there be yet another text in mechanical vibration? My reason for writing is simple. I have great respect and admiration for the many excellent books and monographs on the subject of mechanical vibrations, but none of them fit the view of the field that I have experienced. My only course has been to try to develop it in my own way, with the hope that others see it as I do, and those that do not will tolerate a slightly different approach.

As I see it, there are two phases to the study of mechanical vibration. Mechanical systems are not normally designed to withstand vibration. There are some exceptions, as in the case of aircraft design, but vibration is usually a problem that has to be faced after it occurs. If it doesn't occur, it is ignored. As a consequence, experience with vibration is problem solving, and therefore learning to solve vibration problems is the first phase. The second phase of study is that the vibration engineer must not only solve problems but also explain them. The literature is crammed with case histories of disaster, diagnosis, and ultimate triumph. The question that always arises is "why"? He must explain why the structure or system must be changed, the material changed, or the speed changed, and why the cost must go up. Explaining to others is an important part of the vibration engineer's task. Explanation requires a clear, concise, and simple approach to the problem.

I have three major objectives in organizing an introduction to mechanical vibrations. The first is to stress problems and examples. This book is much more a collection of problems with an accompanying text, than it is the reverse. I would like to have it treated that way. It is my educational conviction that learning is best accomplished by working problems, which I have already emphasized. Most engineering students learn best when they have a good instructor in the classroom, a good text to keep them company in the dark hours of the night, and some good problems that force thinking. These problems must be carried through to an answer. It is not possible to generalize and leave the numbers to someone else. Society is not only holding the engineer responsible for his work but liable for his errors. In fact, the possibility of going to jail for engineering malpractice is becoming very real. This places an additional emphasis on finding answers to real problems.

My second objective is to stress concepts. Mechanical vibration is not a self-evident subject. It has a language of its own, and most engineers are very rigorous in their semantics. Unless concepts are firm, you cannot even discuss mechanical vibration problems, let alone solve them. Such concepts as coordinates, frequency and characteristic frequencies, modes, damping, phase and amplitude, transient and steady state vibration are all very simple, but they are often rushed over for the more sophisticated topics of nonlinearity, instability, random vibration, etc. My experience has been that mechanical and civil engineering seniors hear or read about all these things but do not understand them.

The third is a special consideration of energy dissipation or damping in mechanical engineering systems. Damping is most accepted as being viscous, but in mechanical engineering, there are only two common examples of viscous damping, one is laminar flow of a fluid through a slot and the other is laminar flow through an orifice. All other forms of damping are something else, and why a text should run headlong into a discussion of viscous damping in its first chapter has been particularly bewildering to me. Another problem with damping is the question of whether it should be included when considering more than one degree of freedom. My inclination is that it should not, for it adds little to an explanation and adds much complication. In most structures, damping is extremely light, less than 5% of critical damping, and can easily be neglected except at resonance.

The first eight chapters constitute a reasonable introduction to mechanical vibrations, and they would be a satisfactory basis for a one-quarter, ten-week course. The book is laid out so that allowing one week for each chapter through Chapter 6, and two weeks each for Chapter 7 and Chapter 8, brings the student through free and forced vibration, damped and undamped, and into a discussion of systems with two degrees of freedom. There is no need to discuss more than two degrees of freedom to introduce the concepts of many degrees of freedom. You can learn as much from two degrees of freedom as from ten, and more easily.

Chapters 9 and 10 do consider more than two degrees of freedom and techniques for solving such problems. These chapters were included as an afterthought, at the suggestion of the reviewer and publisher. I would not include them in a one-quarter course for undergraduates, but I would for graduate students or if a longer time were available, such as a 14-week semester. Teaching is a very personal thing, and I know that others think it necessary to include these subjects. With more time available, I would allow at least three weeks for Chapter 9 and one for Chapter 10.

Each of the first eight chapters contains two or three sections. Each section is meant for one class period, and each section is followed by a number of problems based on that section. I have found that five problems in one week is an ample assignment. With a larger assignment, learning is lessened. I have not discriminated between the difficult and the easy problems, mainly because I cannot make that judgment. What is difficult for one student is easy for another. There is a solutions manual, containing the worked solutions to all problems, available to instructors.

I would like to express my gratitude to five people who helped me form my ideas concerning mechanical vibration. The first is my instructor at Columbia University, Professor Dudley Fuller, who introduced me to mechanical vibration. The second is my former instructor and Department Chairman at Berkeley, Clyne F. Garland. In his quiet and efficient way, he taught me the value of his meticulous approach to engineering problems. It is something I have not yet mastered. J. Lathrop Meriam, the former Dean of Engineering at Duke University, was also my instructor, but he was also my colleague in mechanics for ten years. He taught me the value of working problems and understanding concepts. He is also a master of the physical explanation. Some of the problems used in this book have the same generic background as some of those in Dean Meriam's works. One or two are identical and are used with permission. These, I fully acknowledge. The fourth person to whom I am indebted is A. L. Austin, of the Lawrence Radiation Laboratory. Dr. Austin was one of my students, years ago. He is a personal friend, and he has reviewed much of the manuscript. The fifth person is my wife, Jean, who has patiently endured and shared these years of preparation with me. Without her presence and comfort, this book would still be only a good intention.

I am also indebted to the hundreds of mechanical engineering students who have unknowingly served as guinea pigs as I tried out one problem after another. I would like to thank Mrs. Margaret French, who typed much of the original manuscript, and Mrs. Margaret Hansen, Mrs. Barbara von der Meden, and Mrs. Mary Jane Alpaugh who typed what Margaret French did not.

ROBERT F. STEIDEL, JR.

Berkeley, California
August 1971

ABOUT THE AUTHOR

ROBERT F. STEIDEL JR., professor of Mechanical Engineering at the University of California at Berkeley, received his B.S. and M.S. Degrees in Mechanical Engineering from Columbia University. He was appointed to the faculty of Oregon State University in 1949, serving three years before leaving to enroll at the University of California at Berkeley as a National Science Foundation Fellow. In 1955, he received the Doctor of Engineering Degree and became a member of the faculty of Mechanical Engineering at Berkeley. He served as Chairman of the Department of Mechanical Engineering, 1969–1974, and from 1981–1986 as The Associate Dean of Engineering. At various times he has interrupted his long teaching career to serve as consultant and adviser to industry and national laboratories such as The Bonneville Power Administration, The Jet Propulsion Laboratory, and the Lawrence Livermore National Laboratory, where, notable, he was the first project engineer for the Polaris. Regardless of these experiences, when they were over he always returned to his first love, which was engineering teaching.

Professor Steidel is a Fellow of the American Society of Mechanical Engineers and a Fellow of the Institution of Mechanical Engineers, (UK). His ASME involvement includes accreditation, local section activities and the promotion of student-industry interaction. He is the recipient of the ASEE Chester F. Carlson Award, given for innovation in engineering education, and the Western Electric Fund Award. In 1960 he received the Distinguished Teaching Award of the University of California, the second year the award was given.

Professor Steidel is the author or coauthor of three books, 35 journal articles, and numerous technical reports and notes. He is the coauthor, with Professor J. M. Henderson, of the GRAPHIC LANGUAGES OF ENGINEERING, which was written to put the practice of engineering into engineering graphics instruction. His widely used text on Mechanical Vibration is the product of forty years of teaching and professional experience in the field of Mechanical Vibration. Most of the problems he encountered have found their way into the text, where they can be found within its five hundred and fifty-six problems.

Professor Steidel has been the Engineering Advisor to John Wiley & Sons Publishing Company for the past fifteen years, advising the college division on the selection of published texts.

CONTENTS

CHAPTER ONE

DYNAMICS

1.1. INTRODUCTION

Vibration is most simply defined as oscillating motion. It has been implicit in the past and included in most definitions that this oscillating motion is also periodic. This is not as true today as it once was, since we now have an interest in nonperiodic and transient motion and the analytic methods to analyze such motion. There remains the lesser implication that the motion is continuous and has some average or mean value. This is of engineering interest in the prediction of maximum stress, mean displacement, or some measure of reactive force.

Any study of vibration must first turn to a review of *dynamics*, for vibration is first of all motion, and *dynamics* is the term used to denote that portion of mechanics that is devoted to the study of bodies in motion and the forces causing motion. *Kinematics* is that part of dynamics that is the study of the geometry of motion, without reference to force or mass. *Kinetics* is the study of the relation between the motion of bodies and the forces acting upon them.

A logical beginning for the subject of dynamics and vibration is with the work of Galileo (1564–1642). At the age of twenty in 1584, he correctly conceived the principle of the isochronous pendulum. In 1657, the Dutch mathematician, Huygens (1629–1695), applied this principle to a clock. In 1590, Galileo crowned his achievements in experimental physics by discovering and proving the law of falling bodies. His work on motion and acceleration formed the basis for the laws of motion that Sir Isaac Newton (1642–1727) formulated.

For 200 years any interest in vibration was confined to the period of a

1

pendulum, the movement of astronomical bodies and tides, and pertur-bations on these. Toward the end of the nineteenth century, however, high-speed machinery introduced many new problems including the phe-nomena now associated with mechanical vibration. Baron John William Strutt, Lord Rayleigh (1842–1919), organized and developed the theory of mechanical vibration and it is on his contributions that the modern field is founded. In modern times, there have been many men who have contributed to its growth and the subject matter has expanded enor-mously. Special reference will be made to the work of S. Timoshenko and J. P. Den Hartog, who were pioneers in the solution of industrial problems in mechanical vibration as well as being great engineering teachers.

1.2. DISPLACEMENT, VELOCITY, AND ACCELERATION

Kinematics is the study of motion without reference to force or mass. It is primarily concerned with the interrelation of displacement, velocity, acceleration, and time.

Linear displacement is the directed distance that a point has moved on a path from a convenient origin. Since it is a directed quantity, it is a vector and subject to all the laws and characteristics of vectors. If the origin is fixed, the displacement is absolute. If the origin itself is in motion, the displacement is relative. It is essential in kinetics to know whether motion is absolute or relative. Newton's laws of motion formu-late the relation between force, mass, and absolute acceleration.

In Figure 1.1, the vector **s** is the displacement of point P from the origin 0. Any convenient set of coordinates can be selected with which to describe the vector **s**.

$$\mathbf{s} = \mathbf{x} + \mathbf{y} + \mathbf{z} \qquad (1.1)$$

The choice of coordinates should be one of convenience in mathematics. The physical analysis of a problem is independent of the coordinates which are selected.

The instantaneous velocity of point P is the time rate of change of the displacement **s**,

$$\mathbf{v} = \frac{d\mathbf{s}}{dt} = \dot{\mathbf{s}} \qquad (1.2)$$

The dot over the variable is an accepted convention for indicating the first derivative with respect to time. Two dots over the variable indicate the second derivative with respect to time. We will use this convention interchangeably with the written form of the derivative when its use is

Fig. 1.1

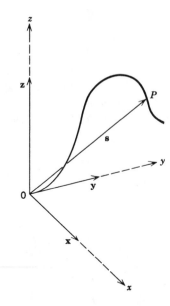

simple and clear. Indiscriminate use of the convention does tend to confuse rather than clarify when the motion is a function of other variables as well as time.

If point *P* moves in a straight line, say along the *x*-axis, as it does in Figure 1.2, the displacement and velocity are functions of the single coordinate *x*, and scalar notation can be used to describe motion,

$$v_x = \frac{dx}{dt} = \dot{x} \tag{1.3}$$

Fig. 1.2

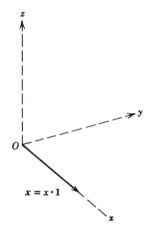

In this case, point P has one degree of freedom, since a single coordinate uniquely determines the position of P.

This also applies to curvilinear motion, such as that of the simple pendulum of Figure 1.3, which moves in the z–y plane. The direction of the vector $\mathbf{v}_\theta$ is tangent to the path of the pendulum and is a function of the manner in which the system is constrained. The magnitude of the vector is proportional to the time rate of change of angular displacement

$$\mathbf{v}_\theta = \dot{\boldsymbol{\theta}} \times \boldsymbol{\rho} \tag{1.4}$$

The vector $\boldsymbol{\rho}$ is the directed distance from A to P, the length of the pendulum. Separately, the velocity can be described in terms of $\dot{z}$ and $\dot{y}$

$$\mathbf{v}_\theta = \dot{\mathbf{z}} + \dot{\mathbf{y}} \tag{1.5}$$

where $\dot{\mathbf{z}}$ and $\dot{\mathbf{y}}$ are component vectors of $\mathbf{v}_\theta$. This equation requires two coordinates z and y and an additional equation of geometric constraint, which is

$$\rho^2 = (\rho - z)^2 + y^2$$

The instantaneous acceleration is the time rate change of velocity. This can also be expressed in terms of component acceleration in the direction of whatever coordinates are convenient,

$$\mathbf{a} = \frac{d\mathbf{v}}{dt} = \dot{\mathbf{v}} = \ddot{\mathbf{s}} \tag{1.6}$$

Fig. 1.3

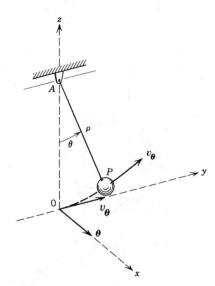

Scalar acceleration can be used if the acceleration is a function of a single coordinate.

The motion discussed is that of a particle or point. If the motion of a rigid body is involved, we must also concern ourselves with its angular displacement.

Angular displacement is the change in angular position of a given line as measured from a convenient reference line. In Figure 1.4, consider the motion of line AB as it moves from its original position to position $A'B'$. The angle between lines AB and $A'B'$ is the angular displacement of line AB, θ_{AB}. It is also a directed quantity. The conventional notation used to designate angular displacement is a vector normal to the plane in which the angular displacement occurs. The length of the vector is proportional to the magnitude of the angular displacement. If the linear displacement of point A is identical with the linear displacement of B, line AB will have no angular displacement. Combining linear displacement with angular displacement, we can completely describe the three-dimensional motion of any rigid body.

Angular velocity is defined as the time rate of change of angular displacement, in a manner similar to the definition of linear velocity.

$$\omega = \frac{d\theta}{dt} = \dot{\theta} \qquad (1.7)$$

Fig. 1.4

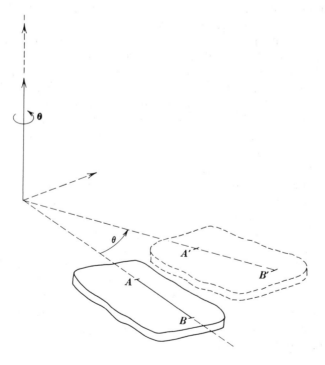

Angular acceleration is the time rate of change of angular velocity

$$\alpha = \frac{d\omega}{dt} = \dot{\omega} = \ddot{\theta} \tag{1.8a}$$

Each of these definitions may be set in scalar form if the angular motion can be expressed in terms of a single coordinate.

1.3. COORDINATES

The selection of coordinates is a very important aspect of the study and solution of problems in dynamics. The mathematics of even the simplest problems can be made difficult through the use of improper or unwise coordinates. This is particularly true in the study of mechanical vibration.

Conventional coordinates may be selected, such as Cartesian coordinates and polar coordinates, but any convenient measure of displacement can be used as a *coordinate*. We are using the term *coordinate* here in its more general form as an independent quantity, which will specify position.

In Figure 1.5, the motion of the simple pendulum, described also in Figure 1.3, can be stated in terms of either the Cartesian coordinates z and y or in terms of the angle θ. The choice of θ as the coordinate is more convenient than the choice of z and y. The pendulum has but a single degree of freedom, provided that we restrict motion to a single plane. Choosing θ as a coordinate makes use of the additional fact that the length of the pendulum is constant. This is a *constraint* on the system. The coordinates z and y are restricted by the equation of constant $z^2 + y^2 = \rho^2$. Constraint plays a very important role in the solution of problems. Recognizing constraint is as important or more important than selecting coordinates. If we recognize constraint as a limitation to motion, we can usually completely describe motion with fewer coordinates. The minimum number of coordinates required to completely describe motion corresponds to the number of degrees of freedom of the system. This implies that if we have a greater number of coordinates than there are degrees of freedom, there must be enough equations of constraint to make up the difference.

A set of coordinates that describe general motion and recognize constraint are called *generalized coordinates*. In Figure 1.5a, the angular coordinate θ is the generalized coordinate that recognizes the fixed length of the pendulum as a system constraint. The linear coordinates z and y do not. For the semicylinder of Figure 1.5b, either the angular coordinate θ or the linear coordinate x can be used to describe motion. They are linearly related, since $rd\theta = dx$, and one coordinate is as convenient as the other.

Fig. 1.5

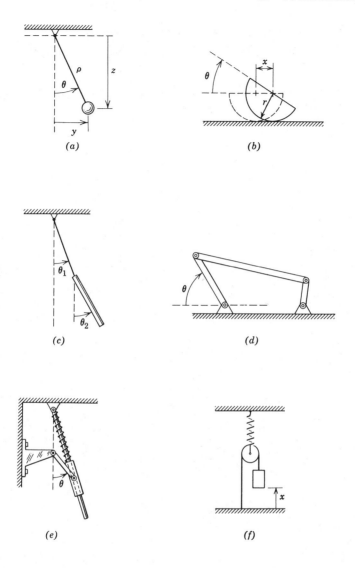

(a)

(b)

(c)

(d)

(e)

(f)

In Figure 1.5c, the motion of the rod swinging as a pendulum can be described in terms of the general coordinates θ_1 and θ_2. Considering that the lengths of the pendula l_1 and l_2 are constant, nothing more is needed to describe motion.

In Figure 1.5d the four-bar linkage has but a single degree of freedom. The motion of any link or any point on these links can be completely described in terms of the angular coordinate θ and its derivatives with respect to time. The same is true for the slider mechanism of Figure 1.5e.

The motion of the suspended mass and pulley of Figure 1.5f can be described in terms of the single coordinate x. The linear displacement of the pulley would have to be determined relative to the linear displacement of the mass, in order to define the spring force, but this is a kinematic exercise.

One further statement on coordinates is needed. For it, we can refer to the pendulum of Figure 1.6, which is a simple example that is easy to

Fig. 1.6

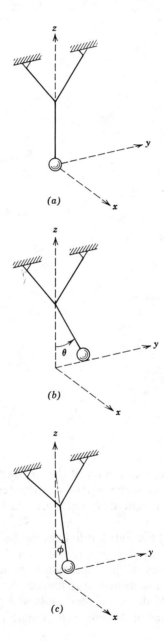

(a)

(b)

(c)

visualize and easier to verify by experiment. The pendulum has two degrees of freedom. If displaced, it will oscillate back and forth in a generally circular pattern. There are two exceptions. If the pendulum is displaced by an angle θ in the z–y plane, the motion will remain in the z–y plane. If it is displaced by an angle ϕ in the z–x plane, the motion will remain in the z–x plane. It is evident that motion in the direction of one coordinate will be independent of the other. The motion in which only one coordinate varies is called a *principal mode of motion*. A characteristic of a principal mode is the common pattern of all motion. All particles pass through maximum and minimum velocities at the same time. Defining the motion of one particle defines the motion of all particles. If only one coordinate is needed to define motion, such a coordinate is called a *principal coordinate*. Analytically, most problems do not require the use of principal coordinates, but the concept of principal coordinates is extremely important. In linear systems, all motion can be described by the superposition of principal modes.

It is interesting to note that generalized coordinates are not necessarily principal coordinates. In Figure 1.6, the angles θ and ϕ are both generalized coordinates as well as principal coordinates. In Figure 1.5c, the angles θ_1 and θ_2 are generalized coordinates but are not principal coordinates. There is no manner of motion that can be described completely in terms of θ_1 or θ_2 without involving the other coordinate. Principal coordinates can be defined, however, and the definitions will involve θ_1 and θ_2. It is not important to make that definition, at the present time, but it is important to recognize that principal coordinates and principal modes do exist.

1.4. KINEMATIC EQUATIONS

The definitions of velocity and acceleration involve the four variables: displacement, velocity, acceleration, and time. If we eliminate the variable of time from each definition, we have a third equation of motion.

$$\mathbf{v} = \frac{d\mathbf{s}}{dt} \tag{1.9a}$$

$$\mathbf{a} = \frac{d\mathbf{v}}{dt} \tag{1.9b}$$

$$\mathbf{a} \cdot d\mathbf{s} = \mathbf{v} \cdot d\mathbf{v} \tag{1.9c}$$

These three equations are the differential equations of motion for the kinematics of a particle. There are three similar differential equations for the kinematics of a line.

$$\omega = \frac{d\theta}{dt} \qquad (1.10a)$$

$$\alpha = \frac{d\omega}{dt} \qquad (1.10b)$$

$$\alpha \cdot d\theta = \omega \cdot d\omega \qquad (1.10c)$$

The first two kinematic equations are vector equations, but since they refer to a single coordinate, scalar equations could be used as well. The third equation involves the dot product of two vectors making it a scalar equation.

Quite frequently, data are obtained in terms of acceleration as a function of time. The integration of these data yields velocity and displacement. Knowing the physical relation between any kinematic quantity and time or any two kinematic quantities is sufficient to obtain a complete kinematic understanding of motion. Table 1.1 is a brief summary of the scalar kinematic equations for the motion of a particle along one coordinate. An analogous set of equations for the motion of a line could be given, but they are less frequently used in mechanical vibration, with the possible exception of torsional vibration.

Table 1.1

Variables	Displacement Time	Velocity Time	Acceleration Time	Displacement Velocity Acceleration
Displacement	—	$\int_{s_1}^{s_2} ds = \int_{t_1}^{t_2} v\, dt$	$\int_{s_1}^{s_2} ds = \int_{t_1}^{t_2}\int_{t_1}^{t_2} a\, dt\, dt$	$\int_{s_1}^{s_2} ds = \int_{v_1}^{v_2} \frac{v}{a}\, dv$
Velocity	$v = \frac{ds}{dt}$	—	$\int_{v_1}^{v_2} dv = \int_{t_1}^{t_2} a\, dt$	$\int_{v_1}^{v_2} v\, dv = \int_{s_1}^{s_2} a\, ds$
Acceleration	$a = \frac{d^2 s}{dt^2}$	$a = \frac{dv}{dt}$	—	$a = v\frac{dv}{ds}$

An understanding of the graphical expression of kinematics is both useful and convenient. If the displacement is expressed as a function of time, the slope of the resulting curve is the instantaneous velocity. If velocity is expressed as a function of time, the slope of the curve is the instantaneous acceleration and the area under the velocity–time curve during an interval t_1 to t_2 is the change in displacement. These are simple relations that are universally understood. Less familiar are the expressions of acceleration. The area under the acceleration–time curve during the interval t_1 to t_2 is the velocity change between these limits, while the area under the acceleration–displacement curve occurring during a displacement from s_1 to s_2 represents half the difference in the square of the velocity at t_1 and t_2.

EXAMPLE PROBLEM 1.1

A particle is moving between two charged parallel plates along a curved path that is in the plane of the diagram. At position P, the radius of curvature of the path is ρ and the vector $\boldsymbol{\rho}$ makes an angle θ with the plates. The velocity of the particle is 240 mm/s, ρ is 80 mm and is increasing at a rate of 40 mm/s. The angle θ is increasing at a rate of 10 rad/s for each second. What is the total acceleration of the particle?

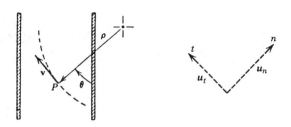

Solution:
Using normal and tangential coordinates, by kinematic definition

$$\mathbf{v} = \boldsymbol{\omega} \times \boldsymbol{\rho}$$

and

$$\mathbf{a} = \frac{d\mathbf{v}}{dt} = \frac{d\boldsymbol{\omega}}{dt} \times \boldsymbol{\rho} + \boldsymbol{\omega} \times \frac{d\boldsymbol{\rho}}{dt}$$

The meaning and value of $\boldsymbol{\rho}$, $\boldsymbol{\omega}$, and $d\boldsymbol{\omega}/dt$ are all easily found. The meaning of $d\boldsymbol{\rho}/dt$ is unclear. How is it directed?

Defining $\boldsymbol{\rho}$, using normal and tangential coordinates, and $\mathbf{u}_n$ and $\mathbf{u}_t$, the unit vectors in the normal and tangential directions,

$$\boldsymbol{\rho} = -\rho\mathbf{u}_n$$

The negative sign simply indicates that the radius of curvature is in a direction opposite to the n coordinate.

Differentiating with respect to time, the vector $d\boldsymbol{\rho}/dt$ is found to have two components, one normal and one tangential.

$$\frac{d\boldsymbol{\rho}}{dt} = -\left(\rho\frac{d\mathbf{u}_n}{dt} + \frac{d\rho}{dt}\mathbf{u}_n\right)$$

$$\frac{d\boldsymbol{\rho}}{dt} = \rho\omega\mathbf{u}_t - \frac{d\rho}{dt}\mathbf{u}_n$$

The full statement for acceleration is

$$\mathbf{a} = \frac{d\boldsymbol{\omega}}{dt} \times \boldsymbol{\rho} + \boldsymbol{\omega} \times \left(\rho\omega\mathbf{u}_t - \frac{d\rho}{dt}\mathbf{u}_n \right)$$

or,

$$\mathbf{a} = \left(\rho\frac{d\omega}{dt} + \omega\frac{d\rho}{dt} \right)\mathbf{u}_t + \rho\omega^2\mathbf{u}_n$$

Some readers may be looking for a factor of 2, $2\omega\, d\rho/dt$! It does not exist! We are using normal and tangential coordinates and not polar coordinates! The radius of curvature is not a radius vector from a fixed set of coordinates. This problem was chosen to make this distinction. Coordinates are selected for your convenience in interpreting the problem, and not for any other reason! Some of these terms are given and others can be found from basic kinematics. The angular velocity ω, is determined since we know the linear velocity and the radius of curvature.

$$v = 240 \text{ mm/s}$$

$$\rho = 80 \text{ mm}$$

$$\omega = \frac{v}{\rho} = \frac{240}{80} = 3 \text{ rad/s}$$

We also are given the rate of change of ω and ρ with time,

$$\frac{d\omega}{dt} = 10 \text{ rad/s}^2 \quad \text{and} \quad \frac{d\rho}{dt} = 40 \text{ mm/s}$$

Substituting each of the proper terms in the vector equation

$$\rho\frac{d\omega}{dt} = 800 \text{ mm/s}^2$$

$$\omega\frac{d\rho}{dt} = 120 \text{ mm/s}^2$$

$$\rho\omega^2 = 720 \text{ mm/s}^2$$

the total acceleration is the vector sum of a normal acceleration of 720 mm/s² and a tangential acceleration of 920 mm/s².

EXAMPLE PROBLEM 1.2

Experimental observations of an earth-penetrating projectile yield the following data. Determine the velocity–time curve by approximating the deceleration–time curve. Estimate the depth of penetration. The projectile entered the earth's surface at a velocity of Mach 1 and was recovered at a depth of 15.5 m. Are the data valid?

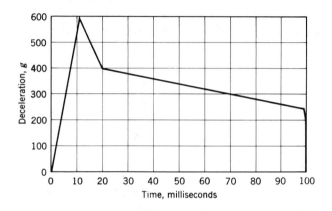

Solution:
Using the kinematic equations of motion

$$v - v_0 = \int_0^t a \; dt$$

and

$$s - s_0 = \int_0^t v \; dt$$

Integrating numerically, with 10-ms intervals, the velocity–time curve can be estimated. The area under the acceleration time curve represents the velocity change of the projectile

$$0 - v_0 = -\sum_0^{100} a \; \Delta t = 329.9 \text{ m/s}$$

Integrating again, the displacement–time curve can be estimated. The area under the velocity–time curve is the change in displacement of the projectile over the 100-ms interval.

$$s - 0 = \sum_0^{100} v \; \Delta t = 15.26 \text{ m}$$

If the projectile were recovered at a depth of 15.5 m, this agrees quite well with observed data and it is reasonable to believe the data to be valid.

EXAMPLE PROBLEM 1.3

A mechanism consists of link *MN*, which moves horizontally in frictionless guides. It carries a collar *A*, in which link *CD* is free to move. Determine the velocity and acceleration of links *MN* and *CD* in terms of generalized coordinates. The coiled spring is unstretched when $\theta = 0$.

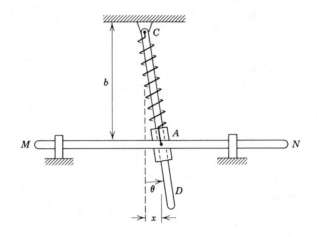

Solution:

This system has a single degree of freedom. That is, only one coordinate is necessary to completely describe the motion of both links *MN* and *CD*. That coordinate is either the horizontal displacement of link *MN*, which is *x*, or the angular displacement θ of link *CD*. Kinematically, they are related. From trigonometry,

$$x = b \tan \theta$$

Differentiating with respect to time, the velocity,

$$\dot{x} = b\dot{\theta} \sec^2 \theta$$

and the acceleration,

$$\ddot{x} = 2b\dot{\theta}^2 \sec^2 \theta \tan \theta + b\ddot{\theta} \sec^2 \theta$$

In terms of the generalized coordinate θ,

$$\dot{\theta} = \frac{\dot{x}}{b}\cos^2 \theta$$

and,

$$\ddot{\theta} = \frac{\ddot{x}}{b}\cos^2 \theta - \frac{2\dot{x}^2}{b^2}\cos^3 \theta \sin \theta$$

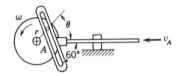

PROBLEM 1.4   Determine the angular velocity ω of the sphere of radius r rolling in a shallow spherical disk of radius R in terms of the single coordinate ϕ.

Answer: $\omega = \dfrac{(R-r)}{r} \cdot \dfrac{d\phi}{dt}$

PROBLEM 1.5   A harmonic motion, $x = A \cos \omega t + B \sin \omega t$, has a frequency of 25 Hz, an initial amplitude of 9 mm, and an initial velocity of 2 m/s. What is the maximum amplitude?

Answer: 15.58 mm

PROBLEM 1.6   The yoke A engages the pin on the wheel, which has a constant angular velocity ω. Determine the velocity of the yoke as a function of time.

PROBLEM 1.7   The amplitude of an oscillation can be described by the relation $s = A \sin(\omega t + \phi)$. At $t = 0$, the velocity is v_0 and the displacement is s_0. Determine expressions for the phase angle ϕ and the maximum amplitude A in terms of s_0, v_0, and ω.

Answer: $A = \sqrt{s_0^2 + \dfrac{v_0^2}{\omega^2}}$; $\tan \phi = \dfrac{s_0 \omega}{v_0}$

PROBLEM 1.8   Determine the angular velocity and angular acceleration of the link AB in terms of the single coordinate x.

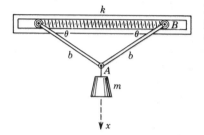

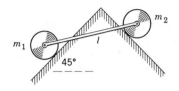

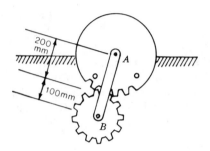

PROBLEM 1.9   Each of two identical solid cylinders rolls without slipping on an inclined plane. They are connected by a weightless bar of length l and each has a radius r. Determine the velocities of m_1 and m_2 in terms of a single generalized coordinate.

Answer: $\omega_1 = \dfrac{\dot{x}_1}{r}; \ \omega_2 = \dfrac{\dot{x}(x_1 + r)}{r\sqrt{l^2 - (x_1 + r)^2}}$

PROBLEM 1.10   Gear A is fixed and cannot move. Gear B meshes with gear A and rotates about gear A with arm AB as a radius. Determine the angular velocity of the gear B in terms of the angular velocity of the arm AB. $r_A = 2r_B$.

Answer: $\omega_B = 3\omega_{AB}$

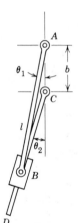

PROBLEM 1.11   The sliding collar B moves without friction on the link CD. Links AB and CD are smooth uniform rods, with the same length and the same mass. How many degrees of freedom does the system have? Define the velocity of each link in terms of generalized coordinates. Limit displacement to small oscillations.

Answer: $\dot{\theta}_2 = \dfrac{l}{l - b}\dot{\theta}_1$

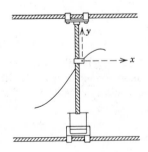

PROBLEM 1.12   An x–y plotter consists of a recording stylus that writes on rectangular coordinate paper. The stylus has a limiting speed of 52 mm/s in the x direction and 30 mm/s in the y direction. The limiting accelerations are 260 mm/s^2 in the x direction and 150 mm/s^2 in the y direction. For the limiting conditions, what is the minimum radius of curvature that the stylus will follow if the x acceleration is positive and the y acceleration is negative. Is the answer the same for a positive y acceleration?

Answer: $\rho = 13.87$ mm; no

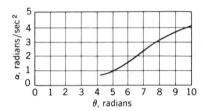

PROBLEM 1.13 Measurements of the angular acceleration α of a flywheel are recorded for various angular displacements θ of the wheel and are plotted as shown. If the angular velocity was 6 rad/s clockwise at $\theta = 5$ rad, estimate with the aid of the graph the angular velocity when $\theta = 10$ rad. The graph refers to clockwise motion.

Answer: $\omega = 7.9 \text{ s}^{-1}$

PROBLEM 1.14 The force needed to compact corrugated cardboard packing increases with the deformation of the cardboard. The voids within the packing disappear when the deformation reaches 75%. Estimate the maximum acceleration for a package dropping from a height of 1 m, if 16 mm of corrugated cardboard packing is used, and the packing is fully compacted during impact. Assume that the force of deformation is directly proportional to deceleration during deformation.

Answer: $a = -263$ g

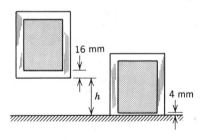

PROBLEM 1.15 A particle moves in a straight line with an acceleration as shown. Determine the final velocity of the particle, if it starts from rest.

Answer: $v = 14.14$ m/s

PROBLEM 1.16 The study of automobile and aircraft crashes shows that complete body support (seatbelts) provides maximum protection against accelerating forces and gives the best chance for survival. In a crash where an automobile is stopped from 80 km/h in 0.09 s, with a linearly decreasing velocity, determine the deceleration and determine the stopping distance, which must be built into the automobile as a crushing of the automobile body. Determine the deceleration and stopping distance for 80 km/h.

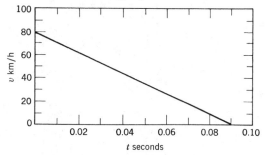

PROBLEM 1.17   The following graph is the acceleration–time record of an experimental solid propellant rocket sled. The sled starts from rest at time $t=0$. Using the dashed lines to simplify the acceleration–time record, construct the velocity–time record, determine the terminal velocity at $t=12$ s, and determine the total distance traveled at that instant.

Answer: 105 m/s, 494 m

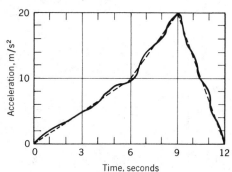

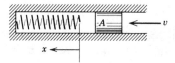

PROBLEM 1.18   The slider A hits the spring with an initial velocity of 40 m/s. Its velocity is reduced to 20 m/s when the compression of the spring is 0.5 m. The deceleration is proportional to x. Determine the deceleration when $x=0.25$ m.

Answer: $a = -1200$ m/s^2

PROBLEM 1.19 A rocket rises vertically with a constant acceleration of 20 m/s² for 20 s. At 20 s the motor shuts off and the rocket continues upward for another 20 s. The tail fins are adjusted to spin the rocket about its own axis. If the angular acceleration of spin is proportional to the upward velocity of the rocket such that $\ddot\theta = V/200$, where $\ddot\theta$ is in radians per seconds², when V is in meters/second, what is the angular spin velocity at $t = 40$ s.

Answer: $\omega = 50.2$ s⁻¹

PROBLEM 1.20 The second stage of a two-stage rocket rises vertically, accelerating linearly from 1 to 4 g in 60 s. If the separation of the second stage takes place at a height of 15 km and a velocity of 1000 m/s, determine the velocity and height when the second stage burns out.

PROBLEM 1.21 A bridge crossing a brook on a country road is approximately sinusoidal in shape. For a car crossing the bridge at a constant speed v, determine the location and magnitude where the car will experience (a) the maximum vertical velocity and (b) the maximum vertical acceleration. The car has zero vertical displacement and zero vertical velocity as it enters the bridge.

Answer: $\dot y = \dfrac{vb\pi}{c}$; $\ddot y = 2b\left(\dfrac{v\pi}{c}\right)^2$

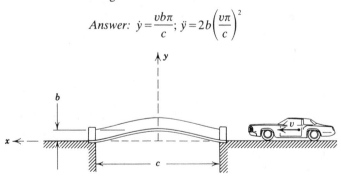

PROBLEM 1.22 The motion of a body can be defined in terms of a single coordinate, x, and moves with a constant acceleration in the direction of that coordinate. At $t = 0$, the displacement is -6 m from a convenient origin. At $t = 2$ s, the

displacement is zero. At $t=4$ s, the velocity is zero. What is the velocity at $t=6$ s?

Answer: $v = -2$ m/s

1.5. KINETICS OF A PARTICLE

Kinetics is the study of the relation between the motion of bodies and the forces that cause motion. It remained a subject of half-philosophy and half-science, steeped in dogma, until well into the eighteenth century. The experimental work of Galileo opened the way for the analysis and reassessment of mechanics. However, one hundred years passed before Sir Isaac Newton (1642–1727) formulated the basic laws of kinetics. This he did in his momentual work, *Philosophiae Naturalis Principia Mathematics*, published in 1687, which stated the fundamental laws of modern physics for the first time.

In essence, the three laws governing the kinetics of a particle are

1. If a balanced force system acts on a particle at rest, it will remain at rest. If a balanced force system acts on a particle in motion, it will remain in motion in a straight line without acceleration.
2. If an unbalanced force system acts on a particle, it will accelerate in proportion to the magnitude and in the direction of the resultant force.
3. When two particles exert forces on each other, these forces are equal in magnitude, opposite in direction, and collinear.

Newton's second law is uniformly accepted as the fundamental principal of classical dynamics. If $\mathbf{R}_1$ is the resultant of the force system, the acceleration will be $\mathbf{a}_1$. If $\mathbf{R}_2$ is the resultant, then the acceleration will be $\mathbf{a}_2$, and

$$\frac{\mathbf{R}_1}{\mathbf{a}_1} = \frac{\mathbf{R}_2}{\mathbf{a}_2} = \frac{\mathbf{R}_n}{\mathbf{a}_n} = \text{constant}$$

This constant of proportionality is called mass. It is a quantitative measure of inertia. It can be stated mathematically as

$$\sum \mathbf{F} = m\mathbf{a}_G \tag{1.11}$$

The resultant of the unbalanced force system is equal to the product of mass, and the acceleration of the mass center, $\mathbf{a}_G$.

In the SI system, the units of force are Newtons, derived by Newton's second law from the product of mass (kilograms, kg) and acceleration (metres per second squared, m/s^2). It is called an absolute system, since force is derived and mass and acceleration are absolute.

In the US–British system, the units of mass (slug, 32.174 lbm) are derived from the ratio of force (pounds, lbf) and acceleration (feet per second squared, ft/s²). A second, but rarely used, system has been to quote force in pounds force (lbf), mass in pounds mass (lbm), and acceleration in feet per second squared (ft/s²). However, equation 1.11 must then be replaced with

$$\sum \mathbf{F} = \left(\frac{1}{32.174}\right) m\mathbf{a}_G$$

The justification for this peculiar version of Newton's second law is to be able to state that a one pound force (lbf) is the force required to hold a one pound mass (lbm) in equilibrium at sea level where $g = 32.174$ ft/s². Both systems are known as gravitational systems, since the value of the slug or the value of the constant $1/32.174$ depend on the acceleration of gravity at the earth's surface. Since the slug has never been widely used, and using the pound for both mass and force is confusing, in dynamics, the SI system is favored. It is also an international system.

While the SI system removes any doubt about units, the vibration engineer will have to be familiar with the common engineering gravitational system quoted because of the enormous and invaluable engineering and scientific literature devoted to structural dynamics, which has accumulated over the years.

The solution of most problems in vibration must begin with an analysis of the force system acting on the particle or body. If the problem is strictly geometrical, the solution may involve only kinematics, but most problems are not so limited. An unbalanced force system acting on a particle or body will cause a particle or body to accelerate, and vibration is the response if the particle or body is part of a system, which resists this acceleration with a force that is linearly or nonlinearly proportional to how far the particle is displaced.

1.6. KINETICS OF A RIGID BODY

The equations of motion for a rigid body are

$$\sum \mathbf{F} = m\mathbf{a}_G = m\dot{\mathbf{v}}_G \tag{1.12a}$$

$$\sum \mathbf{M}_G = \dot{\mathbf{H}}_G \tag{1.12b}$$

The symbol $\mathbf{H}_G$ is the angular momentum of the rigid body. The moment sum $\Sigma \mathbf{M}_G$ is the resultant moment of all external forces and couples about the same axis through the mass center.

Each vector can be expressed in three orthogonal coordinates. Three degrees of freedom and three generalized coordinates describe the motion of a particle, and six degrees of freedom and six generalized coordinates are needed to describe completely the motion of a rigid body. For a rigid body, the first equation describes translation and is identical with the equation of motion for the translation of a particle (equation 1.11). The second equation recognizes that a rigid body may be displaced in rotation as well as translation. The measure of rotation is the displacement of a line through the angular coordinate θ.

For planar motion, which is defined as motion such that the paths of all points on the rigid body lie on parallel planes, the rigid body will have three degrees of freedom. Planar motion involves both translation and rotation and will involve coordinates x, y, and θ. In planar motion, all lines on or within a rigid body have the same angular displacement. This explanation follows from the definition of a rigid body. The second vector equation is not needed for the motion of a particle.

The equations of motion reduce to

$$\sum \mathbf{F} = m\mathbf{a}_G = m\dot{\mathbf{v}}_G \tag{1.13a}$$

$$\sum \mathbf{M}_0 = I_G\ddot{\theta} + \mathbf{r} \times m\dot{\mathbf{v}}_G = I_0\ddot{\theta} + \mathbf{r} \times m\dot{\mathbf{v}}_0 \tag{1.13b}$$

$\Sigma\mathbf{M}_0$ is the resultant moment about an arbitrary axis O. These equations involve both linear acceleration, $\dot{\mathbf{v}}_G$ or $\dot{\mathbf{v}}_0$ and the angular acceleration $\ddot{\theta}$. The term $\mathbf{r}$ is the vector displacement of the mass center from the reference axis. Only the coordinate θ is needed to describe planar motion if the reference axis is either fixed or has zero acceleration. In this event, the moment equation reduces to the kinetic equation of motion for rotation.

$$\sum \mathbf{M}_0 = I_0\ddot{\theta} \tag{1.14}$$

EXAMPLE PROBLEM 1.23

A mass m is held in place by two wires of equal length. Determine the tension in each wire statically and determine the force in the remaining wire if one is cut.

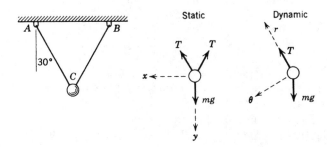

Solution:
This is an excellent example to show the difference between a static and dynamic situation in mechanics.
Static: In static equilibrium, the resultant of the force summation is zero. Cartesian coordinates are convenient

$$\sum F_x = 0; \qquad \sum F_y = 0$$
$$2T \cos 30° = mg$$
$$T = \frac{mg}{\sqrt{3}}$$

Dynamic: With the right wire *BC* cut, the mass *m* swings in an arc about *A* as a center. This is a constraint on the motion. Coordinates that recognize this constraint are displacements normal and tangential to the path.
   In the *r* direction,

$$\sum \mathbf{F}_r = m\mathbf{a}_r$$
$$T - mg \cos 30° = ma_r$$

At the instant the wire is cut, $a_r = \ddot{r} - r\dot{\theta}^2 = 0$ and

$$T = mg \frac{\sqrt{3}}{2}$$

EXAMPLE PROBLEM 1.24

A slender rod *AB*, with a length *l* and a mass *m*, rests with end *A* on a frictionless horizontal surface. End *B* is held. If end *B* were suddenly released, with the rod inclined in the position shown, what is the initial reaction at end *A*?

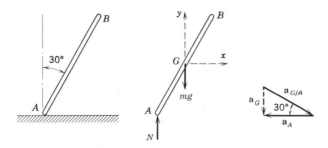

Solution:
The key to the solution of most problems in kinetics is a knowledge of the active forces. The free body diagram shows that there are two active

forces: the normal force between the horizontal surface and end A of the rod and the weight force due to the distributed mass m, acting at the center of gravity G. Since both of these forces are vertical, there can be no horizontal acceleration of the rod. The center of gravity is taken as the origin of our coordinate system.

$$\sum \mathbf{F} = m\mathbf{a}_G$$

$$\sum \mathbf{F}_x = 0 = m\ddot{x} \quad \text{and} \quad \ddot{x} = 0$$

$$\sum \mathbf{F}_y = N - mg = m\ddot{y} \tag{1}$$

This yields our first equation of motion. A second can be formed by taking the moment sum

$$\sum \mathbf{M}_G = \dot{\mathbf{H}}_G$$

$$N\frac{l}{2}\sin \theta = \frac{1}{12}ml^2\ddot{\theta} \tag{2}$$

The angular displacement of the rod AB is positive in the direction of the moment sum.

These two equations have three unknown quantities: N, $\ddot{\theta}$, and $\ddot{y}$. Kinematically, $\ddot{y}$ and $\ddot{\theta}$ are related.

$$\mathbf{a}_G = \mathbf{a}_A + \mathbf{a}_{G/A}$$

The acceleration of end A must be horizontal, and the acceleration of G with respect to A is

$$\mathbf{a}_{G/A} = \frac{l}{2}\alpha \qquad \overset{30}{\searrow}$$

A vector diagram shows that

$$\mathbf{a}_G = \frac{l}{4}\alpha = \ddot{y} \downarrow$$

Substituting, the normal force must be

$$N = \frac{4}{7}mg$$

PROBLEM 1.25 A delicate instrument is packed in a container in order to prevent damage from a sudden drop on a rigid surface. The packing material can be symbolized as elastic springs capable of taking tension and compression. The instrument has a mass of 5 kg and each spring has a constant of 5000 N/m. If the maximum internal displacement of the instrument is

50 mm during impact, what is the maximum deceleration which the instrument experiences?

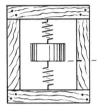

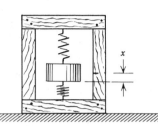

PROBLEM 1.26 A chain of length l and weight μ per unit length is suspended from an elastic cord. Determine the acceleration of the mass center of the remaining chain if the lower quarter is suddenly cut away.

Answer: $a = \dfrac{1}{3}g$

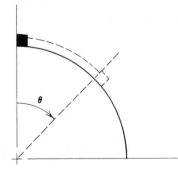

PROBLEM 1.27 A small block slides on a perfectly smooth circular slope. What angle does the block make with the center of the circle when it loses contact with the slope?

Answer: $\theta = 48.2°$

PROBLEM 1.28 A bungee cord is made of 100 rubber strands collected in a cable. The cable supports a mass of 2000 kg and deflects 0.4 m, statically. Determine the additional static deflection, the maximum deflection, and the maximum acceleration if one strand of the cable breaks.

Answer: 4 mm; 0.098 m/s^2

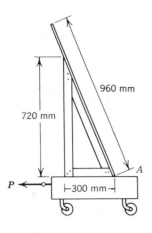

PROBLEM 1.29 Schematically shown here is a 45-kg cart for transporting plate glass. Determine the maximum pull P that can be applied to the car before the glass overturns about end A. The plate glass has a mass of 75 kg.

Answer: $P = 490$ N

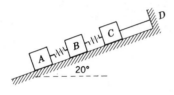

PROBLEM 1.30 Bodies A, B, and C, each with a mass of 5 kg, are connected with springs having moduli of 75 N/m. They are in static equilibrium on a smooth inclined plane. If the cable CD is suddenly cut, determine the initial acceleration of bodies A, B, and C and the initial acceleration of the mass center.

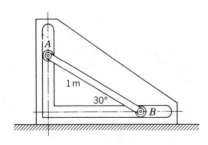

PROBLEM 1.31 The uniform 6-kg link AB is released from rest in the position shown. Determine the reactions at A and B as the link begins to slide if friction in the guides is negligible.

Answer: $A = 19.11$ N
$B = 25.74$ N

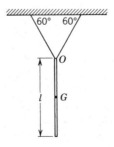

PROBLEM 1.32 A thin rod of mass m and length l is held in place by two wires of equal length. Determine the tension in the remaining wire immediately after one is cut.

Hint: Note that the acceleration of point O will be tangent to an arc of a circle whose radius is the length of the remaining uncut wire. Why?

Answer: $T = 0.495$ mg

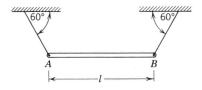

PROBLEM 1.33 A 100-kg uniform beam is supported at its ends by two cables as shown. (a) Determine the tension in the cable at B an instant before and (b) an instant after the cable at A breaks.

Answer: (a) $T = 0.577 \, mg$

(b) $T = 0.266 \, mg$

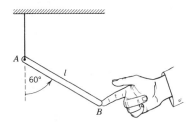

PROBLEM 1.34 A slender rod AB, with a length l and a mass m, is supported at rest in the position shown. If end B were suddenly released, what is the initial tension in the cable at end A?

Answer: $T = \frac{4}{13} mg$

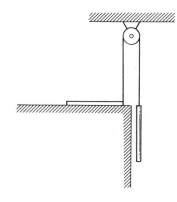

PROBLEM 1.35 Two uniform bars of equal length l and equal weight mg are attached to a wire passing over a pulley. One rests on a horizontal smooth surface and the other hangs free. Determine the acceleration of the free-hanging bar if the system is released from rest in the position shown.

Answer: $a = \frac{3}{8} g$

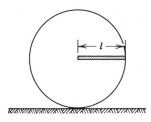

PROBLEM 1.36 A uniform bar of length l and mass m is secured to a circular hoop of radius l as shown. The weight of the hoop is negligible. If the bar and hoop are released from rest in the position illustrated, determine the initial values of the friction force F and the normal force N under the hoop. Friction is sufficient to prevent slipping.

Answer: $F = \frac{3}{8} mg$; $N = \frac{13}{16} mg$

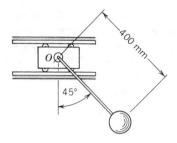

PROBLEM 1.37   The 4-kg pendulum is pivoted on a 2-kg carriage, which moves freely without friction in the horizontal slot. If the pendulum is released from rest in the 45° position, what is the initial acceleration of the carriage?

Answer: 4.9 m/s²

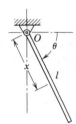

PROBLEM 1.38   The uniform slender bar of mass m and length l is pivoted freely at its end about a horizontal axis through O and released from rest at $\theta = 0$. For a given θ, find the maximum value of the bending moment M in the bar in terms of x and the value of θ at which it occurs.

Answer: $M = \dfrac{mgl}{27}$, at $x = \dfrac{l}{3}$, $\theta = 0$

1.7. WORK AND ENERGY

The concepts of work and energy are also fundamental to the study of dynamics and specifically to mechanical vibration. An increment of work is defined as the product of an increment of displacement and the component of the vector force **F** in the direction of the displacement. It is a vector dot product and therefore a scalar.

$$dU = \mathbf{F} \cdot d\mathbf{s} \tag{1.15}$$

It is evident that an increment of work may be equally well defined as the product of the force and the component of the increment of displacement in the direction of the force. The increment of work done by a couple **M** acting on a body during an increment of an angular rotation $d\boldsymbol{\theta}$ in the plane of the couple is $\mathbf{M} \cdot d\boldsymbol{\theta}$.

Energy is defined as the capacity of a body to do work by reason of its motion or configuration. The configuration of a body within a system is a set of positions for all particles that make up that body. Mechanical energy can be either kinetic or potential. Kinetic energy is the energy that is a direct result of velocity of motion. By definition, the kinetic energy of a particle is

$$T = \tfrac{1}{2}mv^2 \tag{1.16}$$

The kinetic energy of a rigid body is more complicated.

$$T = \tfrac{1}{2}\mathbf{v}_G \cdot m\mathbf{v}_G + \tfrac{1}{2}\dot{\boldsymbol{\theta}} \cdot \mathbf{H}_G \tag{1.17}$$

In this case, $\dot{\boldsymbol{\theta}}$ is the angular velocity and $\mathbf{H}_G$ is the angular momentum of the rigid body referred to an axis through the mass center. In planar motion such as torsional vibration, these terms can be considerably simplified. The angular momentum and the angular velocity vectors are collinear which reduces equation 1.17 to a more familiar form

$$T = \tfrac{1}{2}mv_G^2 + \tfrac{1}{2}I_G\omega^2 \qquad (1.18)$$

Caution should be taken in expressing the kinetic energy of a rigid body. Many times a troublesome problem in mechanical vibrations turns out to be nothing more than a misunderstanding of basic kinetics.

Potential energy may be due to position or deformation. Potential energy of position is the work that must be done against a field force to change the position of a particle. If work must be done on a particle to change its position, the potential energy is increased. If the particle does work in changing its position, the potential energy is decreased. Potential energy may also be produced by elastic members that have been stretched, compressed, twisted, or otherwise deformed. This potential energy is the strain energy of deformation and is sometimes called elastic energy.

If the potential energy of a particle or a system depends only on the position or configuration of the particle or the particles within the system, it is said to be *conservative*. As this implies, the potential energy of a conservative system is independent of the positions or configurations through which the system passes and depends only on the initial and final states. Since it must be computed with respect to some selected position or configuration, which is known as the datum and is arbitrary, potential energy does not have an absolute value in a conservative system. The net work done by a particle that returns to its original position or state of motion in a conservative system is zero.

Energy methods involve an *energy balance* using scalars rather than a *force balance* with vectors. *If no work is done and no energy is lost in friction, any increase in the energy of a conservative system must increase the kinetic energy of the system or be stored as potential or internal energy.* This conservation principle is commonly called the *Conservation of Energy* or the *First Law of Thermodynamics*.

Friction forces are nonconservative, since they are not a function of position or configuration and the dissipated energy is nonrecoverable. Although a real system must include friction and must therefore be nonconservative, it is often convenient to neglect friction losses in the solution of vibration problems.

1.8. IMPULSE AND MOMENTUM

The product of force and time is defined as linear *impulse*. It is a vector quantity that has the direction of the resultant force. Linear *momentum* is the product of mass and linear velocity. Angular momentum is the product of mass moment of inertia and angular velocity. Linear and angular momentum are also vector quantities and can be added and resolved in the same manner as force and impulse.

An alternate statement of Newton's second law of motion is that the resultant of an unbalanced force system must be equal to the time rate of change of momentum of the mass center. For linear momentum, this would be

$$\sum \mathbf{F} = \frac{d}{dt}(m\mathbf{v}_G) \tag{1.19}$$

The collision between two bodies where relatively large forces result over a comparatively short interval of time is called *impact*. During collision, kinetic energy is absorbed as the impacting bodies are deformed. There follows a period of restoration, which may or may not be complete. If complete restoration of the energy of deformation occurs, the impact is elastic. If the restoration of energy is incomplete, the impact is inelastic. After collision, the bodies continue to move with changed velocities. Since the contact forces on one body are equal and opposite to the contact forces on the other, the sum of linear momentum for the two bodies is conserved. The law of *Conservation of Momentum* states that the *momentum of a system of bodies is unchanged if there is no resultant external force on the system.*

For two bodies that impact and separate, the ratio of the velocity of separation, $v_1' - v_2'$ to the velocity of approach, $v_2 - v_1$, is called the coefficient of restitution, e. Figure 1.7 shows such an impact.

$$e = \frac{v_1' - v_2'}{v_2 - v_1} \tag{1.20}$$

Its value will depend on the shape and material properties of the colliding bodies. In elastic impact, the coefficient of restitution is one and there is no energy loss. A coefficient of restitution of zero indicates perfectly inelastic or plastic impact, where there is no separation of the bodies

Fig. 1.7

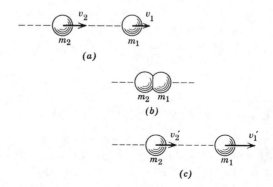

(a)

(b)

(c)

after collision and the energy loss is a maximum. In oblique impact, the coefficient of restitution applies only to those components of velocity along the line of impact or normal to the plane of impact.

EXAMPLE PROBLEM 1.39

As an example of the principles of conservation of energy and momentum, let us consider a helical spring, fixed at its upper end and supporting a mass of 5 kg at its lower end. The spring constant of proportionality is 350 N/m. A collar with a mass of 6 kg can be placed on top of the first mass. Initially, the spring and the 5-kg mass are at rest, with the upper face of the 5-kg mass 168.1 mm above a horizontal platform. The platform has a hole in it through which the 5-kg mass will pass, but the 6-kg collar will not. Describe the motion over one full cycle.

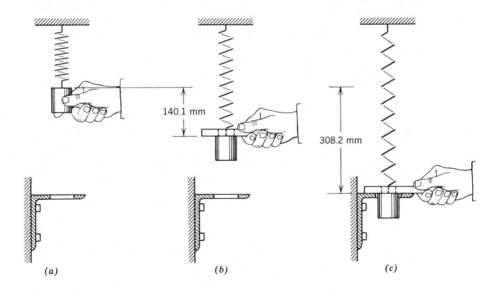

(a)   *(b)*   *(c)*

Solution:
If the spring constant of proportionality is 350 N/m, the spring will be extended 140.1 mm from its free length, supporting only the 5-kg mass. If the 6-kg collar is carefully placed on top of the first mass, and is supported by hand as the helical spring extends to a new equilibrium position, the spring mass and collar will come to rest with the spring extended 308.2 mm from its free length and the upper face of the 5-kg mass exactly even with the upper face of the platform, and the lower face of the 6-kg collar touching, but not being supported by the platform.

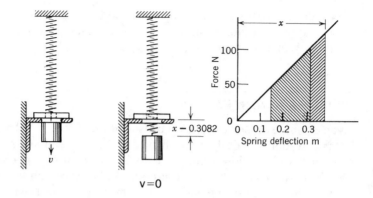

On the other hand, if the collar is placed on the first mass and released, both masses will move down with increasing velocity, the 5-kg mass will pass through the hole and continue down, and the collar will be stopped by the platform. The 5-kg mass will reverse its direction, pass up through the hole in the platform and pick up the collar again. This cycle will be repeated until the entire system comes to rest in the same position it would have if the collar had been put in place carefully.

For the first phase, energy must be conserved

$$\Delta Ve + \Delta Vg + \Delta T = 0$$

$$\Delta T = \tfrac{1}{2}(5+6)(v^2 - 0)$$

$$\Delta Vg = -(5+6)(9.8065)(0.3082 - 0.1401)$$

$$\Delta Ve = \tfrac{1}{2}(350)(0.\overline{3082}^2 - 0.\overline{1401}^2)$$

ΔT and ΔVe are positive since both represent an increase in energy. ΔVg is negative since the potential energy of position for both the 5- and 6-kg masses decreases. The free length of the spring is taken as the original datum.

$$\tfrac{1}{2}(350)(0.\overline{3082}^2 - 0.\overline{1401}^2) - (5+6)(9.8065)(0.3082 - 0.1401)$$
$$+ \tfrac{1}{2}(5+6)(v^2 - 0) = 0$$
$$v = 0.948 \text{ m/s}$$

When the 5-kg mass passes through the hole in the platform, its kinetic energy is conserved. It will reverse direction and return to strike the collar with the same velocity as it had at the instant of separation, $v = 0.948$ m/s. The kinetic energy of the collar is lost in the impact.

Let x be the maximum displacement of the 5-kg mass

$$\Delta Ve + \Delta Vg + \Delta T = 0$$

$$\Delta T = \tfrac{1}{2}(5)(0 - \overline{0.948^2})$$
$$\Delta Vg = -(5)(9.8065)(x - 0.3082)$$
$$\Delta Ve = \tfrac{1}{2}(350)(x^2 - \overline{0.3082^2})$$

$$\tfrac{1}{2}(350)(x^2 - \overline{0.3082^2}) - (5)(9.8065)(x - 0.3082) - \tfrac{1}{2}(5)(\overline{0.948^2}) = 0$$

This reduces to a quadratic equation.

$$x^2 - 0.2802x - 0.02148 = 0$$
$$x = 0.3428 \text{ m}, \ -0.06265 \text{ m}$$

Only the positive value has meaning. The maximum displacement of the 5-kg weight is 34.6 mm below the platform

$$0.3428 - 0.3082 = 0.0346 \text{ m}$$

When the 5-kg mass strikes the collar as it passes upward through the hole in the platform, momentum is conserved. The linear momentum of the 5-kg mass is 4.74 kg·m/s. The collar, before impact, has no linear momentum. After impact the momentum of the two weights moving upward must be 4.74 kg·m/s

$$0 + (5)(0.948) = (5 + 6)v_1$$
$$v_1 = 0.431 \text{ m/s}$$

Since they move together, both masses must have the same velocity. This means that some of the kinetic energy of the 5-kg mass is lost as a result of impact.

$$\Delta U = \tfrac{1}{2}(5)(\overline{0.948^2} - \overline{0.431^2}) = 1.7 \text{ N·m} \qquad \text{is lost.}$$

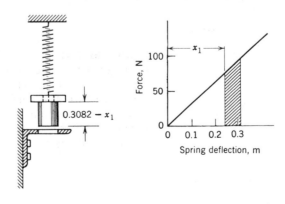

After the 6-kg collar and the 5-kg mass leave the platform, they will move upward until all the kinetic energy is transferred to potential energy of position or stored as elastic energy in the spring. Let the displacement of that position be x_1,

$$\Delta Ve + \Delta Vg + \Delta T = 0$$

$$\Delta T = \tfrac{1}{2}(5+6)(0 - \overline{0.431}^2)$$
$$\Delta Vg = (5+6)(9.8065)(0.3082 - x_1)$$
$$\Delta Ve = \tfrac{1}{2}(350)(x_1^2 - \overline{0.3082}^2)$$

Solving for x_1,

$$x_1^2 - 0.6164 x_1 + 0.08915 = 0$$
$$x_1 = 0.2318 \text{ m}$$

which is 0.0764 m above the platform. The cycle now repeats until all the energy of motion is dissipated and the system comes to rest.

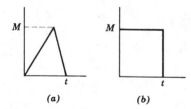

(a)   (b)

PROBLEM 1.40   An angular impulse acting on a flywheel according to Figure (a) increases its revolutions per minute (rpm) from 20 to 40. With the impulse diagram for Figure (b), what would the final revolutions per minute (rpm) have been?

Answer: 60 rpm

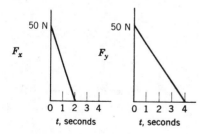

PROBLEM 1.41   A particle with 4 kg mass moves in the horizontal $x-y$ plane under the action of the forces F_x and F_y, which vary with time as shown. If the particle starts from rest at $t=0$, what is the velocity at 4 s?

Answer: 28 m/s

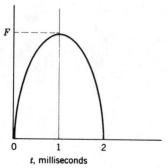

PROBLEM 1.42   A baseball with a mass of 0.145 kg reaches a batter with a velocity of 25 m/s, and after being batted, it leaves the bat with a velocity of 40 m/s in the opposite direction. Find the maximum force of the ball on the bat if the contact force can be approximated by the semiellipse as shown.

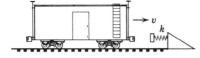

PROBLEM 1.43 A railroad car with a mass of 15,000 kg and rolling along a level track is brought to rest by a bumper spring. The spring is deformed 200 mm and has a modulus of 130 kN/m. With what velocity did the car strike?

Answer: $v = 0.589$ m/s

PROBLEM 1.44 A 2-kg mass rests on a spring-supported platform which is compressed 100 mm in the position shown. After moving 50 mm, the platform is halted by stops. Determine the maximum height h reached by the weight. The spring modulus is 2000 N/m.

Answer: $h = 382.4$ mm

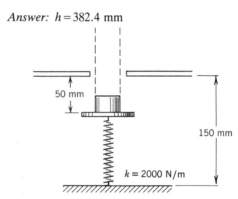

PROBLEM 1.45 The uniform 10-kg bar is designed to snap up when the catch at end A is released. A tightly wound coil spring exerts a moment of 80 N·m on the link at the hinge O. The spring is wound two revolutions from its free position. The moment-displacement diagram for the coil spring is included. Determine the angular velocity of the bar as it passes through the vertical position.

Answer: $\omega = 6.42$ rad/s

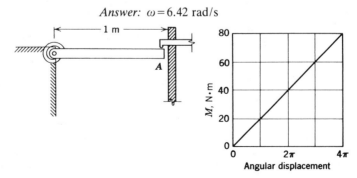

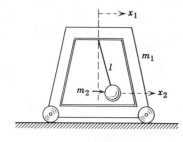

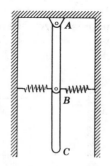

PROBLEM 1.46   The 0.3-kg block fits loosely in a 3-kg barrel and the spring is under an initial compression of 80 mm. Find the final velocities acquired by the block and barrel. The spring modulus is 1750 N/m. Neglect friction and assume that the assembly is placed on a smooth horizontal surface.

Answer: $v_A = 5.826$ m/s; $v_B = 0.583$ m/s

PROBLEM 1.47   A frame supports a pendulum of length l, which swings free in the same plane in which the frame moves. Determine an expression for the kinetic energy and potential energy of the system in terms of the generalized coordinates x_1 and x_2.

PROBLEM 1.48   Determine an expression for the kinetic energy of a thin rod suspended as a pendulum from a light wire in terms of the generalized coordinates θ_1 and θ_2. Make approximations for small oscillations.

Answer: $T = \frac{1}{24} m l^2 (3\dot{\theta}_1^2 + 6\dot{\theta}_1 \dot{\theta}_2 + 4\dot{\theta}_2^2)$

PROBLEM 1.49   Determine an expression for the total kinetic energy of the system, which consists of two slender rods hinged at A and B.

Answer: $T = \frac{1}{6} m l^2 (4\dot{\theta}_1^2 + 3\dot{\theta}_1 \dot{\theta}_2 + \dot{\theta}_2^2)$

PROBLEM 1.50   Determine an expression for the kinetic energy of the thin rigid rod, which is supported as a pendulum at end A, has a mass m, and which is also pinned to a roller and held in place by two elastic springs with constants k.

Answer: $T = \frac{1}{6}ml^2\left(\dot{\theta}^2 + \frac{3\dot{x}\dot{\theta}}{l} + \frac{3\dot{x}^2}{l^2}\right)$

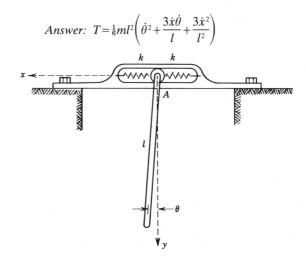

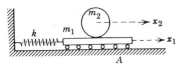

PROBLEM 1.51 A platform of mass m_1 supports a circular cylinder of mass m_2 and is elastically suspended from the wall by a spring with a modulus of k. Determine an expression for the kinetic energy of the system in terms of the generalized coordinates x_1 and x_2, if the cylinder rolls without slipping and $m_1 = m_2 = m$. The platform moves with negligible friction over the surface A.

Answer: $T = \frac{1}{4}m(3\dot{x}_1^2 - 2\dot{x}_2\dot{x}_1 + 3\dot{x}_2^2)$

PROBLEM 1.52 Determine an expression for the total kinetic energy of the system, which consists of a slender rod with a mass m_1, hinged at A and supported by a spring of stiffness k_1 at B, and in turn it supports a spring and mass, k_2 and m_2.

Answer: $T = \frac{1}{6}(m_1\dot{x}^2 + 3m_2\dot{x}_2^2)$

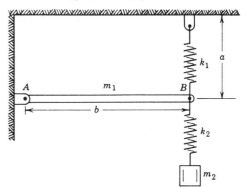

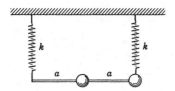

PROBLEM 1.53 Two identical springs support a rigid rod and two identical masses. Choose coordinates that will describe the motion of each mass. Determine an expression for the kinetic energy of the system in terms of your coordinates.

Answer: $T = m\dot{x}^2 + \frac{1}{4}ma^2\dot{\theta}^2$

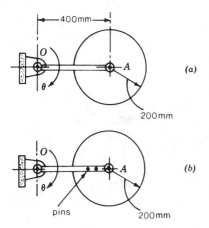

PROBLEM 1.54 The uniform circular disk of 200-mm radius has a mass of 30 kg and is mounted on the rotary bar OA in two ways. In case (a), the disk is not pinned and rotates freely at A. Since there is no friction, the disk has no rigid body rotation. In case (b), the disk is pinned to the bar and has rigid body rotation.

If the systems are released from rest in the positions shown, determine the angular velocity of the bar OA as it passes through its vertical position, in each case.

Answer: (a) $\omega_{OA} = 7$ rad/s
(b) $\omega_{OA} = 6.6$ rad/s

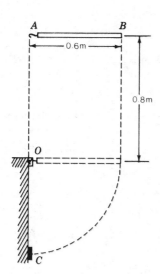

PROBLEM 1.55 The uniform rod AB has a mass of 2 kg and is released from rest from the horizontal position shown. As it falls, the end A becomes hooked at pin O. End B remains free. Determine the speed at which end B strikes the stop at C.

Answer: $v_B = 7.28$ m/s

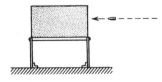

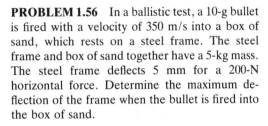

PROBLEM 1.56 In a ballistic test, a 10-g bullet is fired with a velocity of 350 m/s into a box of sand, which rests on a steel frame. The steel frame and box of sand together have a 5-kg mass. The steel frame deflects 5 mm for a 200-N horizontal force. Determine the maximum deflection of the frame when the bullet is fired into the box of sand.

Answer: $x = 7.8$ mm

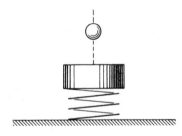

PROBLEM 1.57 A 45-g steel ball is dropped from rest through a vertical height of 2 m. Impact is on a solid steel cylinder with a 0.45-kg mass. The cylinder is supported by a light spring with a modulus of 1600 N/m. The cylinder deflects a maximum of 12 mm from the static equilibrium as a result of the impact. Calculate the height to which the ball will rebound and the coefficient of restitution for the impact.

Answer: $h' = 40.6$ mm; $e = 0.257$

CHAPTER TWO

PERIODIC
MOTION

2.1. INTRODUCTION

Watching the daily ebb and flow of tides and listening to the steady hum of industrial machinery, man is constantly concerned and fascinated with periodic motion. The swaying of a tree in the wind, the pitching and rolling of a ship at sea, and the turning of a generator in its journals all lead to cyclic variations of force and displacement. These motions, visible to the eye or perceptible to the body, have challenged man from the beginning of time. The study of these cyclic or periodic variations is the study of vibration and it is one of the most important aspects of dynamics.

Often, vibration problems are complicated. The motion of a simple pendulum is easy to understand, but the flutter and buffeting of aircraft required years of exhaustive research and study before the problem was solved in any sense of the word. As problems are understood and controlled, new ones arise that were not expected. Often, these are caused by no more than subtle changes in manufacturing procedures, small errors in machining, or a redesign of parts or a system.

Vibration can be classified in several ways. A *free* vibration occurs without externally applied forces. Normally, free vibration arises when an elastic system is displaced or given some initial velocity, as might result from an impact. A *forced* vibration occurs with the application of external forces. Forced vibrations can be *periodic*, *aperiodic*, or *random*. Periodic motion simply repeats itself in regular intervals of time. In aperiodic or random motion, there is no such regular interval. Both free and forced vibrations can be *damped*, which is the term used in the study of vibration to denote a dissipation of energy. Vibrations are also classified by the number of *degrees of freedom* of motion. The number of *degrees of freedom* corresponds to the *number of independent* coordinates needed to completely describe motion.

A particular problem of study may be described with more than one classification. That is, forced damped vibration is motion that is externally forced while energy is dissipated.

If the dissipative forces are proportional to the velocity of motion, the restoring forces proportional to displacement, and the inertial forces proportional to acceleration, a vibration is said to be *linear*. If any of these proportionalities is not satisfied, a vibration is said to be *nonlinear*. The terminology is borrowed from linear and nonlinear differential equations. Linearity is important since linear differential equations can be solved much more easily. Nonlinear vibration can sometimes be linearized by restricting the study of motion to *small oscillations*.

2.2. FREE VIBRATION

Consider an elastic spring stretched by an applied force, **f**. It is called elastic because it obeys Hooke's law, the force **f** varying linearly with displacement, **x**. The constant of proportionality for the spring is the spring constant or spring modulus k, the slope of the force-displacement curve. The units of the spring modulus are force per unit displacement. The spring force is then,

$$\mathbf{f} = k\mathbf{x} \qquad (2.1)$$

Figure 2.1 descriptively shows the variation of the scalar f with x.

If a mass m is attached to the lower end of the spring and the spring and mass are allowed to come to an *equilibrium position*, the spring will deflect statically a distance $\Delta = mg/k$ from its free position, and the spring force will be equal to the weight of the suspended mass.

If the mass is deflected from this new equilibrium, it will oscillate about the equilibrium position. Using Newton's second law of motion, the equation of motion for any displacement **x** is $\Sigma \mathbf{F} = m\ddot{\mathbf{x}}$, which can be written as a linear second-order differential equation in terms of the single coordinate x.

$$mg - (mg + kx) = m\ddot{x}$$
$$m\ddot{x} + kx = 0 \qquad (2.2)$$

Fig. 2.1

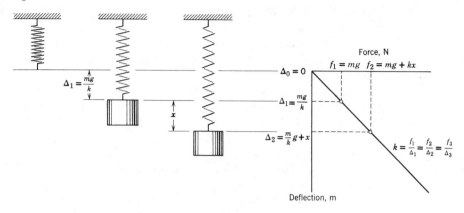

It should be observed that the deflection x is taken from the equilibrium position, that is, static equilibrium occurs at $x=0$, and that the weight force cancels. The weight force will always cancel if it does not restore the system to the equilibrium position. In Section 3.2, this particular point is developed.

For this relatively simple problem, it is not difficult to see that the solution would be harmonic. This can be verified by substituting either an exponential or trigonometric function. For the trial solution of the homogeneous equation, let

$$x = Ce^{rt}$$
$$\dot{x} = rCe^{rt}$$
$$\ddot{x} = r^2Ce^{rt}$$

Substitution shows that the trial solution is an integral of the differential equation if the *characteristic equation* is satisfied.

$$r^2 + \frac{k}{m} = 0 \tag{2.3}$$

The quantity $\sqrt{k/m}$ is the frequency of the harmonic motion, in radians per second, and is generally called the *natural circular frequency*, ω_n.

The characteristic equation has two roots,

$$r_{1,2} = \pm\sqrt{-\frac{k}{m}} = \pm\sqrt{-\omega_n^2} \tag{2.4}$$

These roots are called the *characteristic values* or *eigenvalues* of the characteristic equation.

There are two particular integrals of equation 2.2

$$x_1 = C_1 e^{i\omega_n t}$$

$$x_2 = C_2 e^{-i\omega_n t}$$

If we substitute both solutions into equation 2.2 and add, the sum also satisfies the equation of motion and has the required number of arbitrary constants. The general solution is then

$$x = C_1 e^{i\omega_n t} + C_2 e^{-i\omega_n t} \qquad (2.5)$$

provided that x_1 and x_2 are not linearly dependent. The general integral can also be written in hyperbolic form

$$x = (C_1 + C_2)\cosh i\omega_n t + (C_1 - C_2)\sinh i\omega_n t \qquad (2.6)$$

or in trigonometric form,

$$x = A \cos \omega_n t + B \sin \omega_n t \qquad (2.7)$$

C_1, C_2, A, and B are arbitrary constants that depend on the initial conditions of motion. If the free end of spring instead of the equilibrium position had been chosen as the origin for the coordinate x, a term mg/k would be added to equations 2.6 and 2.7. The term is a static term and is of no importance to dynamics. It can always be eliminated if the equilibrium position is selected as the origin for the coordinate x.

2.3. HARMONIC MOTION

Equations 2.5 to 2.7 are harmonic functions of time. The motion is symmetric about the equilibrium position. The velocity is a maximum and the acceleration is zero each time the mass passes through this position. At the extreme displacements, the velocity is zero and the acceleration is a maximum. This is the simplest form of vibration and it is often called *simple harmonic motion*. This motion is typical of most systems with a single degree of freedom that have been displaced from a position of stable equilibrium by a small amount and released. It models accurately a surprising number of real systems.

It is often convenient to use a vector diagram to visually picture harmonic motion. In Figure 2.2 the displacement x is the sum of the projections on the x-axis of the two vectors **A** and **B**, which rotate about the origin at an angular velocity ω_n and at right angles to one another. The angular displacement of either vector at any time t is $\theta = \omega_n t$. This displacement is measured from the original positions of the vectors **A** and **B** at time $t = 0$, the vertical axis for vector **A**, the horizontal axis for

Fig. 2.2

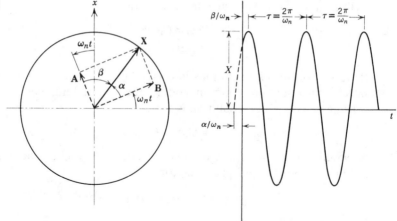

vector **B**. The magnitudes A and B depend on the initial conditions of motion.

The motion completed in any one period is called a *cycle* and conversely the *period* τ is the time necessary to complete one cycle of motion,

$$\tau = \frac{2\pi}{\omega_n} = 2\pi\sqrt{\frac{m}{k}}$$

$$(2.8)$$

$$f_n = \frac{1}{2\pi}\sqrt{\frac{k}{m}}$$

It is sometimes convenient to describe the period of harmonic motion, usually given in seconds, and at other times it is convenient to speak of the frequency of motion, the number of cycles completed in any unit of time. The conventional unit for frequency is the hertz (Hz), which is one cycle per second (cps). The use of hertz is standard practice in many European countries and is widely used in electronics, acoustics, and most scientific fields. The symbol f_n is used for the natural frequency of a vibrating system.

We are limited to two arbitrary constants, but they need not both be amplitudes. At times, it is more convenient to think in terms of an amplitude and a phase angle

$$x = X\sin(\omega_n t + \alpha)$$

$$(2.9a)$$

or

$$x = X\cos(\omega_n t - \beta)$$

$$(2.9b)$$

The displacements $(\omega_n t + \alpha)$ and $(\omega_n t - \beta)$ are measured from the horizontal and vertical axes, respectively, which are again, the general starting positions for the vectors **A** and **B**. X, α, and β are new arbitrary constants that can be described in terms of A and B

$$X = \sqrt{A^2 + B^2} \tag{2.10}$$

$$\tan \beta = \frac{B}{A} = \cot \alpha \tag{2.11}$$

The term X is the *maximum amplitude* or the peak value of the displacement.

EXAMPLE PROBLEM 2.1

Determine the natural frequency of the spring, mass, and pulley as shown. The spring has a modulus of k and the pulley can be assumed to be frictionless and have negligible mass.

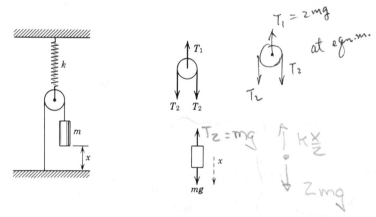

Solution:
A generalized coordinate for this system is the displacement of the mass m. The displacement of the pulley and the deformation of the spring can all be described in terms of x, the displacement of the mass m from its position of static equilibrium. From simple statics,

$$T_1 = 2T_2 = k\left(\frac{x}{2}\right) + 2mg$$

and from Newton's second law of motion, $\Sigma \mathbf{F}_x = m\ddot{x}$

$$mg - T_2 = m\ddot{x}$$

$$mg - \frac{[k(x/2) + 2mg]}{2} = m\ddot{x}$$

$$\ddot{x} + \frac{k}{4m}x = 0$$

This is a linear second-order differential equation similar to equation 2.2. From this equation, the natural frequency is

$$f_n = \frac{1}{4\pi}\sqrt{\frac{k}{m}}$$

EXAMPLE PROBLEM 2.2

Determine the natural frequency for small oscillations of the single-mass system shown. Both springs have a modulus of k and the mass of the pulley is negligible. Friction is also negligible.

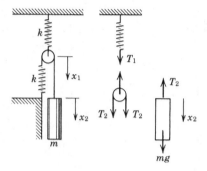

Solution:

This problem is more difficult. This system has only one degree of freedom, but, in order to describe the motion of the system, two geometric coordinates are needed.

Let x_1 be the motion of the pulley and x_2 be the motion of the mass m. Although we need both geometric coordinates, they are kinematically related, and we must find and use that relation.

From simple statics, the tension in the upper spring is

$$T_1 = kx_1 + 2mg$$

and, the tension T_2 in the lower spring is

$$T_2 = k(x_2 - 2x_1) + mg$$

Note that the spring force is a function of the spring stretch, which is the difference, $x_2 - 2x_1$. This is the key to the problem solution, because it relates x_2 to x_1. You can test for this relationship by first holding $x_1 = 0$ and then holding $x_2 = 0$. Do this.

T_1 and T_2 are related from the free body of the pulley. At all times,

$$2T_2 = T_1$$

From which,

$$2k(x_2-2x_1)+2m\ddot{g}=kx_1+2mg$$

This constrains x_1 and x_2 so that

$$2x_2=5x_1$$

Note, also, that the weight forces cancel, as they should.

Now, for the mass m, using Newton's second law of motion,

$$\sum \mathbf{F}=m\ddot{x}_2$$

$$mg-T_2=m\ddot{x}_2$$

or, $$mg-[k(x_2-2x_1)+mg]=m\ddot{x}_2$$

substituting the kinematic relation that $2x_2=5x_1$, yields the linear second order differential equation

$$\ddot{x}_2+\frac{k}{5m}x_2=0$$

From this, the natural frequency is

$$f_n=\frac{1}{2\pi}\sqrt{\frac{k}{5m}}$$

EXAMPLE PROBLEM 2.3. THE MANOMETER

Determine the natural frequency of a U-tube manometer open at both ends and containing a column of liquid mercury of length l and weight density γ.

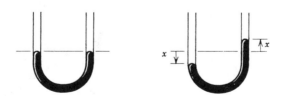

Solution:

The coordinate x established as the displacement of the manometer meniscus from its equilibrium position, completely describes the movement of the manometer fluid and does recognize the constraint of the fluid in the manometer tube. It is therefore a generalized coordinate. The cross-section of the tube is A and it is uniform. The head of pressure acts over the cross-section and acts in a direction opposite to the acceleration $\ddot{x}$. γ is the weight density of the fluid. Using Newton's second law of motion,

$$\sum \mathbf{F}_x = m\ddot{x}$$

$$-2A\gamma x = \frac{A\gamma l}{g}\ddot{x}$$

$$\ddot{x} + \frac{2g}{l}x = 0$$

This is also a linear second-order differential equation analogous to equation 2.2. The frequency of small oscillation is

$$f_n = \frac{1}{2\pi}\sqrt{\frac{2g}{l}}$$

This problem stretches your concepts of mass and coordinate. Here, the mass is a fluid, liquid mercury, and the coordinate x is measured along the manometer tube. Neither affects our use of Newton's second law of motion, which applies to both rigid and nonrigid bodies, and which applies along any coordinate path, so long as the mass is constrained to follow that path. Here, the mercury is constrained to the manometer tube. An equivalent system could have been established, but we should think beyond a vibration world of only blocks and springs. The only limitation on this manometer solution is that the rise and fall of the meniscus must be within the straight portion of the tube for this solution to be valid. Why?

PROBLEM 2.4   The motion of a linear oscillator can be described by the equation $s = A \cos \omega t + B \sin \omega t$, where s is displacement, ω is the natural circular frequency in radians per second, and A and B are arbitrary constants. The natural frequency of the oscillator is 5 Hz. At $t = 0$, the amplitude is 10 mm and the velocity is 0.2 m/s. What is the maximum amplitude displacement, velocity, and acceleration? Plot the motion.

Answer: 11.9 mm; 0.372 m/s; 11.7 m/s²

PROBLEM 2.5   For the particular amplitude response signal shown,

(a) Determine the natural circular frequency ω, (s^{-1}), and period τ, (s).
(b) For the response equation $x(t) = X \cos (\omega t - \beta)$, Find X and β.

(c) For the response equation $x(t) = X$ sin $(\omega t + \alpha)$, Find X and α.

(d) Express $x(t)$ as the sum of sin ωt and cos ωt, $x(t) = A$ cos $\omega t + B$ sin ωt. Find A and B.

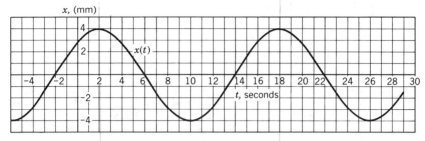

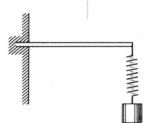

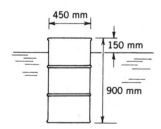

PROBLEM 2.6 A 25-kg mass is suspended from a spring with a modulus of 2 N/mm, which is in turn suspended at its upper end from a thin steel cantilevered beam with a thickness of 3 mm, a width of 20 mm, and a length of 250 mm. Determine the natural frequency of the motion of the weight.

Answer: $f_n = 0.899$ Hz

PROBLEM 2.7 A partially filled oil drum floats in the sea. Determine the frequency of vertical motion as it bobs up and down. Sea water has a density of 1.025.

Answer: $f_n = 0.576$ Hz

PROBLEM 2.8 A heavy table is supported by flat steel legs. Its natural period in horizontal motion is 0.4 s. When a 30-kg plate is clamped to its surface, the natural period in horizontal motion is increased to 0.5 s. What is the effective spring constant and the mass of the table?

Answer: 53.3 kg; 13160 N/m

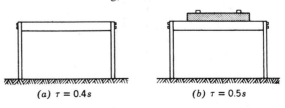

(a) $\tau = 0.4s$ *(b)* $\tau = 0.5s$

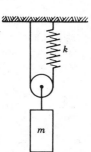

PROBLEM 2.9 Determine the natural frequency for small oscillations of the single mass system shown. The mass of the pulley is negligible.

Answer: $f_n = \dfrac{1}{\pi}\sqrt{\dfrac{k}{m}}$

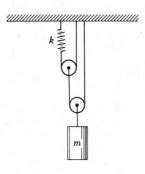

PROBLEM 2.10 Determine the natural frequency for small oscillations of the system shown. The mass of the pulleys and friction are negligible.

Answer: $f_n = \dfrac{2}{\pi}\sqrt{\dfrac{k}{m}}$

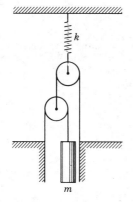

PROBLEM 2.11   Do Problem 2.10 for the following system.

Answer: $f_n = \dfrac{1}{8\pi}\sqrt{\dfrac{k}{m}}$

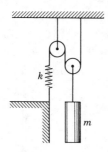

PROBLEM 2.12   Do Problem 2.10 for the following system.

Answer: $f_n = \dfrac{1}{\pi}\sqrt{\dfrac{k}{m}}$

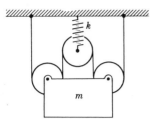

PROBLEM 2.13 The mass m is suspended from two pulleys around which a weightless cord passes. The spring has a modulus of k. Determine the natural frequency of the system.

Answer: $f_n = \dfrac{1}{\pi}\sqrt{\dfrac{k}{m}}$

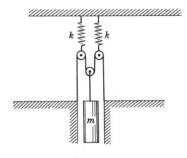

PROBLEM 2.14 The mass m is supported by a system of pulleys and two springs with identical moduli of k. Determine the natural frequency of the system.

Answer: $f_n = \dfrac{1}{2\pi}\sqrt{\dfrac{k}{2m}}$

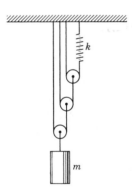

PROBLEM 2.15 The mass m is suspended from three massless pulleys around which massless cords pass. The spring has a modulus of k. Determine the natural frequency of the system.

Answer: $f_n = \dfrac{4}{\pi}\sqrt{\dfrac{k}{m}}$

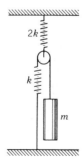

PROBLEM 2.16 Determine the natural frequency for small oscillations of the single mass system shown. The mass of the pulley is negligible.

Answer: $f_n = \dfrac{1}{2\pi}\sqrt{\dfrac{k}{3m}}$

PROBLEM 2.17   Do Problem 2.16 for the following system.

Answer: $f_n = \dfrac{1}{2\pi}\sqrt{\dfrac{4k}{5m}}$

PROBLEM 2.18   The cart is connected to the wall by a rope and pulleys. The flexibility of the rope can be represented by a spring with a modulus k. Determine the natural frequency of the cart, rope, and pulley system. Neglect all friction and the mass of the rope, pulleys, and wheels. What would the natural frequency be if the angle of the incline were changed to $\theta = 60°$?

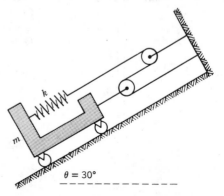

$\theta = 30°$

PROBLEM 2.19   Determine the lowest natural frequency of a thin ring of radius r and mass density ρ vibrating in a radial direction.

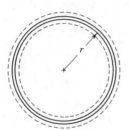

Hint: Circumferential strain $\varepsilon = u/r$, where u is the radial displacement.

Answer: $f_n = \dfrac{1}{2\pi}\sqrt{\dfrac{E}{\rho r^2}}$

PROBLEM 2.20 Solve Problem 2.19 for a thin sphere of radius r.

Answer: $f_n = \frac{1}{2\pi}\sqrt{\frac{2E}{\rho r^2}}$

PROBLEM 2.21 A 200-kg electromagnet lifts 200 kg of iron scrap. The elastic constant is 20 kN/m. Determine the equation of motion of the magnet when the current is turned off and the scrap is dropped.

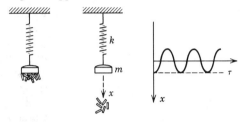

PROBLEM 2.22 The mass m is at rest, partially supported by the spring and partially supported by the stops. In the position shown, the spring force is $mg/2$. At time $t=0$, the stops are rotated, suddenly releasing the mass. Determine the transient equation of motion.

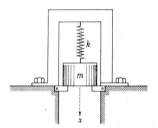

PROBLEM 2.23 A 5-kg fragile glass vase is packed in chopped sponge rubber and placed in a cardboard box that has negligible weight. It is then accidentally dropped from a height of 1 m. This particular sponge rubber exhibits a force deflection curve in bulk as shown. Determine the maximum deformation of the packing within the box and the maximum acceleration of the vase in g's.

Answer: 20.22g

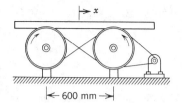

PROBLEM 2.24 A device designed to measure the kinetic coefficient of friction consists of two 90° Vee-grooved pulleys, rotating in opposite directions, across which a cylindrical bar of some known material is placed. When displaced, the bar will perform simple harmonic motion. Derive an expression for the kinetic coefficient of friction μ in terms of the frequency of vibration in cycles per second.

Answer: $\mu = 0.854 f^2$

PROBLEM 2.25 Two sliders are constrained to move within a smooth tube that is rotating in the horizontal plane about the fixed axis 0. Each of the sliders is elastically suspended from identical springs with a modulus k. The ends of the spring are fixed at 0 and the unstretched length of the spring is r_0. Determine the frequency of vibration for a constant angular velocity ω.

Answer: $f_n = \dfrac{1}{2\pi} \sqrt{\dfrac{k}{m} - \omega^2}$

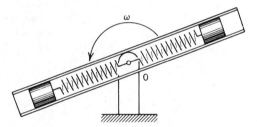

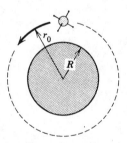

PROBLEM 2.26 A small orbiting body is displaced from its circular orbit a small distance δ. Determine the period and the equation of motion for the disturbance.

Hint: Kepler's law is that the radial line to an orbiting body sweeps equal areas of space in equal times ($r^2\dot{\theta} = $ constant).

Answer: $\tau = 2\pi \sqrt{\dfrac{r_0^3}{gR^2}}$

2.4. TORSIONAL VIBRATION

Torsional vibration refers to vibration of a rigid body about a specific reference axis. In this case, displacement is measured in terms of an angular coordinate. The restoring moment may be either due to the torsion of an elastic member or to the unbalanced moment of a force or a couple.

In Figure 2.3, one end of a long rod supports a disk that has mass, which is large relative to that of the rod, and the other end of the rod is fixed to a rigid foundation. If the rod is elastic, any angular displacement of the disk away from the equilibrium position will create a restoring moment of

$$M = \frac{JG\theta}{l} \qquad (2.12)$$

where J is the second moment of area (polar moment of inertia of area) about the axis of the shaft, G is the modulus of rigidity, l is the length of shaft, and θ is the angular coordinate measure of the displacement of the disk about the axis of the rod. This restoring moment is linearly proportional to the angle θ and the constant of proportionality is defined as the torsional spring constant.

$$K = \frac{M}{\theta} = \frac{JG}{l} \qquad (2.13)$$

The symbol for the torsional spring constant is K and the units for the torsional spring constant are torque per unit of angular displacement, Newton metres per radian (N·m/rad). Taking the moment sum about the axis of the rod, Newton's second law of motion can be stated as

$$\sum \mathbf{M}_0 = I_0 \ddot{\theta}$$
$$-K\theta = I_0 \ddot{\theta}$$

The restoring moment is $K\theta$. The equation of motion is

$$\ddot{\theta} + \frac{K}{I_0}\theta = 0 \qquad (2.14)$$

Fig. 2.3

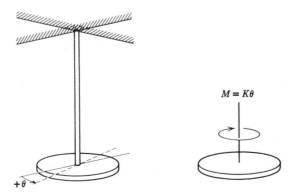

$M = K\theta$

$+\theta$

which is similar to equation 2.2, where θ is in place of the coordinate x, and K/I_0 replaces k/m. The natural frequency is

$$f_n = \frac{1}{2\pi} \sqrt{\frac{K}{I_0}} \qquad (2.15)$$

This system of torsional spring and mass is referred to as a *torsional pendulum*. The most significant application of a torsional pendulum is in a mechanical clock. In a clock, a ratchet and pawl translate the regular oscillation of a small torsional pendulum into the movement of hands across the face of the clock. The principle has not changed since Huygens invented the balance wheel three centuries ago.

2.5. SIMPLE PENDULUM

Another example of torsional vibration is the classic simple pendulum. A small mass or bob m is suspended vertically on a light wire from a hinge at 0, as in Figure 2.4. When the bob is displaced from the vertical it will oscillate about the vertical with a regular periodic motion. All this is a very familiar observation.

If we restrict motion to a single plane, the generalized coordinate which describes motion is the angular displacement from the vertical, θ, measured in that plane. The wire length is a constraint that restricts the bob of the pendulum to move in a circular path about the hinge. Recognition of this constraint makes θ a generalized coordinate.

The free body diagram shows the active forces on the bob when displaced slightly from the equilibrium position. The tensile force in the

Fig. 2.4

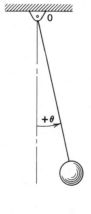

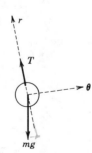

supporting wire and the weight of the bob are the only active forces. Taking a force summation in the θ direction,

$$\sum F_\theta = -mg \sin \theta = ma_\theta = ml\ddot{\theta}$$

The acceleration a_θ is positive in the direction of positive θ.

For small angles of oscillation, the sine can be replaced by the angle, within 1% accuracy for up to 5.5° of motion. Substituting θ for $\sin \theta$,

$$-mg\theta = ml\ddot{\theta}$$

or

$$\ddot{\theta} + \frac{g}{l}\theta = 0 \tag{2.16}$$

This is a linear second-order differential equation, again analogous to equation 2.2, with θ in place of x and g/l in place of k/m. The frequency of small oscillations is

$$f_n = \frac{1}{2\pi}\sqrt{\frac{g}{l}} \tag{2.17}$$

It is dependent only on the length of wire supporting the pendulum, and it is independent of the mass of the bob.

The simple pendulum has also been used as a means of keeping time. The grandfather's clock is a well-known example of a pendulum clock of great accuracy and dependability.

2.6. COMPOUND PENDULUM

A rigid body will oscillate as a pendulum, if it is suspended from some point other than its mass center. As an example, let us consider the rigid body of Figure 2.5, suspended from point 0 and oscillating in the x–y plane. The mass center is at G.

Take the moment sum about an axis through point 0 and normal to the x–y plane,

$$\sum M_0 = I_0\ddot{\theta} = -mgr \sin \theta$$

For small angles, replacing $\sin \theta$ by θ, the restoring moment is a function of the angular displacement, and

$$\ddot{\theta} + \frac{mgr}{I_0}\theta = 0 \tag{2.18}$$

Fig. 2.5

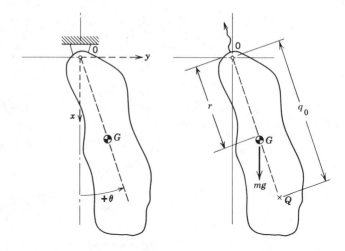

This is also similar to equation 2.2. The natural frequency is

$$f_n = \frac{1}{2\pi}\sqrt{\frac{mgr}{I_0}}$$

(2.19)

I_0 is the mass moment of inertia of the rigid body as measured about the z-axis passing through 0. It is a measure of the resistance of the body to angular acceleration. Mass moment of inertia is an awkward term and, except for certain elementary shapes, it is never calculated directly from the geometry of the rigid body. It can be measured quite accurately from the observed effect of moment of inertia on the dynamic response of a rigid body to angular acceleration, such as observing the frequency of oscillation of a rigid body in free vibration.

At this point, it is convenient to refer to two other arbitrary measures of the mass moment of a rigid body. If all of the mass of a rigid body were concentrated in a thin ring, which had the same inertial properties of resistance to angular acceleration as the rigid body, the radius of that ring would be the *radius of gyration*, k_0. If all of the mass were concentrated at a point, the distance to that point from the fixed axis would be q_0. The point itself is called the *center of percussion*. The location of the center of percussion and the radius of gyration are related to the location of the mass center and the center of rotation.

$$q_0 = \frac{k_0^2}{r}$$

(2.20)

Here, r is again the radial distance of the mass center from the fixed axis.

If the mass moment of inertia is replaced by its equivalent $I_0 = mq_0r$,

Fig. 2.6

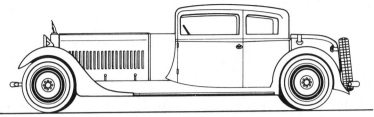

Pre–1934 Automobile design

equation 2.19 becomes

$$f_n = \frac{1}{2\pi}\sqrt{\frac{g}{q_0}}$$

This is analogous to the natural frequency of a simple pendulum, which has the same length as the distance from the axis of oscillation to the center of percussion. A rigid body suspended in this manner is called a *compound pendulum*. The term has a simple meaning, for the center of percussion and the center of oscillation can be interchanged, and the same natural frequency will result in each case.

A good example is an automobile. Considering motion in a profile plane, the automobile is a compound pendulum. If the front wheels strike a bump, a reaction will be felt by passengers unless the center of percussion is located at or near the rear axle. The reverse is true as the rear wheels strike a bump. A reaction will be felt unless the center of percussion is at or near the front axle. As a consequence, good vehicle design places the center of percussion about one axle with the center of oscillation about the other. Most automobile manufacturers changed their designs in 1934, the year this elementary principle of dynamics was first used. Prior to 1934, good automobile esthetics centered the radiator cap over the front wheels. The actual dynamics of a moving automobile are quite complicated, since there are many degrees of freedom, but this simple example explains how the principle of the compound pendulum can be used. Figures 2.6 and 2.7 show the differences in automobile design.

Fig. 2.7

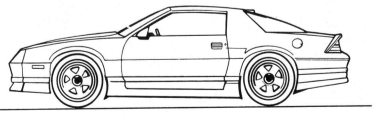

Recent Automobile

2.7. FILAR PENDULUM

A rigid body can also be suspended as a filar pendulum by supporting it so that it will oscillate in the horizontal plane. There are bifilar, trifilar, and quadrifilar pendula, depending on the number of supporting wires. Mass moments of inertia can be found from filar suspensions. Measuring the frequency of a filar pendulum happens to be a very convenient way of finding mass moment of inertia and it is used extensively in practice. Not too many students realize how difficult it is to find the mass moment of inertia of a jet airplane or spacecraft by other means, or how important this quantity is to the guidance and control of the airplane or spacecraft.

Example Problem 2.27 shows how this can be done.

EXAMPLE PROBLEM 2.27. THE BIFILAR PENDULUM

Determine the moment of inertia I_z for an aircraft propeller from observations of the natural frequency of free oscillations for the propeller suspended from two light wires attached to the tips of the propeller blades. The length of each suspension wire is h and the diameter of the propeller is D. The weight of the propeller is mg and it is known.

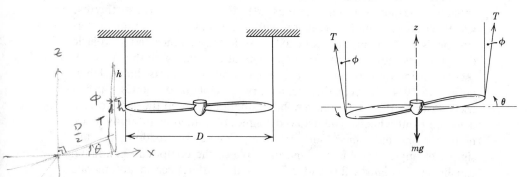

Solution:

As the propeller rotates in the horizontal plane, it moves up and down a small but measurable amount, similar to a simple pendulum. The restoring moments are, for a positive angular displacement,

$$-2T \sin \phi \frac{D}{2}$$

The equation of motion about the geometric z-axis

$$\sum \mathbf{M}_z = I_z \ddot{\boldsymbol{\theta}}$$

$$-TD \sin \phi = I_z \ddot{\theta}$$

I_z is the mass moment of inertia about the z-axis.

$$-TD\left(\frac{D}{2h}\sin\theta\right)=I_z\ddot\theta$$

From geometry, $h \sin \phi = (D/2)\sin \theta$, hence

$$-\frac{TD^2}{2h}\sin \theta = I_z\ddot\theta$$

For small angles of oscillation, $\sin \theta$ can be replaced by the angle θ. Substitution yields the equation

$$\ddot\theta + \frac{TD^2}{2I_z h}\theta = 0$$

$$T \approx \frac{mg}{2}.$$

For small angles of oscillation, T is approximately equal to $mg/2$, and

$$\ddot\theta + \frac{mgD^2}{4I_z h}\theta = 0$$

Using our knowledge of this equation of motion, the natural frequency is

$$f_n = \frac{1}{2\pi}\sqrt{\frac{mgD^2}{4I_z h}}$$

Solving for the mass moment of inertia,

$$I_z = \frac{mgD^2}{16\pi^2 f_n^2 h}$$

Bifilar suspension is used to determine the mass moment of inertia of many objects that have axial symmetry.

EXAMPLE PROBLEM 2.28

A light aircraft is powered by a 9-cylinder radial aircraft engine, which uses a four-stroke cycle. Power is transmitted through a short shaft to a

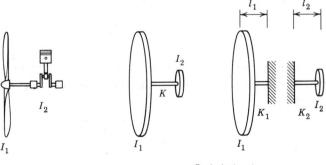

Actual system   Equivalent system

two-bladed aluminum propeller. Determine the natural frequency of the system, if the axial moment of inertia of the propeller is 17.62 kg·m², and the effective moment of inertia of the moving parts of the radial engine is 0.544 kg·m². The torsional spring constant of the propeller shaft is 0.45×10^6 N·m/rad. What would be the expected result at an engine speed of 2000 rpm?

Solution:
For the one principal mode of vibration, the propeller and engine will move out of phase with each other, but with the same frequency. The propeller will move clockwise while the engine rotates counterclockwise and vice versa. This motion is superimposed on the constant rotation of the propeller and engine. A standing node will appear at some point between the two masses and the system can be considered to be two simple torsional pendulums placed end to end. For these two systems, the frequencies f_1 and f_2 are the same as the natural frequency f_n.

$$f_n = f_1 = f_2 = \frac{1}{2\pi}\sqrt{\frac{K_1}{I_1}} = \frac{1}{2\pi}\sqrt{\frac{K_2}{I_2}}$$

This leads to the statement that the ratios K_1/I_1 and K_2/I_2 are identical.

From equation 2.13, the torsional spring constant is inversely proportional to the length of the shaft. I, J, and G are the same,

$$K = \frac{JG}{l}$$

and

$$Kl = K_1 l_1 = K_2 l_2$$

combining

$$\frac{I_1}{I_2} = \frac{K_1}{K_2} = \frac{l_2}{l_1}$$

using the total length of shaft $l = l_1 + l_2$

$$l = l_1\left(1 + \frac{I_1}{I_2}\right)$$

Starting with the expression for f_1, multiplying both the numerator and denominator by l, substituting the above equation for l and substituting $Kl = K_1 l_1$

$$\frac{K_1}{I_1} = \frac{K_1 l}{I_1 l} = \frac{K_1 l_1 \left(1 + \frac{I_1}{I_2}\right)}{I_1 l}$$

$$= \frac{Kl\left(1 + \frac{I_1}{I_2}\right)}{I_1 l} = K\left(\frac{1}{I_1} + \frac{1}{I_2}\right)$$

$$f_n = \frac{1}{2\pi}\sqrt{K\left(\frac{1}{I_1} + \frac{1}{I_2}\right)}$$

using the given values of I_1, I_2, and K,

$$f = \frac{1}{2\pi} \sqrt{0.45 \times 10^6 \left(\frac{1}{17.62} + \frac{1}{0.544} \right)} = 147 \text{ cps}$$

$$= 8820 \text{ cpm}$$

For a 4-stroke cycle radial aircraft engine running at 1960 rpm, there will be

$$\frac{9}{2}(1960) = 8820 \text{ power strokes/min}$$

This corresponds to the natural frequency of the system and resonance will result if the system is run at 1960 rpm. At 2000 rpm, violent shaking of the aircraft would still result.

EXAMPLE PROBLEM 2.29. THE COMPOUND PENDULUM

Find the natural frequency of a long slender rod that is suspended vertically from one end and oscillates in one plane as a pendulum. Locate the center of percussion and determine the radius of gyration.

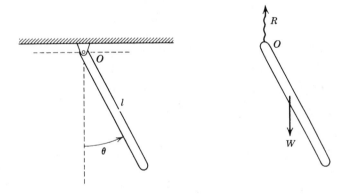

Solution:
The slender rod rotates about point O due to the unbalanced moment of the weight force, W, about point O. The equation of motion for rotation is

$$\sum \mathbf{M}_0 = I_0 \ddot{\theta}$$

$$-mg\frac{l}{2}\sin \theta = \frac{1}{3}ml^2\ddot{\theta} \qquad I_0 = \frac{1}{3}ml^2$$

The moment of inertia for a long slender rod is $\frac{1}{3}ml^2$ about one end.

Replacing sin θ with θ, which is valid for small oscillations, the equation of motion is

$$\ddot{\theta} + \frac{3}{2}\frac{g}{l}\theta = 0$$

This is again similar to equation 2.2, with θ replacing x and $\frac{3g}{2l}$ replacing k/m.

The natural circular frequency is

$$\omega_n = \sqrt{\frac{3}{2}\frac{g}{l}}$$

and the natural frequency, measured in cycles per second, is

$$f_n = \frac{1}{2\pi}\sqrt{\frac{3}{2}\frac{g}{l}}$$

The radius of gyration is

$$I_0 = mk_0^2 = \frac{1}{3}ml^2$$

$$k_0 = \frac{l}{\sqrt{3}}$$

and the distance to the center of percussion is

$$q_0 = \frac{k_0^2}{r} = \frac{2}{3}l$$

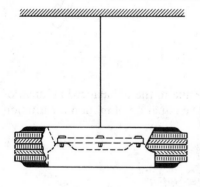

PROBLEM 2.30 A device designed to determine the moment of inertia of a wheel-tire assembly consists of 2-mm steel suspension wire, 2 m long, and a mounting plate, to which is attached the wheel-tire assembly. The suspension wire is fixed at its upper end and hung vertically. When the system oscillates as a torsional pendulum, the period of oscillation without the wheel-tire assembly is 4 s. With the wheel-tire mounted to the mounting plate, the period of oscillation is 25 s. Determine the moment of inertia of wheel-tire assembly.

Answer: 1.01 kg·m²

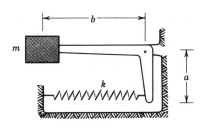

PROBLEM 2.31 Determine the natural frequency for the horizontal pendulum shown. Neglect the mass of the arm.

Answer: $f_n = \dfrac{1}{2\pi} \dfrac{a}{b} \sqrt{\dfrac{k}{m}}$

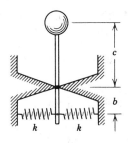

PROBLEM 2.32 A mass m is fixed to one end of a weightless rod which is pivoted a distance c from the weight. What is the natural frequency of vibration for small amplitudes of motion? This is called a vertical pendulum.

Answer: $f_n = \dfrac{1}{2\pi} \sqrt{\dfrac{2kb^2}{mc^2} - \dfrac{g}{c}}$

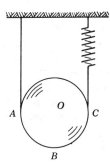

PROBLEM 2.33 The uniform wheel of mass m is supported in the vertical plane by the light flexible band ABC and the spring, which has a stiffness k. The wheel has a moment of inertia I_0 about the geometric center O, and rolls without slipping on the band ABC. Determine the natural frequency of the system.

Answer: $f_n = \dfrac{1}{2\pi} \sqrt{\dfrac{4kr^2}{I_0 + mr^2}}$

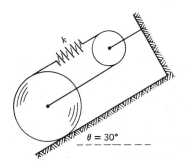

PROBLEM 2.34 A solid circular cylinder is held in place by a cable and a pulley. The flexibility of the cable can be represented by a spring with a modulus k. The cable is wrapped around the cylinder. Friction is sufficient to prevent slipping. Determine the natural frequency of small oscillations. The mass of the pulley may be neglected.

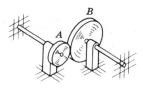

PROBLEM 2.35 Gears A and B mesh with a gear ratio n. They are fixed to circular shafts of equal length and equal diameter. The shafts are built-in at either end. Determine the natural frequency of the system.

Answer: $f_n = \dfrac{1}{2\pi} \sqrt{\dfrac{K(1+n^2)}{I_B + I_A n^2}}$

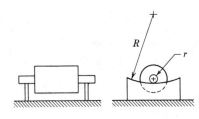

PROBLEM 2.36 A rotor with an axial mass moment of inertia of I is supported by its journals on two curved rails. Its mass is m. The radius of curvature of the rails is R. The journals have a radius of r. For small oscillations, determine the natural frequency of the rotor rolling on the rails without slipping.

Answer: $\omega_n = \sqrt{\dfrac{mgr^2}{(I+mr^2)(R-r)}}$

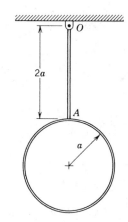

PROBLEM 2.37 A thin, circular ring of mass m is rigidly attached to a slender, vertical rod at A. The rod and ring are pivoted at point 0 and oscillate as a pendulum in a vertical plane. Determine the natural frequency for small amplitude oscillation about equilibrium, if $a = 150$ mm. The slender rod has negligible mass.

Answer: 0.705 Hz

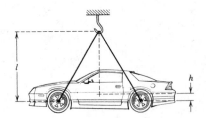

PROBLEM 2.38 A new model automobile is suspended as a pendulum, using cables attached to the front and rear axles. With $l = 4.6$ m the period of oscillation is 4.3 s. For $l = 2.6$ m, the period decreases to 3.3 s. Determine the distance h from the plane containing the geometric center of the axles to the center of gravity.

Answer: 145 mm

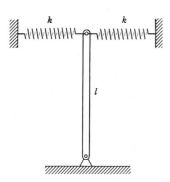

PROBLEM 2.39 A slender, uniform rod of mass m is pivoted at the bottom end and is held in equilibrium by two springs. What is the natural frequency of vibration for small amplitudes?

Answer: $f_n = \dfrac{1}{2\pi} \sqrt{\dfrac{6k}{m} - \dfrac{3g}{2l}}$

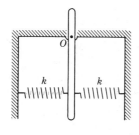

PROBLEM 2.40 A uniform slender bar of mass m and length l hangs vertically at a quarter point of the length as shown. At the bottom quarter point, the bar is attached to two springs. Find the natural frequency (in Hz) of the system, assuming the amplitudes of the motion to be small, if $k = 400$ N/m, $m = 10$ kg, and $l = 800$ mm.

Answer: $f_n = 2$ Hz

PROBLEM 2.41 The front end of a 1000-kg automobile must be raised 100 mm before the front wheels are off the ground. If one front wheel assembly including axle, brake, wheel, and tire as a mass of 30 kg, determine the natural frequency of the assembly if the torsion bar suspension shown below is used. Consider that the weight of the car is evenly distributed on all four wheels.

Answer: $f_n = 4.55$ Hz

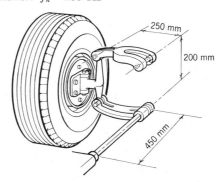

250 mm

200 mm

450 mm

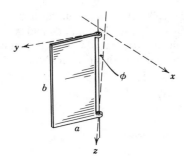

PROBLEM 2.42 Determine the natural frequency of free swing for a rectangular door suspended from a hinge axis that is inclined at a small angle ϕ to the vertical.

Answer: $f_n = \dfrac{1}{2\pi}\sqrt{\dfrac{3g}{2a}\sin\phi}$

PROBLEM 2.43 Repeat Problem 2.29, using the center of percussion, which is $\dfrac{1}{3}l$ from the end of the bar as the center of oscillation, and show that the natural frequency is unchanged. This is the meaning of a compound pendulum.

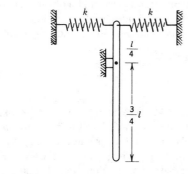

PROBLEM 2.44 A uniform slender bar of mass m and length l hangs *vertically* at a quarter point of the length as shown. The top of the bar is attached to two springs. Find the natural frequency (in Hz) of the system, assuming the amplitudes of the motion to be small, if $k = 4000$ N/m, m $= 10$ kg, and $l = 800$ mm

Answer: $f_n = 3.036$ Hz

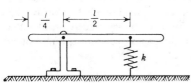

PROBLEM 2.45 The uniform slender bar of mass m is supported in the horizontal position by the spring. Determine the frequency of small oscillations when the bar is set in motion.

Answer: $f_n = \dfrac{1}{2\pi}\sqrt{\dfrac{12\,k}{7\,m}}$

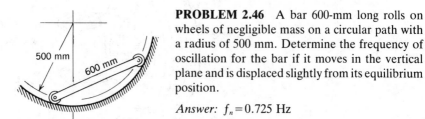

PROBLEM 2.46 A bar 600-mm long rolls on wheels of negligible mass on a circular path with a radius of 500 mm. Determine the frequency of oscillation for the bar if it moves in the vertical plane and is displaced slightly from its equilibrium position.

Answer: $f_n = 0.725$ Hz

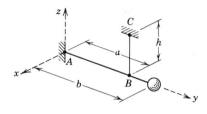

PROBLEM 2.47 A massless arm supports a mass m on the free end, and it is suspended in equilibrium by a vertical string of length h. The equilibrium orientation of the arm is in the horizontal plane and it is pinned to the ground at point A and to the string at point B. What is the natural frequency of small oscillation of the arm about the equilibrium assuming the string is inextensible and always remains in tension?

Answer: $f_n = \dfrac{1}{2\pi}\sqrt{\dfrac{ga}{bh}}$

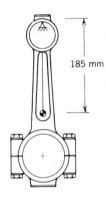

185 mm

PROBLEM 2.48 A connecting rod has a mass of 3.10 kg. It oscillates 59 times in 1 min when suspended on a knife edge about the upper inner surface of the wrist pin bearing. Determine the moment of inertia about the centroid, which is located 185 mm from this surface.

Answer: $I_G = 0.0412$ kg·m²

PROBLEM 2.49 Determine the natural frequency of a solid hemisphere oscillating in planar motion on a horizontal surface, if it rolls without slipping.

Answer: $f_n = \dfrac{1}{2\pi}\sqrt{\dfrac{15g}{26r}}$

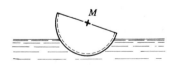

M

PROBLEM 2.50 Determine the natural frequency for small planar oscillations of a hemispherical shell floating half submerged. Assume oscillation about the center of mass of the shell. The metacenter is at M.

Answer: $f_n = \dfrac{1}{2\pi}\sqrt{\dfrac{6g}{5r}}$

PROBLEM 2.51 A hemispherical shell is supported in trunnions on a diameter. Determine the natural frequency for small oscillations of the shell. Note the difference between this problem and Problem 2.50

Answer: $f_n = \dfrac{1}{2\pi}\sqrt{\dfrac{3g}{4r}}$

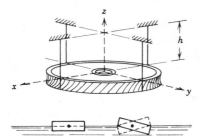

PROBLEM 2.52 A helical gear has a mass of 3.61 kg. It oscillates 89 times in 1 min when suspended as a quadrifilar pendulum. Determine the polar moment of inertia of the gear about its center; $h = 200$ mm and the diameter of the gear is 142.9 mm.

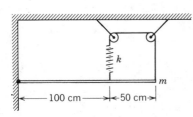

PROBLEM 2.53 A 20- by 240-mm wooden block floats half submerged in water. Determine the frequency of small oscillations of the block rolling from side to side. In this motion, the center of mass remains in the plane of the water surface.

Answer: $f_n = 4.94$ Hz

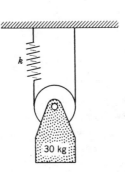

PROBLEM 2.54 Determine the natural frequency of free oscillations for the system. The pulley has a mass of 20 kg and has a radius of gyration of 360 mm. The modulus of the spring is 1.6 N/mm; the diameter of the pulley is 1 m.

Answer: $f_n = 1.639$ Hz

PROBLEM 2.55 Determine the natural frequency for small oscillations of the system shown. The mass of the pulleys and friction are negligible.

Answer: $f_n = \dfrac{5}{6\pi}\sqrt{\dfrac{3k}{m}}$

CHAPTER THREE

ENERGY
METHODS

3.1. ENERGY METHODS

It is often very simple and direct to use energy methods to solve vibration problems. Energy methods involve an *energy balance* using scalars rather than a *force balance* with vectors.

To some, it is easier to conceive of an energy balance than it is to draw free body diagrams and establish vector forces and a force balance. Simply stated, in an energy balance, energy must be conserved. As a principle, this is known as the *Conservation of Energy*. It is a physical law and no violation of it has ever been observed. For the total energy, E,

$$E = T + V \tag{3.1}$$

In this equation, T denotes kinetic energy due to the velocity of the mass of the system, and V is potential energy, either due to the configuration of this mass as measured from some arbitrary datum or due to the stress of the elastic members. The conservation of energy can be expressed in an incremental form, where ΔU is the energy that is added to the total energy as heat or removed as work or friction,

$$\Delta U = \Delta T + \Delta V \tag{3.2}$$

This statement says that the addition or dissipation of energy must also appear as a change in the kinetic or potential energy.

The conservation of energy holds for all systems, whether energy is dissipated or not, but if no energy is added or dissipated, $\Delta U = 0$, and if there is no change in the thermal energy of the system with time,

$$\frac{d}{dt}(T+V) = 0 \qquad (3.3)$$

If T and V are functions of a single generalized coordinate, this expression leads directly to the equation of motion. Equation 3.3 is used extensively for this purpose and using it is referred to as the *energy method* for finding equations of motion and natural frequencies.

For a conservative system, the energy of any specific particle or mass may be either potential or kinetic, but the total energy of the system must remain constant. Cyclic motion is merely one manifestation of the conversion of energy from potential to kinetic energy and back again. Even if some energy is dissipated, energy methods can be used to find the approximate equations of motion. The key word is approximate, for the advantages of using an energy balance, with its inherent simplicity, may outweigh the inaccuracy of ignoring energy dissipation.

3.2. SINGLE DEGREE OF FREEDOM

As an example, let us again examine the simple elastic system of Chapter 2, Figure 2.1, which is repeated in Figure 3.1. At any time, the kinetic energy of the system is expressed as the kinetic energy of the mass, m, with the kinetic energy of the spring being ignored.

$$T = \tfrac{1}{2}m\dot{x}^2 \qquad (3.4)$$

Potential energy is expressed as both elastic potential energy and as potential energy of position. To have meaning, both must be expressed as

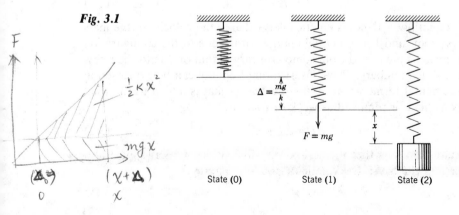

Fig. 3.1

State (0)   State (1)   State (2)

a change from some convenient but arbitrary datum position. For Figure 3.1, if the datum had been taken as the equilibrium position, where the elastic force and weight are balanced, the elastic energy in the spring would be $V_1 = \frac{1}{2}k\Delta^2$. Displacement of the mass from the equilibrium position in the positive x direction would increase the elastic energy stored in the spring and decrease the potential energy due to position. The result would be

$$= \tfrac{1}{2}k\Delta^2 + \tfrac{1}{2}kx^2 + kx\Delta - mgx$$

$$V_2 = \tfrac{1}{2}k(x+\Delta)^2 - mgx = \tfrac{1}{2}k\Delta^2 + \tfrac{1}{2}kx^2 \qquad k\Delta = mg \quad . \tag{3.5}$$

The term $\frac{1}{2}k\Delta^2$ is a constant term for a given weight.

Adding the kinetic and potential energy and differentiating the total energy with respect to time

$$\frac{d}{dt}(T+V) = m\dot{x}\left(\frac{d\dot{x}}{dt}\right) + kx\left(\frac{dx}{dt}\right) = 0$$

$$T = \tfrac{1}{2}m\dot{x}^2$$
$$V = \tfrac{1}{2}kx^2 \quad .$$
$$U = mgx$$

or

$$\frac{d}{dt}(T+V) = dv$$

$$m\ddot{x}(\dot{x}) + kx(\dot{x}) = 0 \qquad m\ddot{x} + kx = mg \quad .$$

Cancelling $\dot{x}$, we have

$$m\ddot{x} + kx = 0 \tag{3.6}$$

which is equation 2.2. Note that if we had taken the unstretched position of the spring as the position where $x = 0$, we would have had a constant term mg/k on the right side of the equation of motion.

$$\frac{mg}{k} \quad .$$

3.3. RAYLEIGH'S ENERGY METHOD

An alternate form of the energy method was devised by Lord Rayleigh for the approximate calculation of the fundamental frequency of a vibrating system, which does not derive and solve the differential equations of motion.

Assuming that the motion is simple harmonic, a basic premise required by this method of solution,

$$x = X \sin \omega_n t$$

Differentiating with respect to time, the velocity is

$$\dot{x} = X\omega_n \cos \omega_n t$$

If the sum total of the kinetic and potential energy is constant, then the average potential energy must be equal to the average (av) kinetic energy over one full cycle for which the period $\tau = 2\pi/\omega_n$

$$T_{av} = \frac{1}{\tau}\int_0^\tau \frac{1}{2}m\dot{x}^2 \, dt = \frac{1}{2}m\frac{X^2\omega_n^2}{\tau}\int_0^\tau \cos^2 \omega_n t \, dt = \frac{1}{4}mX^2\omega_n^2 \qquad (3.7a)$$

$$V_{av} = \frac{1}{\tau}\int_0^\tau \frac{1}{2}kx^2 \, dt = \frac{1}{2}\frac{kX^2}{\tau}\int_0^\tau \sin^2 \omega_n t \, dt = \frac{1}{4}kX^2 \qquad (3.7b)$$

equating

$$T_{av} = V_{av}$$

$$\frac{1}{4}mX^2\omega_n^2 = \frac{1}{4}kX^2 \qquad (3.8)$$

$$\omega_n^2 = \frac{k}{m}$$

This is the square of the natural circular frequency of the system. The same result can be achieved by equating the maximum kinetic energy to the maximum potential energy, bypassing the necessity of integrating over a complete cycle. At the extreme position, the system comes to rest and the energy of the system is entirely potential. Passing through the equilibrium position, the energy is entirely kinetic, provided that we assume the potential energy of the equilibrium position to be zero. Conservation of energy requires that total energy of the system be unchanged or that the change in kinetic energy be equal to the change in potential energy. This means that the change in potential energy at its maximum must be equal to the change in kinetic energy at its maximum.

$$\Delta T_{max} = \Delta V_{max}$$

$$\frac{1}{2}mX^2\omega_n^2 = \frac{1}{2}kX^2 \qquad (3.9)$$

$$\omega_n^2 = \frac{k}{m}$$

Note that the amplitude, X, is eliminated from the expression for ω_n^2 for both equations 3.8 and 3.9. This is not a trivial point, since the independence of natural frequency from amplitude of motion is the basis of *Rayleigh's Principle*.

Finding the equation of motion using energy methods is extremely useful where the simple system has one degree of freedom but geometric or kinematic complexity, such as numerous elastic members. Acceleration can be related directly to the active forces, through the conservation of energy, without considering internal forces that do no work and have no effect on energy changes within the system.

Rayleigh's energy method consists of three important parts. The first is the assumption of a *mode shape*, which is the relation between generalized coordinates. In the case of a single generalized coordinate, it is merely the expression of kinetic and potential energy in terms of the maximum displacement. The matter of mode shape will be discussed at length in the next and in succeeding sections. The second part is the assumption that the motion will be simple harmonic. In an undamped linear system, this is a valid assumption. If damping is light, the distortion is negligible. The third part is equating kinetic energy to potential energy, ignoring heat, work, and friction. With these qualifications, it is remarkable that the natural frequency of harmonic motion can be found with any accuracy, but Rayleigh's energy method is a very powerful method, it does give a very good approximation of the natural frequencies, and it is particularly useful for a system with a single degree of freedom.

In single degree of freedom systems, only one coordinate is involved, and energy can be written as functions of this single coordinate. If multiple degrees of freedom are involved, more than one coordinate will be required, but energy methods provide simple schemes for organizing and ordering a multiplicity of terms. Matrices and matrix algebra can be used effectively when energy methods are extended to problems with many degrees of freedom. The simple energy methods used here are introductory to Lagrangian methods of advanced mechanics.

3.4. THE SELECTION OF THE DATUM

The question of the selection of the most convenient datum position is important. In Section 3.2, for equation 3.5, equilibrium was selected as the datum position. In Figure 3.1, if the datum position had been chosen as the position of the unstretched spring,

$$V_0 = 0$$

Now, in stretching the spring to a new position, a distance Δ such that $\Delta = mg/k$, the potential energy within the spring would be

$$V_1 = \tfrac{1}{2}k\Delta^2$$

Stretching a distance x

$$V_2 = \tfrac{1}{2}k(\Delta + x)^2 - mgx$$

and

$$\Delta V = V_2 - V_1 = \tfrac{1}{2}kx^2$$

Thus, the energy change from state 2 to state 1 is simply $\tfrac{1}{2}kx^2$. In choosing the equilibrium state as the datum position, the potential energy change of the weight forces conveniently cancel. This is true, unless the weight forces change sign with displacement from the equilibrium position. A change in sign refers to the sign of the vector of the generalized coordinate. An example of this exception would be the simple pendulum. Either positive or negative displacement from the equilibrium position results in a positive change in potential energy.

EXAMPLE PROBLEM 3.1. THE MANOMETER

Solve Problem 2.2, using Rayleigh's energy method.

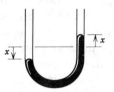

Solution:

The most convenient datum is again the equilibrium position, where the generalized coordinate $x=0$. At any other position, the change in potential energy from the equilibrium position is

$$\Delta V = \frac{A\gamma x^2}{2} - \left(-\frac{A\gamma x^2}{2} \right) = A\gamma x^2$$

One leg of the manometer is raised and the other is depressed. The potential energy of the raised leg is $A\gamma x(x/2)$, since the center of gravity of the segment is displaced $x/2$. The potential energy of the depressed leg is $-A\gamma x(x/2)$, making the total potential energy change from the equilibrium position $A\gamma x^2$.

The kinetic energy change is

$$\Delta T = \frac{1}{2}\frac{A\gamma l}{g}(\dot{x}^2 - 0)$$

Assuming harmonic motion for the generalized coordinate x, where X is the maximum displacement

$$x = X \sin \omega_n t$$

$$\dot{x} = X\omega_n \cos \omega_n t$$

Setting the maximum kinetic energy change to be equal to the maximum potential energy change,

$$\Delta T_{max} = \Delta V_{max}$$

$$\frac{1}{2}\frac{A\gamma l\omega_n^2}{g}X^2 = A\gamma X^2$$

from which,

$$\omega_n^2 = \frac{2g}{l}$$

and

$$f_n = \frac{1}{2\pi}\sqrt{\frac{2g}{l}}$$

This is exactly the same answer as obtained in Problem 2.2.

EXAMPLE PROBLEM 3.2. THE COMPOUND PENDULUM

Repeat Problem 2.29 by solving it using Rayleigh's energy method.

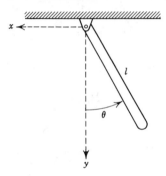

Solution:
The datum position of the pendulum can be chosen as any value of the generalized coordinate θ. The angle $\theta = 0$ is particularly convenient, because it is the equilibrium position of the pendulum, the lowest position of the center of gravity, and all other positions represent an increase in potential energy from the datum position. The equilibrium position is also the position of maximum kinetic energy.

At any position θ, the change in potential energy from the equilibrium is

$$\Delta V = mg\frac{l}{2} - mg\frac{l}{2}\cos\theta$$

The change in kinetic energy of the pendulum is

$$\Delta T = \tfrac{1}{2}(\tfrac{1}{3}ml^2)[(\dot{\theta}^2) - 0]$$

Assuming harmonic motion of the generalized coordinate θ, where θ

is the maximum displacement,

$$\theta = \Theta \sin \omega_n t$$

$$\dot{\theta} = \Theta \omega_n \cos \omega_n t$$

Setting the maximum kinetic energy change to be equal to the maximum potential energy change, which is the most convenient statement of the law of the conservation of energy,

$$\Delta T_{max} = \Delta V_{max}$$

$$\frac{1}{2}\left[\frac{1}{3}ml^{-2}(\Theta\omega_n)^2\right] = mg\frac{l}{2}(1 - \cos \Theta)$$

This is a nonlinear equation. It can be linearized by using two terms of the power series for the cosine function.

$$\cos \Theta = 1 - \frac{\Theta^2}{2!} + \frac{\Theta^4}{4!} - \cdots$$

$$1 - \cos \Theta = \frac{\Theta^2}{2!} - \frac{\Theta^4}{4!} + \cdots$$

All terms after the $\theta^2/2$ term are ignored. Substituting again,

$$\frac{1}{2}\left[\frac{1}{3}ml^2(\Theta\omega_n)^2\right] = mg\frac{l}{2}\frac{\Theta^2}{2}$$

from which,

$$\omega_n^2 = \frac{3g}{2l}$$

and,

$$f_n = \frac{1}{2\pi}\sqrt{\frac{3g}{2l}}$$

which is precisely the same solution as Problem 2.29.

Aside, it is interesting to note that the error in substituting $\sin \Theta$ for Θ is less than 1%, if $\Theta < 5.5°$. The substitution of $\Theta^2/2$ for $1 - \cos \Theta$ is valid with an error less than 1% for $\Theta < 22°$. The natural conclusion is that the limitation to small angles of oscillation is not so restrictive after all.

PROBLEM 3.3   Repeat Problem 2.31 using Rayleigh's energy method.

PROBLEM 3.4   What is the natural frequency of the ballistic test apparatus of Problem 1.56.

Answer: 14.23 Hz

Do the following problems, using Rayleigh's energy method:

PROBLEM 3.5 Do Problem 2.32.

PROBLEM 3.6 Do Problem 2.39.

PROBLEM 3.7 Do Problem 2.43.

PROBLEM 3.8 Do Problem 2.44.

PROBLEM 3.9 Do Problem 2.45.

PROBLEM 3.10 Do Problem 2.47.

PROBLEM 3.11 Do Problem 2.48.

PROBLEM 3.12 Do Problem 2.49.

PROBLEM 3.13 Do Problem 2.50.

PROBLEM 3.14 Do Problem 2.51.

PROBLEM 3.15 Determine the natural frequency of a hemicylindrical shell that oscillates in planar motion on a horizontal surface, if the shell rolls without slipping.

Answer: $f_n = \dfrac{1}{2\pi}\sqrt{\dfrac{0.876g}{r}}$

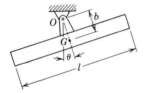

PROBLEM 3.16 The bar of mass m, with its mass center at G, is pivoted about a horizontal axis through O. Determine the natural frequency of oscillation of the bar.

PROBLEM 3.17 The front end suspension of an automobile is shown in the accompanying sketch. If the front coil springs each has a modulus of 50 kN/m and the body has a mass of 1420 kg, determine the natural frequency of vertical oscillation of the front end.

Answer: $f_n = 1.133$ Hz

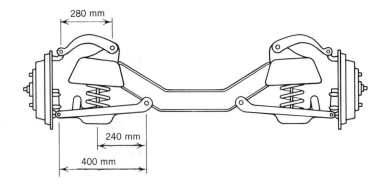

PROBLEM 3.18 The Chilton pendulum consists of a mass m, which is supported by two loose rollers in large holes. Consider the mass to be supported symmetrically, with the center of mass at G. If the radius of the rollers is r and the radius of the holes is R, with $R>r$ determine the natural frequency of the pendulum.

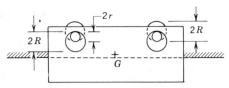

PROBLEM 3.19 The solid circular cylinder rests between two sponge rubber pads. Determine the natural frequency of horizontal oscillations, if the cylinder is displaced from its equilibrium position, rolls without slipping on the flat surface, and does not lose contact with the pads.

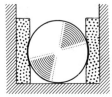

Answer: $f_n = \dfrac{1}{2\pi}\sqrt{\dfrac{4k}{3m}}$

PROBLEM 3.20 Two solid disks are connected by two solid side bars. One side bar is pinned to the disks at their geometric centers. The other is pinned at points halfway from the geometric centers. Determine the natural frequency for small oscillations of the mechanical system about the static equilibrium position. All pinned connections are frictionless. The disks roll without slipping on the horizontal surface.

Answer: $\omega_n = \sqrt{\left(\dfrac{m_1}{6m_2+\frac{5}{2}m_1}\right)\dfrac{g}{r}}$

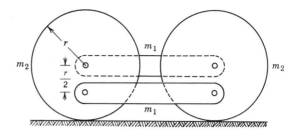

PROBLEM 3.21 An instrument used to count the vertical oscillations of a transmission line consists of a seismic pendulum and escapement mechanism. For the dimensions shown, determine the natural frequency of the instrument.

Answer: $f_n = 1.601$ Hz

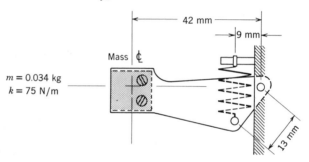

$m = 0.034$ kg
$k = 75$ N/m

PROBLEM 3.22 Determine the natural frequency of small oscillation for the vertical pendulum. Each spring has a modulus of k and is under an initial tension T when the weight is in equilibrium in the vertical position.

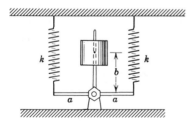

Answer: $f_n = \dfrac{1}{2\pi}\sqrt{\dfrac{2ka^2}{mb^2} - \dfrac{g}{b}}$

PROBLEM 3.23 Derive an expression for the curvature of a spherical surface in terms of the frequency of oscillation of a sphere placed on the surface and disturbed from its equilibrium position.

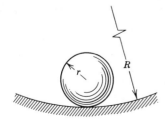

Answer: $R = r + \dfrac{5}{7}\dfrac{g}{(2\pi f_n)^2}$

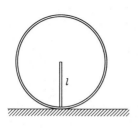

PROBLEM 3.24 A uniform bar of length l and weight mg is secured to a circular hoop of radius l. The weight of the hoop is negligible. Determine the natural frequency of small oscillations if the bar and hoop are disturbed from equilibrium. Friction is sufficient to prevent slipping.

Answer: $f_n = \dfrac{1}{2\pi}\sqrt{\dfrac{3g}{2l}}$

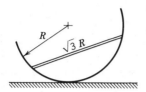

PROBLEM 3.25 A uniform bar of length $\sqrt{3}\cdot R$ and mass m is fixed to a massless ring of radius R. Determine the natural frequency of oscillation, if the bar is offset from its stable horizontal position and the bar and ring roll freely. Friction between the ring and the horizontal surface is sufficient to prevent slipping.

Answer: $f_n = \dfrac{1}{2\pi}\sqrt{\dfrac{g}{R}}$

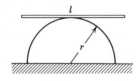

PROBLEM 3.26 The flat plank is placed centrally on a hemispherical surface and displaced from its equilibrium position. For small oscillations about its center of gravity, determine the natural frequency of motion.

Answer: $f_n = \dfrac{1}{2\pi}\sqrt{\dfrac{12gr}{l^2}}$

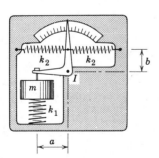

PROBLEM 3.27 An amplitude meter consists of a seismic mass suspended as shown. Determine the natural frequency of the meter in terms of the tension spring k_1, the compression spring k_2, the mass m, and the moment of inertia I.

Answer: $f_n = \dfrac{1}{2\pi}\sqrt{\dfrac{2k_2b^2 + k_1a^2}{I + ma^2}}$

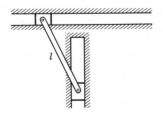

PROBLEM 3.28 Two identical masses are pin-connected to a rigid link of length l. One is free to move in a vertical slot and the other is free to move in a horizontal slot. Both slots are frictionless. Determine the natural frequency of motion for small oscillation.

Answer: $f_n = \dfrac{1}{2\pi}\sqrt{\dfrac{g}{l}}$

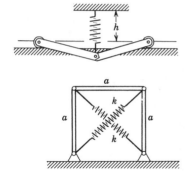

PROBLEM 3.29 The unstretched length of the spring is h and its modulus is k. Each of the identical uniform bars has a mass of m. Determine the natural frequency of the system.

PROBLEM 3.30 Determine the natural frequency of small oscillations for the four-bar frame shown, if it is displaced from the equilibrium position shown. Each of the three links is a uniform bar of mass m.

Answer: $f_n = \dfrac{1}{2\pi}\sqrt{\dfrac{3}{5}\left(\dfrac{k}{m}-\dfrac{2g}{a}\right)}$

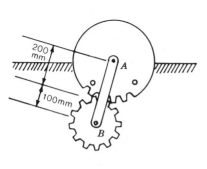

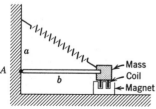

PROBLEM 3.31 The smaller lower gear has a mass of 3 kg and is pinned to the arm AB at B and rotates in the vertical plane about the larger upper stationary gear. The mass of the arm is 2 kg. The small gear has a radius of gyration of 80 mm. Determine the frequency of small oscillations of the gear and arm, if the system oscillates as a pendulum.

Answer: $f_n = 0.71$ Hz

PROBLEM 3.32 A vertical seismometer consists of a large pendulous mass at the end of a massless horizontal boom. The mass and boom are pivoted at end A and suspended by a spring with a constant k. Determine the natural frequency of the seismometer.

Answer: $f_n = \dfrac{1}{2\pi}\sqrt{\dfrac{a^2}{(a^2+b^2)}\left[\dfrac{k}{m}-\dfrac{g}{a}\right]}$

3.5. MODE SHAPE AND THE EFFECT OF THE MASS OF THE ELASTIC MEMBER

Using energy methods, the calculation for the natural frequency can be corrected to include the mass of the elastic member, if its mass is not negligible. It is only necessary to add the kinetic energy of the elastic member, which can be easily done provided that the mode of the vibration is known.

As an example, in Figure 3.2, if the equilibrium position is chosen as the datum, the potential energy change is simply,

$$V_2 - V_1 = \tfrac{1}{2}kx^2 \tag{3.10}$$

Fig. 3.2

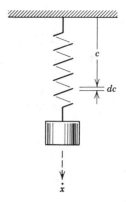

Including the effect of the mass of the spring does not change the potential energy.

The kinetic energy of the system includes the kinetic energy of the mass and the kinetic energy of the spring. The kinetic energy of the mass m is

$$T = \tfrac{1}{2} m \dot{x}^2$$

The kinetic energy of an element of the spring, dc, varies with the distance from the fixed end of the spring. The weight per unit length is μ.

$$dT_s = \frac{1}{2} \frac{\mu \, dc}{g} \dot{c}^2$$

If we assume a linear relation between the velocity $\dot{c}$, and the velocity at the end of the spring, $\dot{x}$, we can integrate to find the total kinetic energy of the spring.

$$\dot{c} = \frac{c}{l} \dot{x}$$

$$\dot{c} = \chi \dot{x}$$

The fraction $c/l = \chi$ is the *mode*, *mode shape*, or *modal fraction*. It describes the displacement at any point on the spring. It is symbolized as χ, and in this case is a linear relation between c and l. It could have been almost anything, but this is the logical shape.

At this point, some remarks about the function χ are in order. It is a function $\chi(c)$ and not of time. The time derivatives $\dot{c}$ and $\ddot{c}$ can be expressed as

$$\dot{c} = \chi \dot{x}$$

$$\ddot{c} = \chi \ddot{x}$$

which make the relation χ quite important.

Substituting for $\dot{c}$, the kinetic energy of the element dc is

$$dT_s = \frac{1}{2g}\frac{\mu c^2}{l^2}\dot{x}^2 dc$$

Integrating over the entire spring,

$$T_s = \int_0^l dT_s = \frac{1}{2}\left(\frac{1}{3}\right)\frac{\mu l}{g}\dot{x}^2$$

Since μl is the weight of the spring, including the effect of the mass of the spring is kinetically equivalent to adding $\frac{1}{3}$ of the mass of the spring to the main mass, m.

$$T = \frac{1}{2}[m + \frac{1}{3}m_s]\dot{x}^2 \tag{3.11}$$

Now let us assume that the oscillation of the system can be expressed as simple harmonic motion of the generalized coordinate, x, the displacement of the main mass at the end of the spring

$$x = X \sin \omega_n t$$

$$\dot{x} = X\omega_n \cos \omega_n t$$

Equating the maximum kinetic energy to the maximum potential energy,

$$\Delta T_{max} = \Delta V_{max}$$

$$\frac{1}{2}(m + \frac{1}{3}m_s)X^2\omega_n^2 = \frac{1}{2}kX^2$$

$$\omega_n^2 = \frac{k}{m + \frac{1}{3}m_s} \tag{3.12}$$

It is obvious that the fraction $\frac{1}{3}$ is consistent with our original assumption of a linear relation between $\dot{c}$ and $\dot{x}$. If we had selected another physical relation for the mode shape such as the parabola,

$$\dot{c} = \frac{c^2}{l^2}\dot{x}$$

this also would have satisfied our knowledge of the boundary conditions, but the result would have been the addition of $\frac{1}{5}m_s$, instead of $\frac{1}{3}m_s$.

3.6. DISTRIBUTED PARAMETERS

Rayleigh's energy method can also be used to determine the lowest or fundamental frequency of systems with distributed parameters, the best practical example being uniform beams.

As an example of such a system, consider a vibrating string stretched between supports with a tensile load, P. This is a continuous system and there are an infinite number of natural frequencies. Referring to Figure 3.3, for the fundamental mode of vibration, the lateral displacement of the string $y(x, t)$ can be expressed in terms of the mode shape $\chi(x)$, which is purely a function of x, and $Y(t)$, which defines the motion of individual elements and is purely a function of time.

Fig. 3.3

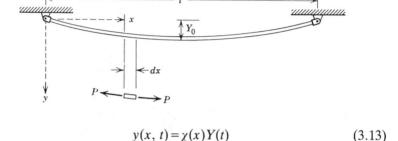

$$y(x, t) = \chi(x)Y(t) \tag{3.13}$$

For the string, the first mode shape $\chi(x)$ is the sinusoid, $\sin \pi x/l$. As a function, it satisfies all the constraints of the system, including the specified boundary conditions. The function $\chi(x)$ is not a function of time.

$$\chi(x) = \sin\frac{\pi x}{l}$$

The displacement, velocity, and acceleration of any point on the string vary according to this mode shape. Y_0 is the displacement at midspan. $Y(t)$ defines the time history of each point. Thus, if we additionally assume that displacement at any point is simple harmonic,

$$y = \chi(x)Y_0 \sin \omega_n t$$
$$\dot{y} = \chi(x)Y_0\omega_n \cos \omega_n t$$

At this point, a comment regarding principal coordinates is in order. Knowing the maximum displacement Y_0 at mid span and the mode shape $\sin \pi x/l$, defines the displacement at any position x. Thus the combination $\chi(x)Y_0$ is a principal coordinate by our definition.

Using these expressions, the kinetic energy of the string can be determined in much the same way as the kinetic energy of the simple elastic spring. If the weight per unit length is μ, the kinetic energy of an element of length dx is

$$dT = \frac{1}{2}\frac{\mu}{g}\dot{y}^2 \, dx$$
$$= \frac{1}{2}\frac{\mu}{g}Y_0^2\omega_n^2 \cos^2 \omega_n t \sin^2\frac{\pi x}{l}dx$$

and, for the entire string, the kinetic energy can be found by integrating from 0 to l.

$$\tfrac{1}{2}\left(\tfrac{x}{2} - \tfrac{\sin \frac{\pi x}{l}}{4}\right)$$

$$T = \frac{1\mu}{2g} Y_0^2 \omega_n^2 \cos^2 \omega_n t \int_0^l \sin^2 \frac{\pi x}{l}\,dx = \frac{1\mu}{4g} l Y_0^2 \omega_n^2 \cos^2 \omega_n t \qquad (3.14)$$

The potential energy stored in an element of string of dl in length would be

$$dV = P\sqrt{1 + \left(\frac{dy}{dx}\right)^2}\,dx - P\,dx$$

Using the first two terms of a binomial expression for the radical,

$$dV = \frac{P}{2}\left(\frac{dy}{dx}\right)^2 dx \qquad \sqrt{1+\delta} = 1 + \tfrac{\delta}{2}$$

For the entire string,

$$\int \cos^2 x\,dx = \tfrac{x}{2} + \tfrac{\sin 2x}{4}$$

$$V = \frac{P}{2}\int_0^l \left(\frac{dy}{dx}\right)^2 dx = \frac{P\pi^2}{2\,l} Y_0^2 \sin^2 \omega_n t \int_0^l \cos^2 \frac{\pi x}{l}\,dx = \frac{P\pi^2}{4\,l} Y_0^2 \sin^2 \omega_n t \qquad (3.15)$$

Setting the maximum potential energy change to be equal to the maximum kinetic energy change,

$$\Delta T_{max} = \Delta V_{max}$$

$$\frac{1\mu l}{4\,g} Y_0^2 \omega_n^2 = \frac{P\pi^2}{4\,l} Y_0^2$$

$$\omega_n^2 = \frac{\pi^2 P}{\mu l^2} g \qquad (3.16)$$

Rayleigh's energy method is useful when the vibrating system is known to be linear and an expression for the fundamental natural frequency is desired. It does require a mode shape, if the kinetic effect of distributed mass is to be included, and if the mode shape is in error, the calculation of the natural frequency will be in error. It should be noted again that natural frequency is independent of amplitude.

3.7. LUMPED SYSTEMS

In the last two sections, the practical problem of modeling the real world has been raised, and it is difficult to avoid. All elasticity is lumped in a spring with an elastic constant, k. All of the inertial properties are lumped in a mass, m. How is this modeling accomplished? When should the elasticity and mass be distributed and not lumped? What numerical values can be placed on the modeled values? The procedure is, of course, largely the art of engineering and can really only be taught by example. Some of the problems suggest how systems can be lumped. We have seen one example. A linear spring, fixed at one end and supporting a mass at the other, can be modeled by lumping one third of the mass of the spring with the supported mass. For a cantilevered beam, this fraction is $\frac{33}{140}$. The inertial effect of the mass of a simply supported beam can be modeled by placing $\frac{17}{35}$ of the beam's mass at its geometric center. For a uniform beam, built in at each end, this fraction is $\frac{3}{8}$. These fractions can be calculated for an ideal beam or spring, but for real beams and springs, these fractions are only approximate.

In each case of modeling, the kinetic characteristics of the model are made equal to the original system. As another example of kinetic equivalence, let us consider the connecting rod of Figure 3.4. The connecting rod is to be modeled as a two-mass system, one mass at the crank pin and one mass at the wrist pin. The total mass and the geometry must be unchanged. The inertial characteristics of the connecting rod and the location of the center of mass must remain unchanged. These lead to four equations.

$$l = r_A + r_B$$
$$m = m_A + m_B$$
$$m_A r_A = m_B r_B$$
$$m_A r_A^2 + m_B r_B^2 = mk_G^2$$

In the last equation, k_G is the radius of gyration with respect to the center of mass. These four statements are usually not compatible. That is, if you are given the location of the center of mass and the total mass, these two statements will uniquely determine r_A, r_B, m_A, and m_B. The fourth, which is the moment of inertia, will follow, but it will probably not be the known moment of inertia of the connecting rod. On the other hand, we could accept all four equations and try to compromise. In actual fact, only a minor compromise is necessary. The connecting rod model is shorter, but the center of mass and the inertial properties are effectively unchanged.

Fig. 3.4

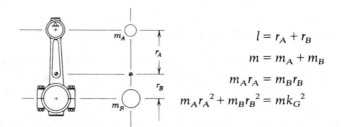

$$l = r_A + r_B$$
$$m = m_A + m_B$$
$$m_A r_A = m_B r_B$$
$$m_A r_A^2 + m_B r_B^2 = mk_G^2$$

EXAMPLE PROBLEM 3.33

A very large luxury liner once had a vibration problem that was critical to its operation until the problem was corrected. The ship had four propellers, each propeller has a mass of 12,200 kg and each was driven by a long hollow shaft, 0.56 m OD and 0.28 m ID, and 71.6 m long. The ship cruised at a speed corresponding to 258 rpm. Determine the natural frequency for longitudinal vibration of the propeller and shaft. What will happen if the ship is equipped (which it was) with four-bladed propellers? How would you correct the problem?

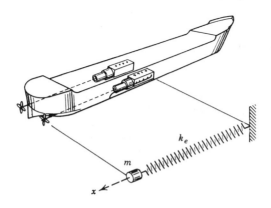

Solution:

For an axial load P, the longitudinal deflection of the propeller shaft will be

$$\delta = \frac{Pl}{AE}$$

The effective spring constant, k_e is

$$k_e = \frac{P}{\delta} = \frac{AE}{l} = \frac{\pi(0.56^2 - 0.28^2)}{4} \frac{}{71.6}(205 \times 10^9) = 528.9 \times 10^6 \text{N/m}$$

The mass of the propeller shaft is 99,200 kg

$$m = \frac{\pi}{4}(0.56^2 - 0.28^2)(71.6)(7.5) = 99{,}200 \text{ kg}$$

Computing the natural frequency without considering the mass of the elastic member, the propeller shaft, would be a gross error, since its effective mass is an order of magnitude larger than the mass of the propeller. Using equation 3.12,

$$\omega_n^2 = \frac{k_e}{m + (m_s/3)}$$

$$f_n = \frac{1}{2\pi}\omega_n = \frac{1}{2\pi}\sqrt{\frac{528.9 \times 10^6}{12,200 + (99,200/3)}} = 17.2 \text{ Hz} = 1032 \text{ cpm}$$

At a cruising speed corresponding to 258 rpm of the propeller shaft, one blade of the four-bladed propeller will pass by the restricted area between the propeller and the hull every 0.058 s, 4 times each revolution or 1032 times each minute, matching the natural frequency.

To eliminate unwarranted vibration, a three-bladed propeller was used. The entire problem is chronicled in the *Journal of the Society of Naval Architects and Marine Engineers*.

EXAMPLE PROBLEM 3.34

A uniform cantilevered beam carries a mass m at one end and the other end is fixed into a vertical wall. Determine the kinetic effect on the natural frequency of considering the mass of the beam.

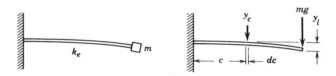

Solution:
The static deflection y_c at any point along the beam is

$$y_c = \frac{mg}{EI}\left(\frac{lc^2}{2} - \frac{c^3}{6}\right)$$

where, I in this case is the second moment of area of the beam cross-section, and the deflection y_l of the end of the beam under the static load mg is

$$y_l = \frac{mgl^3}{3EI}$$

The exact mode shape for the dynamic deflection of a vibrating beam involves hyperbolic functions, but let us make an assumption for the fundamental mode that the static and dynamic mode shapes are identical

$$\chi(c) = \frac{y_c}{y_l} = \frac{1}{2l^3}(3lc^2 - c^3)$$

The velocity of any beam element can be found in terms of the velocity at the end of the beam and the mode shape

$$y_c = \chi y_l$$

$$\dot{y}_c = \chi \dot{y}_l$$

The kinetic energy change of the system, including both the kinetic energy of the beam and the kinetic energy of the concentrated mass at the end of the beam is

$$\Delta T = \tfrac{1}{2}m\dot{y}_l^2 + \int_0^l \frac{\mu dc}{2g}\dot{y}_c^2$$

The mass of the uniform beam is m_s and $m_s g = \mu l$. The weight of the beam per unit length is μ. The mass at the end of the beam is m. Substituting for the velocity $\dot{y}_c$,

$$\Delta T = \tfrac{1}{2}m\dot{y}_l^2 + \int_0^l \frac{\mu \dot{y}_l^2}{8l^6 g}(3lc^2 - c^3)^2 \, dc$$

integrating,

$$\Delta T = \tfrac{1}{2}m\dot{y}_l^2 + \tfrac{1}{2}\tfrac{33}{140}m_s\dot{y}_l^2$$

Including the effect of the vibrating beam is kinetically equivalent to adding $\tfrac{33}{140}$ of the mass of the beam to the mass m.

The equivalent spring constant for a cantilevered beam is $k_e = 3EI/l^3$, in terms of deflection at the end of the beam. The potential energy change in terms of deflection at the end of the beam is

$$\Delta V = \frac{3EI}{2 \, l^3}y_l^2$$

Assuming harmonic motion for the beam,

$$y_l = Y_l \sin \omega_n t$$

where Y_l is the maximum deflection of the tip of the beam.

Setting the maximum kinetic energy change to be equal to the maximum potential energy change,

$$\Delta T_{max} = \Delta V_{max}$$

$$\tfrac{1}{2}m Y_l^2 \omega_n^2 + \tfrac{1}{2}\tfrac{33}{140}m_s \, Y_l^2 \omega_n^2 = \frac{3EI}{2 \, l^3} \, Y_l^2$$

and

$$\omega_n^2 = \frac{3EI}{(m + \frac{33}{140}m_s)l^3}$$

For the sake of comparison, let us consider what happens when $m \to 0$, which is simply the fundamental frequency of a uniform cantilevered beam.

$$\omega_n^2 = \sqrt{\frac{3EI}{\frac{33}{140}m_s l^3}} = 3.567\sqrt{\frac{EIg}{\mu l^4}}$$

The exact value is $\omega_n = 3.515\sqrt{EIg/\mu l^4}$, or about $1\frac{1}{2}\%$ error. Note also that the approximate mode shape gives a natural frequency which is higher than the actual. This is a characteristic trait, to which we will make reference later.

PROBLEM 3.35   As in Example Problem 3.34, determine the lowest natural frequency of a uniform cantilevered beam using the mode shape,

$$\chi(c) = 1 - \cos\frac{\pi c}{2l}$$

which also satisfies the boundary conditions for a cantilevered beam.

Answer: $\omega_n = 3.64\sqrt{\dfrac{EIg}{\mu l^4}}$

PROBLEM 3.36   Determine the lowest natural frequency of a simply supported beam of uniform cross-section, with a central mass m. Assume a curve of the form

$$y = \frac{mg}{48EI}(3xl^2 - 4x^3)$$

which is valid for $0 \leqslant x \leqslant l/2$.

Answer: $\omega_n = \sqrt{\dfrac{48EI}{l^3(m + \frac{17}{35}m_s)}}$

PROBLEM 3.37   Solve the previous problem using the mode shape

$$\chi = \sin\frac{\pi x}{l}$$

Answer: $\omega_n = \sqrt{\dfrac{48EI}{l^3(m + \frac{1}{2}m_s)}}$

PROBLEM 3.38 Determine the lowest natural frequency of a beam with clamped ends, of uniform cross-section, with a central mass m. Assume a curve of the form,

$$y = \frac{mg}{48EI}(3lx^2 - 4x^3)$$

which is valid for $0 \leqslant x \leqslant l/2$.

Answer: $f_n = \dfrac{1}{2\pi}\sqrt{\dfrac{192EI/l^3}{m + \frac{13}{35}m_s}}$

PROBLEM 3.39 Solve the previous problem using the mode shape

$$\chi = \frac{1}{2}\left(1 - \cos\frac{2\pi x}{l}\right)$$

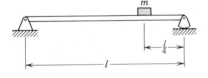

PROBLEM 3.40 A frame building vibrates laterally, as shown. Using Rayleigh's energy method, determine the fraction of the walls which can be considered kinetically to be moving with the roof truss.

Hint: Consider the walls as fixed to the foundation and fixed to the roof structure.

Answer: 0.375

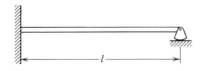

PROBLEM 3.41 Determine the lowest natural frequency of a simply supported beam of uniform cross-section with a mass m located at one of the quarter points.

Hint: Use the mode shape.

$$\chi = \sin\frac{\pi x}{l}$$

Answer: $f_n = \dfrac{1}{2\pi}\sqrt{\dfrac{54.31EI}{l^3(m + m_s)}}$

PROBLEM 3.42 Determine the lowest natural frequency of the beam of uniform cross-section. Use the parabolic mode shape

$$y = b - \frac{4b}{l^2}\left(x - \frac{l}{2}\right)^2$$

Compare with Problem 3.43.

Answer: $f_n = \dfrac{10.95}{2\pi}\sqrt{\dfrac{EIg}{\mu l^4}}$

PROBLEM 3.43   Determine the lowest natural frequency of the beam in Problem 3.42. Use a sinusoid as the mode shape

$$\chi = b \sin \frac{\pi x}{l}$$

Compare with Problem 3.42.

Answer: $f_n = \frac{\pi}{2}\sqrt{\frac{EIg}{\mu l^4}}$

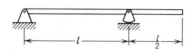

PROBLEM 3.44   Determine the natural frequency of an overhanging simply supported uniform beam. The parabolic equation

$$y = \left[\frac{4b}{l^2}\left(x - \frac{l}{2}\right)^2 - b \right]$$

will satisfy the boundary conditions for the beam.

Answer: $f_n = \frac{2.322}{2\pi}\sqrt{\frac{EIg}{\mu l^4}}$

PROBLEM 3.45   A ship at sea can vibrate in several modes. One mode is laterally as a free–free uniform beam. Using Rayleigh's energy method, determine the fundamental frequency of lateral vibration for a free–free uniform beam with a flexural rigidity of EI and a uniform weight of μ per unit length. The lateral deformation can be described by the equation

$$y = b\left(3 \sin \frac{\pi x}{l} - 2\right)$$

where b is the maximum lateral deflection at mid ship. *Hint:* The strain energy of a uniform beam in bending is

$$V = \int_0^l \frac{EI}{2} \left(\frac{d^2 y}{dx^2} \right)^2 dx$$

Answer: $f_n = \dfrac{11.3}{\pi} \sqrt{\dfrac{EIg}{\mu l^4}}$

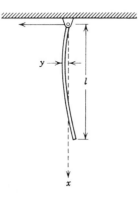

PROBLEM 3.46 A long rod that is pivoted about the upper end hangs vertically. To determine the fundamental natural frequency of lateral vibration, a mode shape is needed for the rod. The equation

$$y = \sqrt{2} b \sin \frac{5\pi x}{4 l}$$

does satisfy the boundary conditions. Determine the kinetic energy, the potential energy and the natural frequency of the rod for the mode, using this mode shape. Remember that for a uniform rod,

$$V = \int_0^l \frac{EI}{2} \left(\frac{d^2 y}{dx^2} \right)^2 dx$$

Answer: $f_n = \dfrac{15.42}{2\pi} \sqrt{\dfrac{EIg}{\mu l^4}}$

PROBLEM 3.47 Determine the natural frequency of the rod in Problem 3.46 using the mode shape

$$y = b \left(\sin \frac{\pi x}{l} - \frac{x}{l} \right)$$

Answer: $f_n = \dfrac{15.73}{2\pi} \sqrt{\dfrac{EIg}{\mu l^4}}$

CHAPTER FOUR

FORCED

PERIODIC

MOTION

4.1. INTRODUCTION

An elastic system that is subjected to externally applied forces is said to be *forced*, and the oscillating motion that results from externally applied forces is *forced vibration*. If energy dissipation or *damping* is present, the motion is *forced damped vibration*. When part of the motion disappears after a period of time, that part is known as the *transient*. The part that remains after the transient has disappeared is called the *steady state vibration*.

Transient vibration is of immense concern where shock, impact, and moving loads are involved. This motion is not necessarily periodic and mechanical failure from transient vibration is generally attributed to the mechanical strength of some component being exceeded. Transient motion will be studied in some detail, in the next chapter.

Steady state vibration exists long after transient vibration has died away. It is generally associated with the continuous operation of machinery, and if mechanical failure occurs, it is usually through the mechanism of fatigue after a long period of time. In some cases, the same motion phenomenon may present differing aspects to different structures. A car moving over a bridge is a transient load on the bridge structure. To the car chassis, however, movement over the bridge is just a brief interval of steady state operation. Our definitions of transient and steady state vibration overlap.

Although steady state vibration never occurs practically, without the presence of some energy dissipation in damping, the effect of damping is small unless the amplitude of motion is large. Since damping does complicate a study of forced vibration, and the various types of damping do not produce similar effects, a study of forced damped vibration will be deferred in favor of concentrating on the physical concepts of *resonance*, *vibration isolation*, and *response*.

4.2. UNDAMPED FORCED HARMONIC VIBRATION, $\mathbf{F}(t) = \mathbf{F}_1 \sin \omega t$

Let us consider the motion of a single degree of freedom spring and mass system subjected to a harmonically varying force, $\mathbf{F}(t) = \mathbf{F}_1 \sin \omega t$. F_1 is the maximum value of the impressed force and ω is the frequency with which the force $F(t)$ varies in radians per second. Referring to Figure 4.1, from Newton's second law, the equation of motion, $\Sigma \mathbf{F} = m\ddot{\mathbf{x}}$ becomes

$$-k\mathbf{x} + \mathbf{F}(t) = m\ddot{\mathbf{x}} \qquad (4.1)$$

As can be seen from the free body diagram, the acceleration is in the direction of the impressed force, and the spring force is opposite to the impressed force. Rearranging terms, and using the scalar component in the direction of the coordinate x,

$$m\ddot{x} + kx = F_1 \sin \omega t \qquad (4.2)$$

$$\ddot{x} + \frac{k}{m}x = \frac{F_1}{m} \sin \omega t$$

Mathematically, this is a particular form of equation 2.2. The solution must contain the general integral for the differential equation of motion

 Fig. 4.1

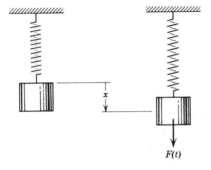

$$F(t)$$

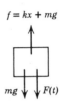

$$f = kx + mg$$

$$mg \quad F(t)$$

including the necessary arbitrary constants, as well as a particular integral that satisfies the particular form of equation 4.2.

$$x = A \cos \omega_n t + B \sin \omega_n t + \frac{F_1}{m(\omega_n^2 - \omega^2)} \sin \omega t \qquad (4.3)$$

The first two terms are called transient, even though damping is absent in this particular case. They are dependent on initial conditions and after a period of time, with any measurable amount of damping, the influence of these terms is small. In the steady state, we need only to direct our attention to the last term, which is called the steady state term. It is not affected by initial conditions and remains as long as the forcing function is applied. Omitting the transient terms,

$$x = \frac{F_1}{m(\omega_n^2 - \omega^2)} \sin \omega t \qquad (4.4)$$

This is similar to the harmonic displacement $x = X \sin \omega t$, where the maximum displacement,

$$X = \frac{F_1}{m(\omega_n^2 - \omega^2)}$$

Harmonically varying forcing functions cause a harmonically varying displacement, with the maximum value of the force, F_1, related to the maximum value of the displacement X. Substituting $k = m\omega_n^2$,

$$\frac{X}{F_1/k} = \frac{1}{1 - \dfrac{\omega^2}{\omega_n^2}} \qquad (4.5)$$

Equation 4.5 is plotted in Figure 4.2 in nondimensional form, the dimensionless *amplitude ratio* $X/(F_1/k)$ plotted as a function of the dimensionless *frequency ratio* ω/ω_n. At low frequencies, when $\omega \ll \omega_n$, the amplitude of motion is approximately F_1/k, which is the deflection that the elastic system would have if F_1 were a static force. It is sometimes referred to as the "static deflection" Δ_{st}. At high frequencies, when $\omega \gg \omega_n$, the motion will be very small, becoming less and less, as the frequency ratio is increased. When the forcing frequency and the natural frequency are nearly equal, it is evident that very large vibration amplitudes can result from very small forces. The condition at $\omega = \omega_n$, when the amplitude ratio is infinite, is known as *resonance*.

The variation of the amplitude or amplitude ratio with frequency is called the *response* of the system. For values of $\omega < \omega_n$, the amplitude ratio is positive. For $\omega > \omega_n$, the amplitude ratio is negative. This is another way of saying that the motion is in phase with the applied force for $\omega < \omega_n$, and the motion and applied force are out of phase for $\omega > \omega_n$.

Fig. 4.2

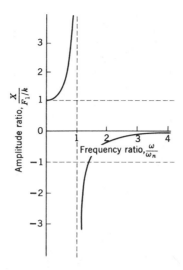

4.3. FORCED VIBRATION CAUSED BY $F(t)=F_1e^{i\omega t}$

If the forcing function had been the exponential function $F(t)=F_1e^{i\omega t}$, the equation of motion would be

$$\ddot{x}+\frac{k}{m}x=\frac{F_1}{m}e^{i\omega t} \tag{4.6}$$

for which the solution is

$$x=A\cos\omega_n t+B\sin\omega_n t+\frac{F_1}{m(\omega_n^2-\omega^2)}e^{i\omega t} \tag{4.7}$$

In the steady state, the displacement is $x=Xe^{i\omega t}$, where the maximum displacement X is exactly the same as in equation 4.5.

$$\frac{X}{F_1/k}=\frac{1}{1-\dfrac{\omega^2}{\omega_n^2}}$$

4.4. FORCED VIBRATION CAUSED BY ROTATING UNBALANCED FORCES, $F(t)=m_0\omega^2 e\sin\omega t$

An obvious source of forced vibration is the unbalance of rotating parts. If the center of gravity of an unbalanced mass m_0 has a radial eccentricity e from the geometric axis of rotation, the applied force is $F(t)=$

$m_0\omega^2 e \sin \omega t$. The displacement x is a function not only of the eccentricity but of the mass ratio m_0/m, where m is the entire mass supported by the elastic system, including the unbalanced rotating mass m_0. Figure 4.3 shows an unbalanced mass m_0 rotating about a geometric axis at O. The entire mass m, which includes the rotor, is constrained to move in one direction only. Lateral motion is ignored. If present, it could be considered in the same way as vertical motion, but it would add another degree of freedom.

Substituting the unbalanced force $m_0\omega^2 e$ for F_1, the steady state displacement becomes

$$x = \frac{m_0\omega^2 e}{m(\omega_n^2 - \omega^2)} \sin \omega t$$

The maximum displacement X, can be expressed as

$$X = \frac{m_0\omega^2 e}{m(\omega_n^2 - \omega^2)}$$

$$\frac{mX}{m_0 e} = \frac{\left(\dfrac{\omega^2}{\omega_n^2}\right)}{\left(1 - \dfrac{\omega^2}{\omega_n^2}\right)} \tag{4.8}$$

$mX/m_0 e$ is the *magnification ratio* of the system and it is also dimensionless.

Figure 4.4 is the nondimensional plot of $mX/m_0 e$ as a function of the frequency ratio ω/ω_n. At low frequencies, where $\omega \ll \omega_n$ the magnification factor is near zero. The unbalanced rotating force is insignificant and there is no vibration. At resonance, $\omega = \omega_n$ and the magnification

Fig. 4.3

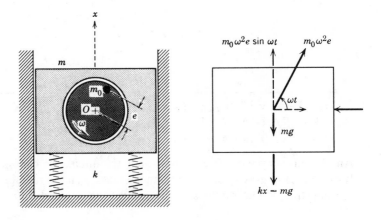

Fig. 4.4

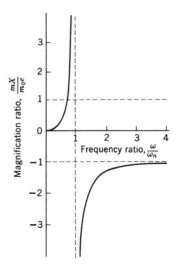

factor and the amplitude of motion are infinite. At high frequencies, when $\omega >> \omega_n$, $X \rightarrow -(m_0/m)e$. If the rotating mass, m_0, is a small fraction of the total mass m, the amplitude of vibration is proportionally small. This is the reason that some machines have very heavy frames or are attached to large base blocks of concrete. Decreasing the ratio m_0/m by increasing m is one way of decreasing the amplitude of vibration.

4.5. TRANSMITTED FORCES AND VIBRATION ISOLATION

The condition of mechanical resonance is clearly something to be avoided, if long life and quiet operation are objectives. At resonance, the amplitude of motion becomes very large and the spring and mass system literally tears itself apart. There are, of course, circumstances when mechanical resonance is desired. Mechanical shakers do have numerous industrial applications. The question of how far from the resonant condition represents safe operation is important and can be answered by considering the force transmitted through the spring.

The disturbing force can be transmitted through the spring to the foundation or the ground only if the spring is compressed or extended. If X is the maximum displacement of the spring, then the maximum transmitted force is kX. This is, of course, added to whatever static force already exists within the spring or supporting structure. The *transmission ratio* (TR) is defined as the absolute value of the fraction of the maximum disturbing force that is actually transmitted to the base of the spring or foundation

$$\text{Transmission Ratio} = \left| \frac{1}{\left(1 - \dfrac{\omega^2}{\omega_n^2}\right)} \right| \tag{4.9}$$

The transmission ratio is shown in Figure 4.5. It is similar to Figure 4.2, except that only a positive value is shown where $\omega/\omega_n > 1$. The transmitted force can be less than the disturbing force only if the forcing frequency exceeds the natural frequency of the elastic system by at least a factor of $\sqrt{2}$. This means that for smooth operation, the natural frequency of the supporting structure must be considerably below the frequency of excitation. As shown in Figure 4.5, for the entire region where $\omega/\omega_n < \sqrt{2}$, the transmission ratio >1. In this case, an elastic supporting structure would do more harm than good. It is conventional to specify the transmission ratio as a positive number, with the phase relation being ignored.

Fig. 4.5

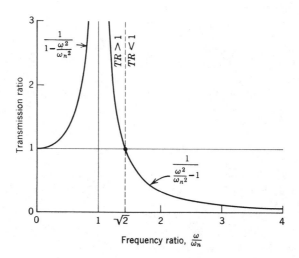

4.6. FORCED VIBRATION CAUSED BY HARMONIC GROUND MOTION

The movement of the base or the foundation may very well result in unwanted vibration of an elastic structure resting on the base. This is the reverse of the previous problem. As an example, consider the elastically supported system of Figure 4.6, in which the spring and mass system is subjected to the movement $u = b \sin \omega t$.

The displacement of the mass m is the generalized coordinate x, but the deflection of the elastic spring is the relative displacement between the mass m and point A. Point A is displaced $u = b \sin \omega t$.

Fig. 4.6

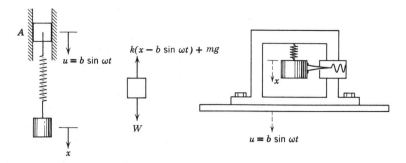

The equation of motion is

$$-k(x-u)=m\ddot{x}$$

or

$$m\ddot{x}+kx=ku=kb \sin \omega t \qquad (4.10)$$

This equation is similar to equation 4.2. The problem of the vibrating foundation is analogous to the problem of the harmonically varying force, substituting kb for F_1. Using this analogy, the steady state solution is

$$x=\frac{b}{1-\dfrac{\omega^2}{\omega_n^2}} \sin \omega t \qquad (4.11)$$

The fraction X/b where $x=X \sin \omega t$ is identical with the amplitude ratio of Figure 4.2.

The relative motion between the mass m and the end of the elastic spring, point A, is $z=x-u$. The equation of motion, equation 4.10, can be expressed in terms of relative motion, z, since $x=z+u$, and $\ddot{x}=\ddot{z}+\ddot{u}$,

$$-k(x-u)=m\ddot{x}$$
$$-kz=m(\ddot{z}+\ddot{u})$$

and

$$m\ddot{z}+kz=-m\ddot{u}=mb\omega^2 \sin \omega t \qquad (4.12)$$

This equation, is also similar to equation 4.2, with $mb\omega^2$ substituted for F_1. The steady state expression for the relative motion z, where $z=Z \sin \omega t$

$$\frac{Z}{b}=\frac{\left(\dfrac{\omega^2}{\omega_n^2}\right)}{\left(1-\dfrac{\omega^2}{\omega_n^2}\right)} \qquad (4.13)$$

The fraction Z/b is identical with the *magnification ratio* of Figure 4.4.

4.7. VIBRATION MEASURING INSTRUMENTS

The motion of undamped seismic vibration measuring instruments can be described by the previous simple dynamic principles. The word "seismic" and its variations, seismometer and seismograph, are derived from the Greek word "seismos," meaning earth.

The second sketch of Figure 4.6 shows a vibration-measuring instrument recording a sinusoidal record of the relative motion between the mass m and the frame. If the natural frequency is very low, $\omega/\omega_n \gg 1$, the relative motion will be nearly identical with the motion of the foundation, $Z \to -b$, and the recorded vibration trace will be a record of the motion of the foundation or base. An instrument that records vibration displacement is called a vibrometer. Its accuracy is strictly a function of how low the natural frequency is in relation to the excitation frequency, ω. For seismographs, which record earthquakes where the periods may be long, natural frequencies are required to be in the order of 5 to 10 cpm.

An accelerometer is a vibration measuring instrument that measures acceleration. Going back to equation 4.13,

$$z = Z \sin \omega t = \frac{b\left(\dfrac{\omega^2}{\omega_n^2}\right)}{\left(1 - \dfrac{\omega^2}{\omega_n^2}\right)} \sin \omega t$$

$$= \frac{b\omega^2}{\omega_n^2\left(1 - \dfrac{\omega^2}{\omega_n^2}\right)} \sin \omega t = \frac{\ddot{u}}{\omega_n^2\left(\dfrac{\omega^2}{\omega_n^2} - 1\right)}$$

For $\omega/\omega_n \ll 1$ the denominator is approximately a constant and the relative motion is proportional to $\ddot{u}$. The vibration record will be a record of the acceleration of the foundation or base. Unfortunately, higher harmonics may be in resonance with the natural frequency. The presence of higher harmonics would destroy the meaning of an accelerometer record unless the instrument were damped. The reverse is true for a displacement-measuring instrument. Damping is not needed and indeed, if damping were present, it could lower the accuracy of the displacement measurement. It is also possible to integrate the output of an accelerometer twice and obtain a displacement-time record. This is often done, since accelerometers are smaller and more readily adaptable than are the larger, more bulky seismic vibrometers.

EXAMPLE PROBLEM 4.1

An electric motor has a mass of 10 kg and is set on four identical springs, each with a spring modulus of 1.6 N/mm. The radius of gyration of the motor assembly about the shaft axis is 100 mm. If the running speed of the motor is 1750 rpm, determine the transmission ratio for vertical vibration and torsional vibration.

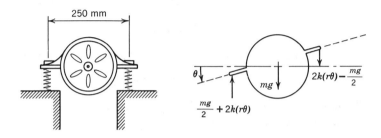

Solution:

The natural frequency for vertical motion can be found from the system constants quite easily. There are four springs, each with a known constant of proportionality. The total spring constant will be four times that of a single spring. The weight of the motor is known and the operating speed is known.

$$\omega_n^2 = \frac{k_e}{m}$$

$$\omega_n^2 = \frac{4(1.6)10^3}{10} = 640 \text{ s}^{-2}$$

The transmission ratio in vertical motion follows,

$$TR = \left| \frac{1}{\left(1 - \frac{\omega^2}{\omega_n^2}\right)} \right|$$

or,

$$TR = \frac{1}{\left\{ \frac{\left[\frac{2\pi}{60}(1750)\right]^2}{640} - 1 \right\}} = -0.0194$$

For torsional vibration, the torsional spring constant must be found in terms of the known constant of the linear springs. Each spring deflects $r\theta$ for an angular displacement of θ. The radius $r = 125$ mm.

The geometric axis is also the center of mass in this problem, and taking the moment sum about the center of mass,

$$\sum \mathbf{M}_G = I_G \ddot{\theta}$$

$$-\left(\frac{mg}{2} + 2kr\theta\right)r - \left(2kr\theta - \frac{mg}{2}\right)r = I_G\ddot{\theta}$$

$$I_G\ddot{\theta} + 4kr^2\theta = 0$$

This is similar to equation 2.14, where the torsional spring constant is

$$K = 4kr^2$$

Note that the static force of the weight cancels. From these equations, it is easy to conclude that

$$\omega_n^2 = \frac{4kr^2}{I_G}$$

$$\omega_n^2 = \frac{4(1.6)(0.125)^2 10^3}{10(0.1)^2} = 1000 \text{ s}^{-2}$$

The transmission ratio in torsional vibration is

$$TR = \left|\frac{1}{\left(1 - \dfrac{\omega^2}{\omega_n^2}\right)}\right|$$

$$TR = \frac{1}{\left\{\dfrac{\left[\dfrac{2\pi}{60}(1750)\right]^2}{1000} - 1\right\}} = 0.031$$

The torsional mode has a higher natural frequency, which is closer to the operating speed, and more torsional vibration is transmitted.

If the center of gravity were not located in the plane of support, the vertical and torsional modes would be coupled and the isolation problem would be more complex. To avoid exciting the vertical mode through torsional vibration and to avoid exciting the torsional mode with vertical vibration, it is good practice to place the plane of support so that it contains the mass center.

EXAMPLE PROBLEM 4.2

A steel frame supports a turbine-driven exhaust fan. At a speed of 400 rpm, the horizontal amplitude of motion is 4.5 mm, measured at the floor level of the fan. At a speed of 500 rpm, the amplitude is 10 mm. No resonant condition is observed in changing speed from 400 to 500 rpm. To decrease this intolerable vibration, it is proposed to add a slab of concrete beneath the turbine. What would be the effect of this added mass? Estimate the amplitude of motion at 400 rpm, if this slab doubles the effective mass of the structure. What happens at 500 rpm?

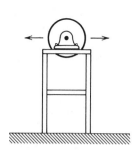 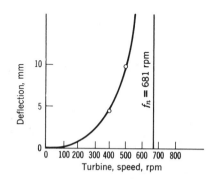

Solution:

Since amplitude increases with increasing turbine speed, the resonant condition is somewhere above 500 rpm, $\omega_n > 500(2\pi)/60$. If we assume that these two bits of observed data fall exactly on the expected curve, we can solve for both the resonant frequency and the parameter m/m_0.

$$\frac{mX}{m_0 e} = \frac{\left(\dfrac{\omega^2}{\omega_n^2}\right)}{\left(1 - \dfrac{\omega^2}{\omega_n^2}\right)}$$

We have two equations,

$$0.0045 = \frac{m_0 e}{m}\left[\frac{(400/f_n)^2}{1-(400/f_n)^2}\right]$$

and,

$$0.010 = \frac{m_0 e}{m}\left[\frac{(500/f_n)^2}{1-(500/f_n)^2}\right]$$

Solving these two equations simultaneously,

$$\frac{m}{m_0 e} = \frac{1}{4.5}\left[\frac{(400/f_n)^2}{1-(400/f_n)^2}\right] = \frac{1}{10}\left[\frac{(500/f_n)^2}{1-(500/f_n)^2}\right]$$

from which, $f_n = 681$ rpm and,

$$\frac{m}{m_0 e} = 117 \text{ m}^{-1}$$

If the effective mass were doubled, $m/m_0 e = 234$ m^{-1} and $f_n = 483$ rpm. The amplitude at 400 rpm would be much greater, since the added concrete slab would lower the natural frequency of the structure, bringing the natural frequency much closer to the operating speed.

$$\frac{mX}{m_0 e} = \frac{\left(\dfrac{\omega^2}{\omega_n^2}\right)}{\left(1-\dfrac{\omega^2}{\omega_n^2}\right)}$$

$$x = \frac{1}{234}\left[\frac{(400/483)^2}{1-(400/483)^2}\right] = 9.3 \text{ mm}$$

The structure would probably not survive a speed transition from 400 to 500 rpm, since the resonant condition would occur at 483 rpm.

This problem actually occurred. The structure has been modified, but the slab was poured and a lower natural frequency did result. As a matter of record, the vibration was so severe that the turbine speed never was raised above 250 rpm, which made the exhaust fan useless.

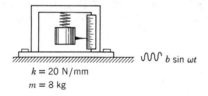

$k = 20$ N/mm
$m = 8$ kg

PROBLEM 4.3 The pointer of the vibration measuring instrument is observed to move between the 0.20 and 0.30 marks on the vertical scale, when subjected to a displacement alternating at a frequency of 100 rad/s. What would the excursion be if the forcing frequency were doubled?

Answer: 0.21 to 0.29

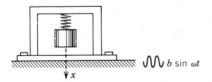

PROBLEM 4.4 The 0.5-kg mass is attached to the end of a light spring, which deflects 10 mm when a 5-N force is statically applied to the mass. If the frame is given a vertical harmonic movement with a frequency of 4 Hz and an amplitude of 2 mm, find the amplitude x of vertical vibration of the mass.

Answer: $x = 5.43$ mm

PROBLEM 4.5 An elastically mounted assembly vibrates at an amplitude of 10 mm at 1200 rpm and 2 mm at 2400 rpm. What is the resonant (natural) frequency for the assembly?

PROBLEM 4.6 A 1570-kg automobile is supported on four-coil springs, each with a spring modulus of 25 N/mm. In traveling over an elevated expressway, a resonant vertical motion is excited. Each span of the expressway is 25 m in length and sags 15 mm at midspan. Determine the maximum vertical excursion of the automobile traveling at 100 km/h over the expressway. Damping is neglected.

Answer: 64 mm

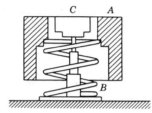

PROBLEM 4.7 The hand vibrometer is a simple seismic device used in field work for approximate measurements. The seismic mass *A* is suspended on a spring *B*. The double amplitude of vibration of a frame, casing, or foundation is indicated by the band swept out by the dial-gage pointer. The natural frequency of the instrument is 4.5 cps. If the indicated double amplitude is 0.95 mm for a known frequency of 20 cps, what is the true double amplitude of motion.

Answer: 0.90 mm

PROBLEM 4.8 A 350-kg gasoline engine driven air compressor operates at 800 power strokes per minute. When mounted on rubber pads, the transmitted vibration is reduced to one fourth of the value without rubber pads. What was the static deflection of the rubber pads?

Answer: 7 mm

PROBLEM 4.9 If the engine˙ of Problem 4.8 is mounted on a 350-kg concrete block, which is in turn mounted on the same rubber pads, what is the percentage decrease in the dynamic deflection?

Answer: 55.6% decrease

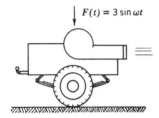

$F(t) = 3 \sin \omega t$

PROBLEM 4.10 A portable shredder is used to shred bark, tree branches, and shrub clippings into 10- to 30-mm pieces. The shredder and trailer have a mass of 200 kg. The tires and support system have an elastic constant of 460 N/mm. Determine the vertical motion of the shredder, if the shredder exerts a vertical force of $3 \sin \omega t$, in kN, and the frequency of the excitation is 1200 rpm.

Answer: 1.1 mm

PROBLEM 4.11 In a remote sugar mill, a processing machine is mounted on a base that is in turn mounted on an elastic cork and rubber pad. It is impossible to measure the static deflection of the pad, so the natural frequency is unknown. There are two operating speeds of the machine, one at 750 rpm and one at 1500 rpm. At 1500 rpm the amplitude of the casing of the machine is 40% of the amplitude at 750 rpm. There is no resonant frequency between 750 and 1500 rpm. These amplitudes are only comparative, since there is no means of calibrating the amplitude measuring instrument. From these two values, estimate the natural frequency of the machine and base.

Answer: $\omega_u = 63.9 \text{ s}^{-1}$

PROBLEM 4.12 During the installation of a 60-cycle induction motor with a mass of 200 kg, it is determined by means of a level that the deflection of the floor under the motor is 0.13 mm. The rated speed of the induction motor is 1800 rpm. Would you recommend vibration isolation? Explain your answer.

PROBLEM 4.13 A self-contained textile machine with a 4500-kg mass must be isolated to prevent the transmission of force to the foundation. The forcing function can be represented as $F(t) = 100 \sin \omega t$ and the frequency of the fundamental is 1800 rpm. Isolators are rated by the measured static deflection when installed. They come in two ratings, (a) 3 mm, and (b) 6 mm. The rating of an isolator is the static deflection of the isolator under its static load, which would reduce the transmission of force below 10%?

Answer: (b)

PROBLEM 4.14 An electric motor and pump have a combined mass of 50 kg and are supported on a welded steel structure. When the motor and pump are in place, it is noted that the support structure deflects 3 mm from its own static position and the resonant frequency is 546 rpm. At the running speed of 600 rpm the amplitude of motion is 5 mm. To avoid resonance, vibration isolators that have a rated deflection of 6 mm are placed between the motor and the support. What will be the amplitude of motion at 600 rpm with the isolators in place?

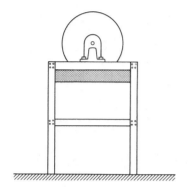

PROBLEM 4.15 A steel frame supports a 1000-kg concrete mass, which in turn supports a turbine driven exhaust fan. At a speed of 400 rpm the amplitude of motion is 4 mm and at a speed of 500 rpm the amplitude is 11.6 mm. Determine, as a factor, the change in spring stiffness necessary to allow operation at the maximum turbine speed of 520 rpm with a maximum deflection of 2 mm.

Answer: $k' = 2.62k$

PROBLEM 4.16 The rotating parts of the right front wheel assembly of a 1979 automobile have a mass of 35 kg and are measured to be out of balance. This unbalance can be corrected by attaching a 150 g lead weight to the 375-mm diameter wheel rim. The tire has a diameter of 700 mm. The car springs have a modulus of 50 N/mm and the tire can be considered to have a modulus of 600 N/mm, if the tire deflection is small. If the balance correction is not made, at what speed will resonant vibration of the wheel assembly occur? What would the amplitude of motion be for a speed of 130 km/h?

Answer: 1.1 mm

PROBLEM 4.17 A delicate aircraft instrument is mounted on four rubber isolators that have a rated deflection of 5 mm. This means that the static deflection of the isolator at equilibrium is 5 mm. What is the transmission ratio of vibration transmitted to the instrument for 1800 rpm?

PROBLEM 4.18 A high-speed diesel engine is mounted on four rubber pads such that the deflection is 5 mm. If the engine and coupling has a mass of 300 kg, above what speed must the motor run for 95% isolation? The engine is a four-cylinder, two-stroke cycle diesel.

Answer: 485 rpm

PROBLEM 4.19 A motor operating at a speed of 950 rpm is supported on four identical rubber pads, and vibrates with an amplitude of 3.8 mm. In starting, it passes through a resonant frequency at 500 rpm. Determine the amplitude of vibration if the motor were mounted on four pairs of pads, each pair consisting of two of the original pads in series.

Answer: 3.2 mm

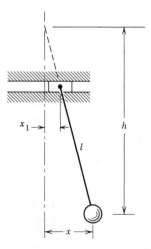

PROBLEM 4.20 A 100-kg motor and generator set is mounted on rubber pads that deflect 3 mm under the static weight of the set. The rated speed of the motor is 1800 rpm, and at that speed the set vibrates with an amplitude of 0.05 mm. What would be the amplitude of motion if the motor generator set is mounted on a 300-kg block of concrete, which is then mounted on the same rubber pads?

Answer: 0.012 mm

PROBLEM 4.21 Determine the error in an accelerometer reading if the natural frequency of the accelerometer is four times the frequency of the observed motion.

Answer: 6.7% high

PROBLEM 4.22 The point of support of a simple pendulum is given a horizontal harmonic oscillation of $x_1 = X_1 \sin \omega t$. Determine an expression for the motion of the pendulum, assuming small deflections. Determine the distance h from the mass to the node in terms of ω and ω_n using the coordinates x and x_1.

Answer: $h = l(\omega_n^2/\omega^2)$

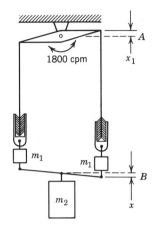

PROBLEM 4.23 A machine designed to test the bond between rubber and wire is shown schematically. The wire is immersed and bonded to a rubber compound, which completely fills the cylindrical sleeve. For a double amplitude of motion at end A of 2 mm, what is the motion of end B? What is the maximum force on the wire at B tending to break its rubber bond? Consider the wires as weightless and rigid ($m_1 = 5$ kg, $m_2 = 50$ kg, $k = 175$ N/mm).

Answer: 1.11 mm; 488 N

PROBLEM 4.24 A suspension system for a four-cylinder inboard-outdrive marine engine consists of four rubber isolators spaced in a rectangle 240 mm wide by 320 mm high. Each isolator consists of a shear type rubber fitting 50 mm in diameter and has a linear stiffness of 300 N/mm, determined by a simple static load. The mass of the engine is 150 kg. Polar radius of gyration of the engine is 175 mm. What is the torsional transmission ratio (TR) at the rated speed of 2200 rpm? Below what revolutions per minute would the $TR>1$? The engine has the four-stroke cycle.

Answer: 0.052; 690 rpm

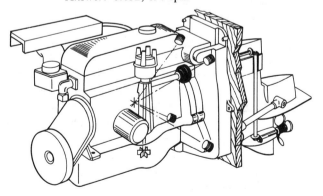

PROBLEM 4.25 Automobile engines are conventionally mounted on three isolators, one on each side of the engine block and one at the rear. For simplicity, assume that the rear isolator is on the centerline of the crankshaft. Each isolator

has a linear stiffness of 200 N/mm. The engine has six cylinders and has a mass of 125 kg. The radius of gyration for the engine is 150 mm.

(a) What is the transmission ratio for vertical motion at an operating speed of 2200 rpm?

(b) At what minimum distance from the centerline of the crankshaft must the side isolators be mounted so that the transmission ratio for torsional vibration is no greater than that for vertical vibration? The engine operates on the four-stroke cycle.

Answer: (a) 0.0102; (b) 184 mm

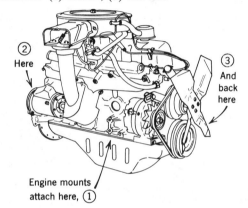

4.8. HARMONIC ANALYSIS

A forcing function or disturbing motion is often periodic but not simple harmonic. An unbalance in rotating machinery almost always causes harmonic motion which is very close to being a pure sinusoid, but there can be many other sources of vibration. If the motion or forcing function is periodic, we can always represent the function by a *Fourier series* of sine and cosine functions of time, each representing some multiple of the fundamental frequency. In a linear system, each harmonic then acts as if it alone were exciting the spring and mass system, and the system response is the sum total of the excitation of all harmonics.

A Fourier series can be written

$$y(\omega t) = \tfrac{1}{2}A_0 + A_1 \cos \omega t + A_2 \cos 2\omega t + A_3 \cos 3\omega t + \cdots$$
$$+ A_n \cos n\omega t + B_1 \sin \omega t + B_2 \sin 2\omega t$$
$$+ B_3 \sin 3\omega t + \cdots + B_n \sin n\omega t \tag{4.14}$$

The function $y(\omega t)$ can represent a force or a displacement if the excitation comes from ground motion. If $y(\omega t)$ represents a force, the amplitudes $\frac{1}{2}A_0, A_1, A_2, \ldots, B_1, B_2, \ldots$ also represent forces. If $y(\omega t)$ is displacement, the amplitudes represent displacements. The fundamental frequency is ω, the frequency at which the forcing function repeats itself, and $A_1 \cos \omega t + B_1 \sin \omega t$ is then known as the fundamental forcing function. The frequency of the second harmonic is 2ω, and $A_2 \cos 2\omega t + B_2 \sin 2\omega t$ is the second harmonic. The frequency of the third harmonic is 3ω, and $A_3 \cos 3\omega t + B_3 \sin 3\omega t$ is the third harmonic, and so on.

Without too much concern for the details of Fourier analysis with which students are intimately familiar, the arbitrary coefficients, A_n and B_n can be found by either direct or numerical integration. For the period $\omega t = 0$ to $\omega t = 2\pi$,

$$A_n = \frac{1}{\pi} \int_0^{2\pi} y(\omega t)\cos n\omega t \, d(\omega t) \qquad (4.15)$$

$$B_n = \frac{1}{\pi} \int_0^{2\pi} y(\omega t)\sin n\omega t \, d(\omega t) \qquad (4.16)$$

The introduction of the term $\frac{1}{2}A_0$ permits the use of the equation for A_n for all values of the coefficients $A_1, A_2, A_3 \ldots, A_n$ including A_0. The magnitude $\frac{1}{2}A_0$ is nothing more than the average value of $y(\omega t)$ over the full period.

$$2\pi\left(\frac{A_0}{2}\right) = \int_0^{2\pi} y(\omega t) \, d(\omega t) \qquad (4.17)$$

With a Fourier analysis of the forcing function and knowing the response of the system to sinusoidal forcing, the total response of all harmonics can be obtained by adding the individual responses of each harmonic. For the system response, let us presume that the periodic forcing function $y(\omega t)$ can be represented by the Fourier series and all the arbitrary constants are known.

$$y(\omega t) = \tfrac{1}{2}F_0 + F_1 \sin(\omega t + \alpha_1) + F_2 \sin(2\omega t + \alpha_2) + \cdots$$
$$+ F_n \sin(n\omega t + \alpha_n) \quad (4.18)$$

This is a general form of the Fourier series in which one trigonometric function and a phase angle are used instead of both sine and cosine terms. F_n and α_n are the arbitrary constants, instead of A_n and B_n. Here, $F_n = \sqrt{A_n^2 + B_n^2}$ and $\tan^{-1} \alpha_n = A_n/B_n$.

The steady state response to the excitation $F_1 \sin(\omega t + \alpha_1)$ will be

$$x_1 = \frac{F_1}{k\left(1 - \dfrac{\omega^2}{\omega_n^2}\right)} \sin(\omega t + \alpha_1)$$

And, the steady state response to the excitation $F_2 \sin(2\omega t + \alpha_2)$ will be similarly

$$x_2 = \frac{F_2}{k\left(1 - \dfrac{4\omega^2}{\omega_n^2}\right)} \sin(2\omega t + \alpha_2)$$

As a general expression, the steady state response to the excitation $F_n \sin(n\omega t + \alpha_n)$ will be

$$x_n = \frac{F_n}{k\left(1 - \dfrac{n^2\omega^2}{\omega_n^2}\right)} \sin(n\omega t + \alpha_n) \qquad (4.19)$$

At this point, we can no longer ignore our use of n in two different ways. Do not confuse the harmonic $\sin n\omega t$ with the natural frequency ω_n. The use of the symbol n in both cases is confusing but conventional, and it is better to familiarize yourself with the convention as it is, rather than to conjure up something that you will never see again. By addition, the total response of the single degree of freedom system will be $x = x_1 + x_2 + x_3 + \cdots + x_n$ or

$$x = \sum_{n=1}^{\infty} \frac{F_n}{k\left(1 - \dfrac{n^2\omega^2}{\omega_n^2}\right)} \sin(n\omega t + \alpha_n) \qquad (4.20)$$

It is clear that the harmonic which most closely approximates the natural frequency of the system will disproportionately influence the system response.

4.9. NUMERICAL SOLUTIONS FOR HARMONIC COEFFICIENTS

In most cases, direct integration of a periodic function is impossible. With hand calculation, harmonic analysis is difficult and time consuming, if more than the third harmonic is required. With the advent of computers and machine computation, it is quite easy to obtain higher order harmonics.

Referring to Figure 4.7 the periodic function $y(\omega t)$ repeats after a period of τ. This period can be divided arbitrarily into N equal parts, $\tau/N = \Delta t$ and $\omega\tau/N = \Delta\omega t$. The frequency of the fundamental is $\omega_1 = 2\pi/\tau$. Replacing the integrals for A_n and B_n, in equations 4.15 and 4.16, by the

Fig. 4.7

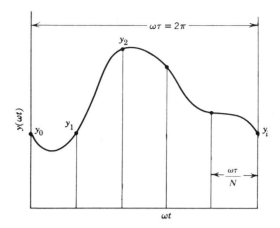

finite sums,

$$A_n = \frac{2}{\omega\tau} \sum_{i=1}^{N} y(\omega t_i) \cos\frac{2n\pi t_i}{\tau} \Delta(\omega t)$$

$$= \frac{2}{N} \sum_{i=1}^{N} y(\omega t_i) \cos\frac{2n\pi t_i}{\tau}$$

$$B_n = \frac{2}{\omega\tau} \sum_{i=1}^{N} y(\omega t_i) \sin\frac{2n\pi t_i}{\tau} \Delta(\omega t) \qquad (4.21)$$

$$= \frac{2}{N} \sum_{i=1}^{N} y(\omega t_i) \sin\frac{2n\pi t_i}{\tau}$$

The subscript i denotes the ith interval. y_i is the value of the function $y(\omega t_i)$ at the ith interval. The angle $2\pi t_i/\tau$ is that fraction of a circle, which the fraction t_i/τ represents, for $n=1$. For higher harmonics, there will be n multiples of 2π radians. In calculating the coefficients of higher harmonics, it must be remembered that the small subscript n indicates the nth harmonic, while the large N is the number of equal intervals into which the period τ is divided. In calculating the Fourier coefficients numerically, they need not be found successively. That is, the value of the nth harmonic may be found independently of the other harmonics.

It is also possible to use a specific truncated series to represent $y(\omega t)$. As an example,

$$y(\omega t) = \tfrac{1}{2}A_0 + A_1 \cos \omega t + A_2 \cos 2\omega t + A_3 \cos 3\omega t$$
$$+ B_1 \sin \omega t + B_2 \sin 2\omega t \qquad (4.22)$$

This series uses six terms. If we use the value of $y(\omega t)$ at six points, we can solve for $\tfrac{1}{2}A_0$, A_1, A_2, A_3, B_1, and B_2, solving six equations, simultaneously. Six points are used because the angles appearing in the sine and cosine terms will be multiples of $\pi/3$. The trigonometric terms will be ± 1, $\pm\sqrt{3}/2$, $\pm\frac{1}{2}$, or zero.

Needless to say, this is a tedious calculation, but matrix methods do simplify the solution of a set of simultaneous equations.

$$
\begin{bmatrix} y_0 \\ y_1 \\ y_2 \\ y_3 \\ y_4 \\ y_5 \end{bmatrix} =
\begin{bmatrix}
1 & 1 & 1 & 1 & 0 & 0 \\
1 & \frac{1}{2} & -\frac{1}{2} & -1 & \frac{\sqrt{3}}{2} & \frac{\sqrt{3}}{2} \\
1 & -\frac{1}{2} & -\frac{1}{2} & 1 & \frac{\sqrt{3}}{2} & -\frac{\sqrt{3}}{2} \\
1 & -1 & 1 & -1 & 0 & 0 \\
1 & -\frac{1}{2} & -\frac{1}{2} & 1 & -\frac{\sqrt{3}}{2} & \frac{\sqrt{3}}{2} \\
1 & \frac{1}{2} & -\frac{1}{2} & -1 & -\frac{\sqrt{3}}{2} & -\frac{\sqrt{3}}{2}
\end{bmatrix}
\begin{bmatrix} \frac{1}{2}A_0 \\ A_1 \\ A_2 \\ A_3 \\ B_1 \\ B_2 \end{bmatrix}
$$

inverting the matrix, we can find each term directly.

$$
\begin{bmatrix} \frac{1}{2}A_0 \\ A_1 \\ A_2 \\ A_3 \\ B_1 \\ B_2 \end{bmatrix} = \frac{1}{6}
\begin{bmatrix}
1 & 1 & 1 & 1 & 1 & 1 \\
2 & 1 & -1 & -2 & -1 & 1 \\
2 & -1 & -1 & 2 & -1 & -1 \\
1 & -1 & 1 & -1 & 1 & -1 \\
0 & \sqrt{3} & \sqrt{3} & 0 & -\sqrt{3} & -\sqrt{3} \\
0 & \sqrt{3} & -\sqrt{3} & 0 & \sqrt{3} & -\sqrt{3}
\end{bmatrix}
\begin{bmatrix} y_0 \\ y_1 \\ y_2 \\ y_3 \\ y_4 \\ y_5 \end{bmatrix}
$$

Similar matrices can be set up for any number of equations, but it is convenient to use submultiples of 2π, which repeat, such as $\pi/2$, $\pi/3$, $\pi/4$, $\pi/6$, $\pi/12$, $\pi/24$. This requires 6, 12, 24, and 48 intervals and a like number of equations. With 6 intervals and 6 ordinates, the coefficients of the first three harmonics can be found. With 12 intervals, the first six harmonics can be found. A 12 interval matrix is included. Normally, more intervals will give greater accuracy.

These devices have no advantage or disadvantage over a conventional Fourier series, but they do force a truncated series to go through a specific set of data points. This may or may not be important. For example, there may be one or more points that the series must pass through exactly.

$$
\begin{bmatrix}
\tfrac{1}{2}A_0 \\
A_1 \\
A_2 \\
A_3 \\
A_4 \\
A_5 \\
A_6 \\
B_1 \\
B_2 \\
B_3 \\
B_4 \\
B_5
\end{bmatrix}
= \frac{1}{12}
\begin{bmatrix}
1 & 1 & 1 & 1 & 1 & 1 & 1 & 1 & 1 & 1 & 1 & 1 \\
2 & \sqrt{3} & 1 & 0 & -1 & -\sqrt{3} & -2 & -\sqrt{3} & -1 & 0 & 1 & \sqrt{3} \\
2 & 1 & -1 & -2 & -1 & 1 & 2 & 1 & -1 & -2 & -1 & 1 \\
2 & 0 & -2 & 0 & 2 & 0 & -2 & 0 & 2 & 0 & -2 & 0 \\
2 & -1 & -1 & 2 & -1 & -1 & 2 & -1 & -1 & 2 & -1 & -1 \\
2 & -\sqrt{3} & 1 & 0 & -1 & \sqrt{3} & -2 & \sqrt{3} & -1 & 0 & 1 & -\sqrt{3} \\
1 & -1 & 1 & -1 & 1 & -1 & 1 & -1 & 1 & -1 & 1 & -1 \\
0 & 1 & \sqrt{3} & 2 & \sqrt{3} & 1 & 0 & -1 & -\sqrt{3} & -2 & -\sqrt{3} & -1 \\
0 & \sqrt{3} & \sqrt{3} & 0 & -\sqrt{3} & -\sqrt{3} & 0 & \sqrt{3} & \sqrt{3} & 0 & -\sqrt{3} & -\sqrt{3} \\
0 & 2 & 0 & -2 & 0 & 2 & 0 & -2 & 0 & 2 & 0 & -2 \\
0 & \sqrt{3} & -\sqrt{3} & 0 & \sqrt{3} & -\sqrt{3} & 0 & \sqrt{3} & -\sqrt{3} & 0 & \sqrt{3} & -\sqrt{3} \\
0 & 1 & -\sqrt{3} & 2 & -\sqrt{3} & 1 & 0 & -1 & \sqrt{3} & -2 & \sqrt{3} & -1
\end{bmatrix}
\begin{bmatrix}
y_0 \\
y_1 \\
y_2 \\
y_3 \\
y_4 \\
y_5 \\
y_6 \\
y_7 \\
y_8 \\
y_9 \\
y_{10} \\
y_{11}
\end{bmatrix}
$$

Solving for the set of coefficients using simultaneous equations will do this, at the cost of some inaccuracy elsewhere. These are just different ways of finding harmonics numerically.

4.10. WORK PER CYCLE

If the energy put into a vibrating mechanical system exceeds that which is dissipated in some form of damping, the amplitude of vibration builds. It is important at this state of our study of vibration to consider the work put into a vibrating system by a harmonic force.

Work is defined as the scalar product of force and the displacement in the direction of the force

$$dU = F(t) \cdot dx \tag{4.23}$$

For the period $\omega t = 0$ to $\omega t = 2\pi$, the work per cycle is

$$\Delta U = \int_0^{2\pi/\omega} F(t) \left(\frac{dx}{dt}\right) dt = \int_0^{2\pi} F(\omega t) \frac{\dot{x}}{\omega} d(\omega t) \tag{4.24}$$

Now, if we assume that the force $F(\omega t)$ is a harmonic of order n, $F(\omega t) = F_n \sin(n\omega t + \alpha_n)$ and the displacement is a simple harmonic with a frequency ω, $x = X \sin \omega t$, the work per cycle can be found by integrating

$$\Delta U = F_n X \int_0^{2\pi} \sin(n\omega t + \alpha_n) \cos \omega t \, d(\omega t)$$

$$= \pi F_1 X \sin \alpha_1 \quad \text{for} \quad n = 1 \tag{4.25}$$

$$= 0, \quad \text{for} \quad n \neq 1$$

Remember, n as used in $n\omega t$ is an integer which is the order of the harmonic. The value of the integral is zero, except where $n = 1$. This simply means that the harmonic force $F(\omega t)$ and the displacement $x(\omega t)$ must be of the same frequency, ω, for energy to be added to the system. The work done by a harmonic force over a complete cycle is a function of the maximum displacement, the maximum force, and the phase angle between the forcing function and the displacement. Referring to the vector diagram of Figure 4.8, $F_n \sin \alpha_n$ is the component of the rotating vector force in the direction of the velocity vector. A natural conclusion is that the force must be in phase with velocity of the same frequency to do work on the system.

This does not mean that ω is necessarily the fundamental. It could be any of the harmonics. If we had started with $F(\omega t) = F_n \sin(n\omega t + \alpha_n)$ and $x = X \sin 2\omega t$, no energy could have been added to the system unless $n = 2$. If the harmonic force and the displacement or velocity are of different frequencies, no work is done by the forcing function over an integral number of cycles.

Fig. 4.8

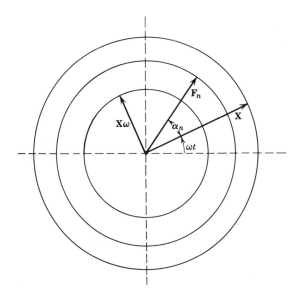

EXAMPLE PROBLEM 4.26

A 3.68-g slider is placed in a smooth tube with an internal diameter of 40 mm. The spring has a stiffness of 284 N/m. One side of the tube is open to the atmosphere. The pressure on the other side varies periodically as shown in the graph. Determine the response of the slider.

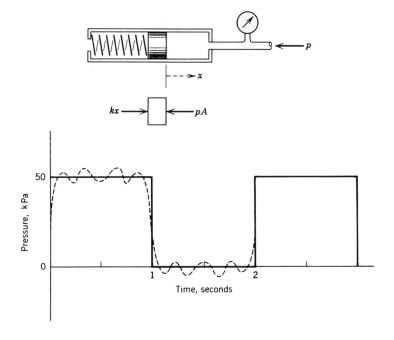

Solution:
The forcing function $p(\omega t)$ may be expressed in a Fourier series

$$p(\omega t) = \frac{A_0}{2} + A_1 \cos \omega t + A_2 \cos 2\omega t + \cdots$$

$$+ B_1 \sin \omega t + B_2 \sin 2\omega t + B_3 \sin 3\omega t + \cdots$$

for a period of $\tau = 2$ s, $\omega\tau = 2\pi$, and $\omega = \pi$ s^{-1}. Solving for amplitudes $A_0, A_1, A_2, \ldots$

$$A_0 = \frac{1}{\pi}\int_0^{2\pi} F(\omega t)\ d(\omega t) = \frac{1}{\pi}\int_0^{\pi} PA\ d(\omega t) = PA$$

$$A_1 = \frac{1}{\pi}\int_0^{2\pi} F(\omega t)\cos \omega t\ d(\omega t) = \frac{1}{\pi}[PA \sin \omega t]_0^{\pi} = 0$$

Likewise, $A_2 = A_3 = A_4 = \cdots = 0$. A without a subscript is the frontal area of the piston.

It is evident that there are no cosine terms in the series. This reflects our second thoughts, since the function $p(\omega t)$ is odd.

Solving for amplitudes $B_1, B_2, B_3, \ldots$

$$B_1 = \frac{1}{\pi}\int_0^{2\pi} F(\omega t)\sin(\omega t)\ d(\omega t) = \frac{PA}{\pi}[-\cos \omega t]_0^{\pi} = 2\frac{PA}{\pi}$$

$$B_2 = \frac{1}{\pi}\int_0^{2\pi} F(\omega t)\sin 2\omega t\ d(\omega t) = \frac{PA}{2\pi}[-\cos 2\omega t]_0^{\pi} = 0$$

$$B_3 = \frac{1}{\pi}\int_0^{2\pi} F(\omega t)\sin 3\omega t\ d(\omega t) = \frac{PA}{3\pi}[-\cos 3\omega t]_0^{\pi} = 2\frac{PA}{3\pi}$$

$$B_n = \frac{2PA}{n\pi} \qquad \text{for } n \text{ odd,}$$

$$B_n = 0 \qquad \text{for } n \text{ even.}$$

The graph $p(\omega t)$ can be expressed by the series

$$p(\omega t) = \frac{PA}{2} + 2\frac{PA}{\pi} \sin \omega t + \frac{2}{3}\frac{PA}{\pi} \sin 3\omega t + \frac{2}{5}\frac{PA}{\pi} \sin 5\omega t + \cdots$$

A plot of the series shows how closely these terms approximate the square wave.

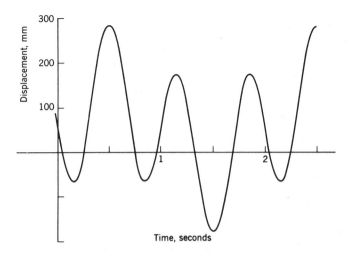

The natural frequency of the spring and mass system is

$$\omega_n^2 = \frac{k}{m} = \frac{284}{3.68} = 77.2 \text{ s}^{-2}$$

For the response to the static term,

$$X_0 = \frac{PA}{2k} = \frac{50000}{2}\left(\frac{\pi}{4}\right)\frac{(0.04)^2}{284} = 0.1106 \text{ m}$$

For the response to the first harmonic,

$$X_1 = 2\frac{PA}{\pi k}\left[\frac{1}{\left(1-\dfrac{\omega^2}{\omega_n^2}\right)}\right]$$

$$= \frac{2}{\pi}(50000)\left(\frac{\pi}{4}\right)\frac{(0.04)^2}{284}\left[\frac{1}{\left(1-\dfrac{\pi^2}{77.2}\right)}\right] = 0.1612 \text{ m}$$

For the response to the third harmonic,

$$X_3 = \frac{2}{3}\frac{PA}{\pi k}\left[\frac{1}{\left(1-\dfrac{9\omega^2}{\omega_n^2}\right)}\right]$$

$$= \frac{2(50000)}{3\pi}\left(\frac{\pi}{4}\right)\frac{(0.04)^2}{284}\frac{1}{\left(1-\dfrac{9\pi^2}{77.2}\right)} = -0.310 \text{ m}$$

And the fifth harmonic,

$$X_5 = \frac{2}{5}\frac{PA}{\pi k}\frac{1}{\left(1 - \frac{25\omega^2}{\omega_n^2}\right)}$$

$$= \frac{2(50000)}{5\pi}\left(\frac{\pi}{4}\right)\frac{(0.04)^2}{284}\left[\frac{1}{\left(1 - \frac{25\pi^2}{77.2}\right)}\right] = -0.0128 \text{ m}$$

The third harmonic is very large since the natural frequency is closer to the third harmonic than it is to the fundamental or any other harmonic. The response is shown in the graph.

EXAMPLE PROBLEM 4.27

Determine the first three harmonics of the triangular wave that has a period of 0.24 s, using numerical methods to determine a Fourier expansion.

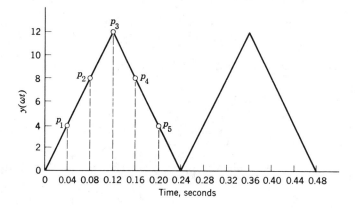

Solution:
The circular frequency of the fundamental is 2π radians every 0.24 s, or $\omega = 26.2 \text{ s}^{-1}$. Dividing the wave into six intervals, each interval will be 0.04 s or $\pi/3$ rad of the full period.

The average value of the function $p(t)$ will be the sum of all six values, $p_1, p_2, p_3, \ldots, p_6$, divided by six, or

$$A_0 = \frac{2}{6}\sum_{i=1}^{6} p_i = 12$$

$$\tfrac{1}{2}A_0 = 6,$$

which is the average value. The other coefficients can be found from equations 4.21.

$$A_n = \frac{2}{6} \sum_{i=1}^{6} p_i \cos \frac{2n\pi t_i}{\tau}$$

$$B_n = \frac{2}{6} \sum_{i=1}^{6} p_i \sin \frac{2n\pi t_i}{\tau}$$

Setting up a table for the tabular computation of the fundamental $A_1 \cos \omega t + B_1 \sin \omega t$, $n=1$:

i	p_i	$\cos \dfrac{2\pi t_i}{\tau}$	$\sin \dfrac{2\pi t_i}{\tau}$	$p_i \cos \dfrac{2\pi t_i}{\tau}$	$p_i \sin \dfrac{2\pi t_i}{\tau}$
1	4	0.50	0.866	2	3.464
2	8	-0.50	0.866	-4	6.928
3	12	-1.00	0.000	-12	0.000
4	8	-0.50	-0.866	-4	-6.928
5	4	0.50	-0.866	2	-3.464
6	0	1.00	0.000	0	0.000
				$\Sigma = -16$	$\Sigma = 0$

$$A_1 = \frac{2}{6} \sum_{i=1}^{6} p_i \cos \frac{2\pi t_i}{\tau} = \frac{-16}{3} = -5.333$$

$$B_1 = 0$$

Setting up the tabular computation of the second harmonic, $A_2 \cos 2\omega t + B_2 \sin 2\omega t$, $n=2$:

i	p_i	$\cos \dfrac{4\pi t_i}{\tau}$	$\sin \dfrac{4\pi t_i}{\tau}$	$p_i \cos \dfrac{4\pi t_i}{\tau}$	$p_i \sin \dfrac{4\pi t_i}{\tau}$
1	4	-0.50	0.866	-2	3.464
2	8	-0.50	-0.866	-4	-6.928
3	12	1.00	0.000	12	0.000
4	8	-0.50	0.866	-4	6.928
5	4	-0.50	-0.866	-2	-3.464
6	0	1.00	0.000	0	0.000
				$\Sigma = 0$	$\Sigma = 0$

$$A_2 = 0$$

$$B_2 = 0$$

And, in the third harmonic, $A_3 \cos 3\omega t + B_3 \sin 3\omega t$, $n = 3$:

i	p_i	$\cos \dfrac{6\pi t_i}{\tau}$	$\sin \dfrac{6\pi t_i}{\tau}$	$p_i \cos \dfrac{6\pi t_i}{\tau}$	$p_i \sin \dfrac{6\pi t_i}{\tau}$
1	4	-1	0	-4	0
2	8	1	0	$+8$	0
3	12	-1	0	-12	0
4	8	1	0	8	0
5	4	-1	0	-4	0
6	0	1	0	0	0
				$\Sigma = -4$	$\Sigma = 0$

$$A_3 = \frac{2}{6} \sum_{i=1}^{6} p_i \cos \frac{6\pi t_i}{\tau} = \frac{-4}{3} = -1.333$$

$$B_3 = 0$$

The approximate expression is therefore,

$$p(t) = 6 - 5.333 \cos 26.2t - 1.333 \cos 78.5t$$

If a large number of intervals are used, the Fourier expansion will be a more accurate expression of the function, and higher harmonics will be determined. As an example, using 24 intervals, each interval will be 0.01s or $\pi/12$ rad of a full period.

Doing the same for the second and third harmonics, an approximate expression using 24 intervals is

$$p(t) = 6 - 4.891 \cos 26.2t - 0.5691 \cos 78.5t$$

The exact expression is

$$p(t) = 6 - \sum_{n=1,3,5}^{\infty} \frac{1}{n^2} \cos \frac{n\pi t_i}{0.12}$$

$$= 6 - 4.05 \cos 26.2t - 0.45 \cos 78.5t$$

Since $p(t)$ can be expressed as a continuous function of time, the expressions for the Fourier coefficients are integrable.

The absence of sine terms could have been predicted, since the wave form is an even function, that is $p(t)$ is symmetric about the origin.

PROBLEM 4.28 Repeat Example Problem 4.27, forming the function to pass through six points, starting with $y_0 = 0$ at $t = 0$.

PROBLEM 4.29 Repeat Example Problem 4.27, using 12 points, starting with $y_0 = 0$ at $t = 0$.

PROBLEM 4.30 Repeat Example Problem 4.27, using 24 intervals, and determine an expression that includes the first, second, and third harmonics.

PROBLEM 4.31 Determine the response of the slider of the Example Problem 4.26 if the pressure varies periodically as below. Plot the response.

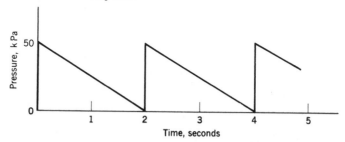

PROBLEM 4.32 Make a harmonic analysis, which includes the first three harmonics, of the forcing function shown.

$\omega t(°)$	$F(\omega t)$	$\omega t(°)$	$F(\omega t)$	$\omega t(°)$	$F(\omega t)$	$\omega t(°)$	$F(\omega t)$
15	1.15	105	4.35	195	5.60	285	4.80
30	1.10	120	5.50	210	4.80	300	4.60
45	1.20	135	6.25	225	4.40	315	4.00
60	1.55	150	6.65	240	4.40	330	3.10
75	2.20	165	6.65	255	4.60	345	2.15
90	3.20	180	6.35	270	4.75	360	1.50

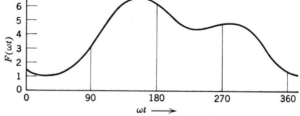

PROBLEM 4.33 Make a harmonic analysis of the wave form shown in the following graph.

Answer: $y = 40.0 \sin \dfrac{\pi t}{0.24} + 6.0 \sin \dfrac{3\pi t}{0.24}$

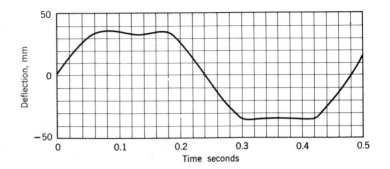

PROBLEM 4.34 The variation of pressure with time in a hydraulic line is shown in the following table. The data were taken directly from recorded pressures at 0.01-s intervals. What is the frequency of the fundamental? Using harmonic analysis, determine a Fourier expansion which will express the data.

Time (s)	Pressure (N/m^2)	Time (s)	Pressure (N/m^2)	Time (s)	Pressure (N/m^2)	Time (s)	Pressure (N/m^2)
0.01	2.9	0.07	28.7	0.13	54.3	0.19	28.6
0.02	10.4	0.08	31.5	0.14	46.8	0.20	25.7
0.03	19.2	0.09	38.0	0.15	38.0	0.21	19.2
0.04	25.7	0.10	46.8	0.16	31.5	0.22	10.4
0.05	28.6	0.11	54.3	0.17	28.7	0.23	2.9
0.06	28.6	0.12	57.2	0.18	28.6	0.24	0

PROBLEM 4.35 The turning effort of a typical six-cylinder four-stroke cycle gasoline engine is described by the following table and shown in the torque–time diagram. Make a harmonic analysis of the turning effort. Find the amplitude of at least the first three harmonics.

Time (s)	Torque $(N \cdot m)$	Time (s)	Torque $(N \cdot m)$	Time (s)	Torque $(N \cdot m)$	Time (s)	Torque $(N \cdot m)$
0.00075	410	0.00525	740	0.00975	750	0.01425	500
0.00150	420	0.00600	860	0.01050	700	0.01500	470
0.00225	440	0.00675	980	0.01125	650	0.01575	440
0.00300	480	0.00750	920	0.01200	600	0.01650	420
0.00375	540	0.00825	860	0.01275	560	0.01725	410
0.00450	640	0.00900	800	0.01350	530	0.01800	400

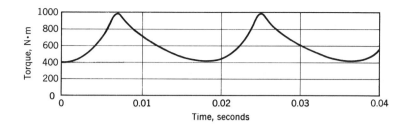

PROBLEM 4.36 The velocity pressure at the outlet of a ram jet, which is a measure of the thrust, varies as shown below as a function of time. Make a harmonic analysis of the pressure as a function of time. Include the first three harmonics. Check the maximum value of the first harmonic with the peak of the curve.

Time (s)	Pressure (MPa)	Time (s)	Pressure (MPa)	Time (s)	Pressure (MPa)	Time (s)	Pressure (MPa)
0.001	− 12.5	0.007	5.0	0.013	15.0	0.019	70.0
0.002	− 12.5	0.008	2.5	0.014	20.0	0.020	65.0
0.003	− 7.5	0.009	0	0.015	25.0	0.021	50.0
0.004	0	0.010	2.5	0.016	35.0	0.022	25.0
0.005	5.0	0.011	5.0	0.017	50.0	0.023	10.0
0.006	7.5	0.012	10.0	0.018	60.0	0.024	− 2.5

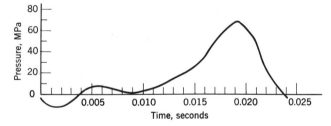

PROBLEM 4.37 A 200-kg vertical shaper is placed on rubber pads which deflect 2 mm under the weight of the shaper. The shaper has a maximum cutting speed of 225 cpm. Determine the vertical motion of the shaper if the vertical shaking force is $200 \cos \omega t + 75 \cos 2\omega t$, expressed in Newtons. The strong second harmonic originates from the quick return mechanism of the shaper.

Answer: $x(t) = 0.224 \cos \omega t + 0.119 \cos 2\omega t$, mm

$$F(t) = 200 \cos \omega t + 75 \cos 2 \omega t$$

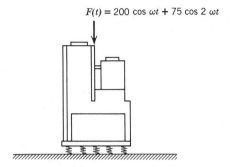

PROBLEM 4.38 A high-speed internal combustion engine has a bore of 92 mm and a stroke of 100 mm. The wrist pin is set laterally within the body of the piston. One end of the 200-mm connecting rod is attached to the wrist pin and the other to the crank shaft. The piston assembly is quite rigid and is ordinarily considered as a single mass. If we consider the wrist pin as a flexible spring, between the piston and connecting rod, the natural frequency of this system is 300 cps. Determine the maximum displacement of the piston relative to the connecting rod for an operating speed of 3600 rpm.

Answer: 2.68 mm

PROBLEM 4.39 A 1000-kg punch operates at 86 rpm and is known to transmit a force to the foundation which can be expressed as

$$F(t) = 25 + 20 \sin 9t - 3 \sin 18t - 2 \sin 27t$$

where t is in seconds. The punch can be mounted on a cork and rubber pad with a rated deflection of 2 mm, or on commercial isolators which deflect from 20 mm to 100 mm under load, depending on the isolator selected. If these are taken as three representative choices, which is preferred and why is it preferred?

Answer: cork and rubber pad

PROBLEM 4.40 A self-contained textile machine with a mass of 5000 kg must be isolated against the transmission of force to the foundation. The forcing function can be represented as

$$F(t) = 100 \sin \omega t + 5 \sin 2\omega t$$

and the frequency of the fundamental is 1200 rpm. Vibration isolators are rated by the static deflection with the isolators under load; isolators come in ratings of 4, 6, and 8 mm. Which would reduce the transmission of force below 10%?

Answer: 8 mm rating

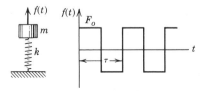

PROBLEM 4.41   A single degree of freedom linear oscillator with mass m and spring constant k is excited in steady state vibration by the square wave, $f(t)$, shown here.

The square wave is to be tuned by adjusting the oscillator period τ. Under what conditions would resonance be expected to occur?

PROBLEM 4.42   A single degree of freedom, linear oscillator with a natural circular frequency $\omega_n = 2\omega$ is driven by the force

$$f(t) = \cos(\omega t) + \cos(3\omega t + \phi)$$

Determine the phase angle ϕ so that the response of the oscillator is maximized. What is the maximum response amplitude?

PROBLEM 4.43   Determine the response x of a simple pendulum to the cam, which moves at a constant angular velocity. The movement of point A is plotted as shown. The spring is stiff enough to maintain contact between the cam and follower at all times.

Answer: $x = 1.09 \sin \omega t + 0.0694 \sin 3\omega t$
$- 0.006 \sin 5\omega t$

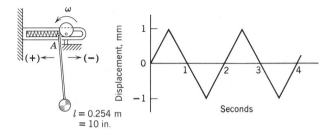

PROBLEM 4.44   Determine the response of the simple pendulum to the cam that imparts a periodic saw to the displacement to point A of

the pendulum of Problem 4.43. The spring is stiff enough to maintain contact between the cam and follower.

Answer: $x = 0.5 - 0.428 \sin \pi t + 7.07 \sin 2\pi t + 0.0815 \sin 3\pi t + 0.0258 \sin 4\pi t + 0.0118 \sin 5\pi t + \cdots$

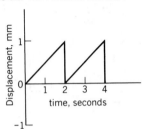

CHAPTER FIVE

INITIAL CONDITIONS AND TRANSIENT VIBRATION

5.1. INTRODUCTION

A temporary component of motion is called a *transient*. In general, transient motion or transients accompany any change in the amount or form of energy stored in a vibrating system. In many cases, transient vibration can be ignored, considering only steady state vibration. However, large margins of safety, in which ignorance of transient conditions can be buried are not always possible, and today, the study of transient phenomena is one of the largest areas of concern in mechanical vibration. The analysis of transient motion requires extensive mathematical treatment and modern computational methods.

In mechanical vibration, the study of transient motion centers around the problems of overstress and the possibility of catastrophic failure or perhaps, low cycle fatigue. Although our mathematics may be similar to other problems in engineering, such as control theory, our perspectives of the physical problem are different. We are not as interested in the length of time during which a transient diminishes as we are in the number of cycles of large amplitude or high stress that the system may have. The presence of the transient condition is not as important as is the maximum amplitude or maximum stress.

The exact solution of transient vibration problems is difficult, requiring advanced mathematical methods, but, there are three basic forcing functions that can be used to approximate transient phenomena. These are the rectangular step forcing function, or suddenly applied load, the exponentially decaying step, and the ramp function, where force increases linearly with time. Since our primary interest is to determine if mechanical failure occurs, many times we can ignore a complete mathematical solution in favor of an approximate solution that uses one of these three basic functions. The approximate solutions are often sufficient to satisfy questions of design.

5.2. THE RECTANGULAR STEP FORCING FUNCTION

If a constant load F_0 is suddenly applied to a simple spring and mass system, the spring will extend a distance F_0/k, and the mass will oscillate about this new equilibrium position. The amplitude of the vibration will depend on the initial conditions of motion.

Consider the free body diagram of Figure 5.1, after the load F_0 is applied. The equation of motion is

$$m\ddot{x} + kx = F_0 \qquad (5.1)$$

and the displacement x, is

$$x = A \cos \omega_n t + B \sin \omega_n t + (F_0/k) \qquad (5.2)$$

If the system were initially at rest, $x(0) = 0$ and $\dot{x}(0) = 0$. The arbitrary constants A and B are no longer arbitrary; substitution of the initial conditions will show that the constants are $A = -F_0/k$ and $B = 0$, and the response would be

$$x = (F_0/k)(1 - \cos \omega_n t) \qquad (5.3)$$

Fig. 5.1

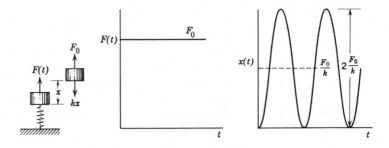

A displacement–time curve shows this motion in Figure 5.1. Note that the maximum amplitude of motion is $2(F_0/k)$. This is a simple statement that is quite valuable in design. The maximum displacement for a suddenly applied load is twice the displacement for the same load applied statically.

If the mass m were not initially at rest, the amplitude of motion would be larger. If $x(0) = x_0$ and $\dot{x}(0) = v_0$, $A = x_0 - (F_0/k)$ and $B = v_0/\omega_n$

$$x = \left(x_0 - \frac{F_0}{k}\right)\cos \omega_n t + \frac{v_0}{\omega_n} \sin \omega_n t + \frac{F_0}{k}$$

which can be rearranged algebraically into the more instructive form,

$$x = \left(\sqrt{x_0^2 + \frac{v_0^2}{\omega_n^2}}\right)\sin(\omega_n t + \phi) + \frac{F_0}{k}(1 - \cos \omega_n t) \tag{5.4}$$

If the system were initially at rest, the first term in equation 5.4 would be zero and equation 5.4 would be identical with equation 5.3.

5.3. THE RAMP OR LINEARLY INCREASING FORCING FUNCTION

In Figure 5.2, the forcing function $F(t)$ linearly increases with time. Naturally, it cannot increase indefinitely, but our interest is confined to the regime where the force does increase linearly and the elastic limit of the spring has not been reached. The equation of motion is

$$m\ddot{x} + kx = Ct \tag{5.5}$$

for which

$$x = A \cos \omega_n t + B \sin \omega_n t + (Ct/k) \tag{5.6}$$

Again, the arbitrary constants A and B depend on the initial conditions. If the system were initially at rest, $x(0) = 0$ and $\dot{x}(0) = 0$, the arbitrary constants $A = 0$ and $B = -C/k\omega_n$

$$x = (C/k\omega_n)(\omega_n t - \sin \omega_n t) \tag{5.7}$$

Fig. 5.2

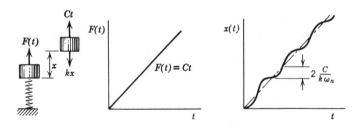

The displacement oscillates at the natural frequency ω_n about a middle position that linearly increases with time. The amplitude of this oscillation is $C/k\omega_n$. Catastrophic failure is possible, and the maximum displacement from the original equilibrium position, at any time t, is

$$X = (C/k\omega_n)(1 + \omega_n t) \tag{5.8}$$

5.4. THE EXPONENTIALLY DECAYING FORCING FUNCTION

A third basic forcing function is the suddenly applied load that decays exponentially with time. It typifies many blastlike impulses and can be adapted to a variety of problems by changing the value of the exponent a, which has the same units as $\omega(\text{rad/s})$.

In the equation of motion,

$$m\ddot{x} + kx = F_0 e^{-at} \tag{5.9}$$

for which the solution is

$$x = A \cos \omega_n t + B \sin \omega_n t + \frac{F_0 e^{-at}}{ma^2 + k} \tag{5.10}$$

The mathematical similarity between this forcing function and the forcing function $F(t) = F_1 e^{i\omega t}$ can be used as an analogy. It is the same as equation 4.6, with a substituted for $i\omega$ and F_0 substituted for F_1.

Assuming rest conditions initially,

$$x = \frac{F_0}{k\left(1 + \frac{a^2}{\omega_n^2}\right)} \left[\frac{a}{\omega_n} \sin \omega_n t - \cos \omega_n t + e^{-at}\right] \tag{5.11}$$

If $a \to 0$, the forcing function decays very slowly, and the displacement-time curve approaches $x = (F_0/k)(1 - \cos \omega_n t)$, the response to a rectangular step forcing function.

If $a \to \infty$ the forcing function immediately decays to zero, the area under the force–time curve (which is the net impulse), approaches zero, and the force imparts no velocity change to the mass m. If $a \to \infty$, the system would remain motionless.

These are the two extremes. Between these, the displacement oscillates, with the amplitude decreasing with time. The steady state amplitude, which is reached for $at > 5$ is

$$X = \frac{F_0}{k\sqrt{1 + \frac{a^2}{\omega_n^2}}} \tag{5.12}$$

Fig. 5.3

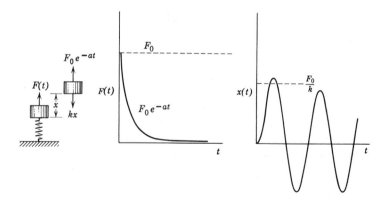

The maximum amplitude depends on the ratio a/ω_n but it is always less than $2(F_0/k)$. This again is an important fact to remember. The only way that the displacement can exceed $2(F_0/k)$ is for the initial conditions to have an initial velocity, imparting kinetic energy to the system. Figure 5.3 shows the response of a simple system to an exponentially decaying step function.

5.5. COMBINATIONS OF FORCING FUNCTIONS

The three basic forcing functions can be combined in a variety of ways to approximate just about any possible transient forcing function. In a linear system the response to a combination of two or more forcing functions, all acting at the same time, can be found by superposing the response from each forcing function, one on the other. In linear mathematics, this is known as the principle of superposition.

Rectangular Step—Exponential Decaying Step. As examples of combinations of practical importance, an exponentially decaying step forcing function can be subtracted from a rectangular step.

$$F(t) = F_0(1 - e^{-at}) \tag{5.13}$$

This function is familiar since it is analogous to the response of a glass thermometer immersed in a liquid bath. It is evident that $F(t)$ could reproduce the transient response to thermal stress or a thermal pulse due to sudden heating.

By combining equation 5.2 with equation 5.10, the solution to such a transient forcing function would be

$$x = A \cos \omega_n t + B \sin \omega_n t + \frac{F_0}{k} \left[1 - \frac{e^{-at}}{\left(1 + \dfrac{a^2}{\omega_n^2}\right)} \right] \tag{5.14}$$

If the system were initially at rest, $x(0) = 0$ and $\dot{x}(0) = 0$

$$x = \frac{F_0}{k} - \frac{F_0}{k\left(1 + \dfrac{a^2}{\omega_n^2}\right)} \left[\frac{a}{\omega_n} \left(\frac{a}{\omega_n} \cos \omega_n t + \sin \omega_n t \right) + e^{-at} \right] \quad (5.15)$$

The response to this motion is shown in Figure 5.4. Note that the motion becomes simple harmonic for $at > 5$, oscillating about a new equilibrium position, which is displaced by a distance F_0/k. Dropping the exponential e^{-at}, for $at > 5$, would constitute an error of less than 0.67%.

Fig. 5.4

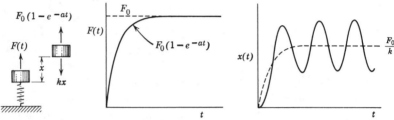

$$x = \frac{F_0}{k} \left[1 - \frac{\left(\dfrac{a}{\omega_n}\right)}{\sqrt{1 + \dfrac{a^2}{\omega_n^2}}} \sin (\omega_n t + \phi) \right] \qquad at > 5 \qquad (5.16)$$

The amplitude of the simple harmonic motion is

$$\frac{F_0}{k} \cdot \frac{\left(\dfrac{a}{\omega_n}\right)}{\sqrt{1 + \dfrac{a^2}{\omega_n^2}}}$$

and the maximum possible excursion is

$$\frac{F_0}{k} \left[1 + \frac{\left(\dfrac{a}{\omega_n}\right)}{\sqrt{1 + \dfrac{a^2}{\omega_n^2}}} \right] \qquad (5.17)$$

Two Exponential Decaying Step Functions. A second example of a practical importance is to subtract one exponentially decaying step from another.

$$F(t) = F_0(e^{-at} - e^{-bt}) \qquad (5.18)$$

Fig. 5.5

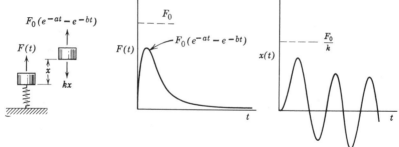

For $F(t)$ to be positive $b>a$. A variety of blastlike impulses can be approximated by changing the time constants a and b. In Figure 5.5 the impulse shown is for $b=2a$. The displacement x is

$$x = A \cos \omega_n t + B \sin \omega_n t + \frac{F_0}{k}\left[\frac{e^{-at}}{\left(1+\dfrac{a^2}{\omega_n^2}\right)} - \frac{e^{-bt}}{\left(1+\dfrac{b^2}{\omega_n^2}\right)}\right] \qquad (5.19)$$

For rest conditions initially,

$$x = \frac{F_0}{k\left(1+\dfrac{a^2}{\omega_n^2}\right)}\left[\frac{a}{\omega_n}\sin\omega_n t - \cos\omega_n t + e^{-at}\right]$$

$$-\frac{F_0}{k\left(1+\dfrac{b^2}{\omega_n^2}\right)}\left[\frac{b}{\omega_n}\sin\omega_n t - \cos\omega_n t + e^{-bt}\right] \qquad (5.20)$$

Successive Functions. It is also possible to add or subtract one function after another, using the previous solutions in succession as building blocks for more complicated forcing functions. As an example, in Figure 5.6a, one rectangular step function is subtracted from another of equal magnitude, but after a time interval τ has passed. The result is a square impulse. A triangular impulse of Figure 5.6b is the result of having a linearly increasing forcing function which has a duration of a period τ, and a maximum force F_0, added to a linearly decreasing forcing function, which is mirror image of the first.

In the same way, in Figure 5.6c, a ramp function is added to a rectangular step, giving the forcing function a finite rise time. A ramp function could be combined with an exponentially decaying function, shown in 5.6d. This last example is an excellent approximation of a shock. If the forcing function is expressed in pressure, it would simulate a shock wave. In each instance, Figure 5.6a to d, we are not superposing forcing functions, but are applying them successively. The terminal condition for one

Fig. 5.6

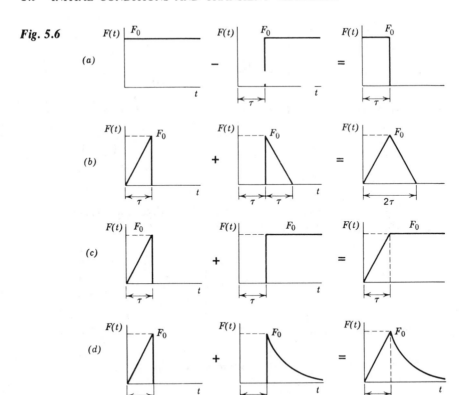

transient becomes the initial condition for the next. This is not an easy way to solve transient problems, but it is direct. It is physically simple and algebraically difficult, unless the simplest of mathematical terms are used. But, it is very easy for an engineer to see what is happening. As an example, if the rise time, τ, is equal to one half the natural period of the single degree of freedom system, the system will be at maximum displacement when the transition occurs. Simlarly, if the rise time is equal to the natural period of motion, or some multiple of the natural period, the system will have a minimum displacement when transition occurs.

EXAMPLE PROBLEM 5.1

A 20-kg piston slides in a smooth cylinder with an internal diameter of 125 mm and is supported on a spring with an elastic modulus of 965 N/m. At time $t=0$, the valve between the gas reservoir and the cylinder is suddenly opened, raising the pressure in the cylinder to 20 kPag. The gas is then bled through an orifice B, the pressure falling exponentially with time so that after 0.2 s, the pressure is reduced to 10 kPag. Determine the transient response of the piston as a function of time, and the steady state amplitude of the piston.

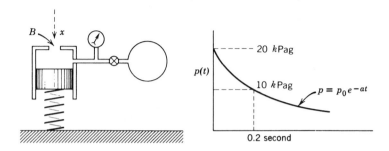

Solution:
The pressure is known to have an initial value of 20 kPag, to decrease exponentially with time, passing through the point where $p(t) = 10$ kPag at $t = 0.2$ s. The physical constants are

$$p = p_0 e^{-at}$$

$$\frac{p}{p_0} = e^{-at}$$

solving for the constant a,

$$\frac{1}{2} = e^{-a(0.2)}$$

$$a = 3.47 \text{ s}^{-1}$$

and the natural frequency, ω_n,

$$\omega_n^2 = \frac{k}{m} = \frac{965}{20} = 48.25 \text{ s}^{-2}$$

$$\omega_n = 6.94 \text{ s}^{-1}$$

for the equation of motion,

$$\sum F = m\ddot{x}$$

$$m\ddot{x} + kx = F_0 e^{-at} = p_0 A_r e^{-at}$$

A_r is the projected area. The solution is

$$x = A \cos \omega_n t + B \sin \omega_n t + \frac{p_0 A_r e^{-at}}{(ma^2 + k)}$$

for rest conditions, initially $x(0) = 0$, and $\dot{x}(0) = 0$

$$x = \frac{p_0 A_r}{(ma^2 + k)} \left(\frac{a}{\omega_n} \sin \omega_n t - \cos \omega_n t + e^{-at} \right)$$

A plot of this equation, as a function of time, shows that the maximum deflection occurs after 0.36 s, and is 0.312 m.

As $e^{-at} \to 0$, the motion approaches a steady state.

$$x = \frac{p_0 A_r}{(ma^2 + k)} \cdot \left(\frac{a}{\omega_n} \sin \omega_n t - \cos \omega_n t \right)$$

$$= \frac{p_0 A_r}{(ma^2 + k)} \left(\sqrt{1 + \frac{a^2}{\omega_n^2}} \right) \sin(\omega_n t + \phi)$$

$$= \frac{p_0 A_r}{k \sqrt{1 + \frac{a^2}{\omega_n^2}}} \sin(\omega_n t + \phi)$$

For this case, $a/\omega_n = \dfrac{1}{2}$ and

$$X = \frac{p_0 A_r}{k \sqrt{1 + \left(\frac{1}{2} \right)^2}} = \sqrt{\frac{4}{5}} \frac{[(20,000)(\pi/4)(0.125)^2]}{965} = 0.227 \text{ m}$$

PROBLEM 5.2 The piston of Example Problem 5.1 is subjected to a pressure that slowly leaks in, instead of out. p_s is the final static pressure. Determine the response of the system and estimate the maximum displacement of the piston in the cylinder.

Answer: $x = 2.54 - 2.19 \, (0.159 \cos 695t + 0.399 \sin 6.95t + e^{-2.77t})$

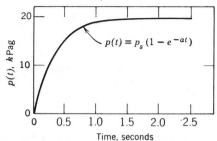

PROBLEM 5.3 Determine an expression for the transient response of a linear single degree of freedom system to the versed sine impulse

$$\frac{F_0}{2} (1 - \cos \omega t)$$

for $t < \tau$, assuming the system to be initially at rest.

Answer:

$$x = \frac{F_0}{2k\left(1 - \dfrac{\omega^2}{\omega_n^2}\right)} \times \left[1 - \cos \omega t - \frac{\omega^2}{\omega_n^2}(1 - \cos \omega_n t)\right]$$

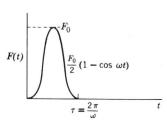

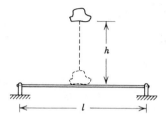

PROBLEM 5.4 A falling mass m drops from a height h and strikes a simply supported beam which has an equal mass and remains attached to the beam without rebound. Determine the maximum dynamic deflection of the beam in terms of the static deflection and the modulus of elasticity, the flexural rigidity and the length of span, provided that the deflection remains elastic. At impact, the velocity of the mass is v. After impact, the deflection of the beam is Δ_{st}.

$$\textit{Answer: } \delta = \Delta_{st}\left[1 + \sqrt{1 + \left(\frac{m}{m + \frac{17}{35}m_b}\right)\frac{v^2}{g\Delta_{ST}}}\right]$$

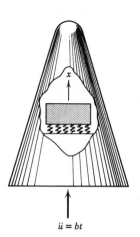

PROBLEM 5.5 An instrument package installed in the nose of a rocket is cushioned against vibration. The rocket is fired vertically from rest with an acceleration which increases linearly with time $\ddot{u} = bt$, where b is a constant. Derive the equation of motion for the instrument package if it has a mass, m. Find expressions for the relative displacement of the package with respect to the rocket and its absolute acceleration as a function of time. Assume initial conditions, $z(0) = 0$, $\dot{z}(0) = 0$, where $z = x - u$.

$$\textit{Answer: } \ddot{x} = bt\left[1 - \frac{1}{\omega_n t}\sin \omega_n t\right]$$

$\ddot{u} = bt$

PROBLEM 5.6 If the rocket of Problem 5.5 were to be fired vertically from rest with a constant acceleration $\ddot{u} = c$, find an expression for the relative displacement of the package with respect to the rocket, and the acceleration of the rocket.

PROBLEM 5.7 In ejection seat experiments, the torso is modeled as a spring and mass system. The head is a single mass weighing 5.44 kg. It is supported by the spinal column, with an elastic modulus of 87.56 N/mm. If ejection follows the acceleration–time curve shown, determine the peak acceleration of the head. Does it match experiment?

Answer: 33 g; yes

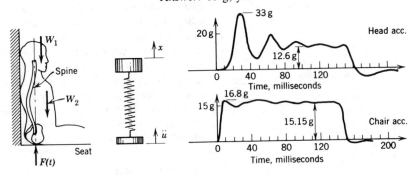

PROBLEM 5.8 1000 kg of sand are loaded from a hopper into a dump truck in 2 s. The dump truck and its load have a mass of 10 000 kg before the 1000 kg of sand are loaded. The springs of the truck have a modulus of 350 kN/m. Estimate the vertical displacement of the truck at the end of the loading process. You may assume that the sand is loaded at a uniform rate during those 2 s.

Answer: 30.4 mm

PROBLEM 5.9 In the firing of the second or third stage of a rocket, the acceleration increases slowly according to

$$\ddot{u} = ae^{bt}$$

Determine an expression for the acceleration $\ddot{x}$ that an astronaut feels. Assume initial conditions $z(0) = 0$, $\dot{z}(0) = 0$, where $z = x - u$.

Answer:

$$\ddot{x} = \frac{a}{(b^2 + \omega_n^2)} \left(\omega_n^2 \cos \omega_n t + b\omega_n \sin \omega_n t - \omega_n^2 e^{bt} \right)$$

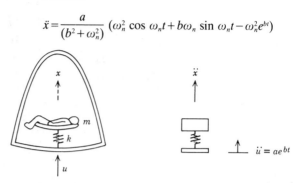

PROBLEM 5.10   A thin wire of length l, linear coefficient of thermal expansion α, cross-section A and elastic modulus E, supports a mass m. Electrical continuity is maintained by a mercury pool that offers little resistance to the motion of the mass. When the switch is closed the capacitor discharges through the wire. The rate of temperature rise for the wire is $\phi - \phi_0 = (\phi_m - \phi_0)(1 - e^{-t/t_1})$, where ϕ_0 is the ambient temperature and ϕ_m is the maximum temperature, and $t_1 = L/R$. Determine the equation of motion and the deflection x if the system were initially at rest.

PROBLEM 5.11   Two identical pistons, each with a mass of 2.5 kg and elastically connected with a constant of 2000 N/m fit in a smooth tube that is 50 mm in diameter. At rest, the spring is unextended and uncompressed and 0.25 m separate the two pistons. The pistons and tube are connected through valve B to a large storage tank containing a dry gas stored at a pressure of 100 kPag. At some instant, the valve B is opened and the pressure in the tube rises to the pressure in the storage tank as shown. Determine the maximum and minimum separation of the pistons after equilibrium has been restored.

Answer: 0.372 m; 0.324 m

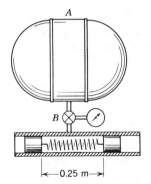

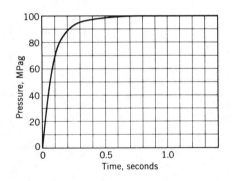

PROBLEM 5.12 An atmospheric pressure pulse can be approximated by the equation

$$p(t) = 500[e^{-100t} - e^{-1000t}]$$

where $p(t)$ is in Pascals and t is in seconds. Estimate the lateral displacement of a small car that has a projected area of 5 m² and has a mass of 800 kg. The four rubber tires have a total resistance to lateral movement measured as 618 N/mm.

Answer: 0.96 mm

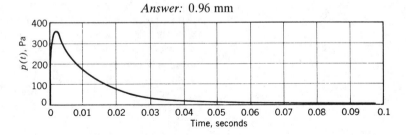

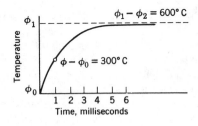

PROBLEM 5.13 A 2-kg target is suspended by a single steel cylindrical strut 1.06 m long and 5 mm diameter and placed within an atomic pile, where it is subjected to neutron bombardment. This induces a rapid rise in temperature in both the target and the strut and induces a thermal stress, $E\alpha(\phi - \phi_0)$ in the strut, where E is the modulus of elasticity, α is the coefficient of expansion, and $(\phi - \phi_0)$ is the temperature differential. If the temperature differential is 300°C after 1 ms, and reaches a steady state value of 600°C exponentially, $(\phi - \phi_0) = 600(1 - e^{-at})$, what is

(a) the amplitude of motion for the thermally induced vibration, and

(b) the maximum displacement of the target from its original position?

Hint: Superpose the elastic stress due to target displacement and the induced thermal stress.

Answer: (a) 3.3 mm; (b) 10.8 mm

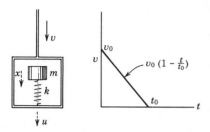

PROBLEM 5.14 An elevator moves downward at a constant velocity v_0. At a time $t=0$, the elevator stops, the velocity decreasing linearly to zero in a time t_0. Determine the displacement x of the mass m for $t \le t_0$

Answer: $x = \dfrac{v_0}{\omega_n t_0}(1 - \cos \omega_n t) + v_0 t\left(1 - \dfrac{t}{2t_0}\right)$

PROBLEM 5.15 Do Problem 5.14, but in this case, let the velocity of the elevator decrease parabolically with time, $v = v_0(1 - t^2/t_0^2)$

Answer: $x = \dfrac{2v_0}{\omega_n^2 t_0^2}(\omega_n t - \sin \omega_n t) + v_0 t\left(1 - \dfrac{t^2}{3t_0^2}\right)$

PROBLEM 5.16 Do Problem 5.14, but in this case, let the velocity of the elevator decrease with time, $v = v_0(1 - t/t_0)^2$

Answer: $x = \dfrac{2v_0}{\omega_n t_0}(\cos \omega_n t - 1)$

$\qquad + \dfrac{2v_0}{\omega_n^2 t_0^2}(\omega_n t - \sin \omega_n t) + v_0 t\left(t - \dfrac{t}{t_0} + \dfrac{t^2}{3t_0^2}\right)$

5.6. STATE SPACE AND THE PHASE PLANE

For a simple elastic system, the equation of motion is described by the linear, second-order differential equation,

$$\ddot{x} + \omega_n^2 x = 0 \qquad (5.21)$$

where ω_n^2 is determined by the system constants. Let us introduce a *state variable*, y, which is defined as

$$y = \frac{\dot{x}}{\omega_n} \qquad (5.22)$$

It is called a state variable because it is a function of the time derivative of the variable x. Differentiating,

$$\dot{y} = \frac{\ddot{x}}{\omega_n}$$

Substituting this expression in the equation of motion reduces the equation from a single second-order equation to two first-order equations, 5.22 and

$$\dot{y}\omega_n + \omega_n^2 x = 0$$
$$x = -\frac{\dot{y}}{\omega_n} \tag{5.23}$$

Dividing $\dot{y}$ by $\dot{x}$, the frequency ω_n cancels.

$$\frac{\dot{y}}{\dot{x}} = \frac{dy}{dx} = -\frac{x}{y} \tag{5.24}$$

The motive behind this mathematical ploy is clear. Equation 5.24 describes a circle where $x^2 + y^2 = X^2$. We can plot $y = f(x)$ as a circle of radius X, which is a constant. Figure 5.7 is such a plot.

Fig. 5.7

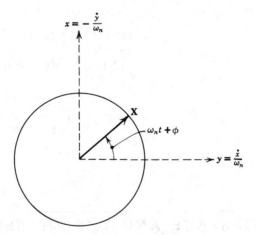

The rotation of the radius vector **X** describes both the harmonic displacement and the velocity. The vector **X** rotates at angular velocity ω_n and the position of the vector is located by the angle $(\omega_n t + \phi)$. The physical meaning is unchanged. For a simple elastic system, the displacement is harmonic, and

$$x = X \sin(\omega_n t + \phi)$$
$$\dot{x} = y\omega_n = X\omega_n \cos(\omega_n t + \phi)$$

Cancelling ω_n,

$$y = X \cos(\omega_n t + \phi)$$

Adding $\sin^2(\omega_n t + \phi) + \cos^2(\omega_n t + \phi) = 1$, or $x^2/X^2 + y^2/X^2 = 1$, we have a graphical description of $y = f(x)$, where y and x are the ordinate and abscissa, respectively, known as the *phase-plane*. Each point in the phase-plane represents a unique displacement and a unique velocity. The rest state is the origin, where $x = 0$ and $y = \dot{x}/\omega_n = 0$. All points that satisfy the equation of motion $\ddot{x} + \omega_n^2 x = 0$ lie on the circle of radius X. The magnitude of the vector $\mathbf{X}$ is determined by the initial conditions of motion.

5.7. THE STATE SPACE RESPONSE TO A RECTANGULAR STEP IMPULSE

If a constant load is suddenly applied to a simple spring and mass at rest, the equilibrium position is changed by a displacement of F/k. The equation of motion is $\ddot{x} + \omega_n^2 x = F/m$. Since the system was at rest, the initial velocity is zero, but the initial displacement from equilibrium is now F/k. A circle with a radius of $X = F/k$, centered at the new equilibrium position, will be the locus of all the state points in the phase-plane or state space, which represent displacement and velocity.

After a period t_1, the suddenly applied load is just as suddenly removed. The state variables at that time are y_1 and x_1. These become the initial state conditions for motion which is now described by the rotation of the vector $\mathbf{X}$ about the original origin. Both the displacement and the velocity may be found by projection of the respective vectors $\mathbf{X}_1$ on either the x or y axis, respectively. Figure 5.8 shows graphically the response to such a suddenly applied load.

If several rectangular step impulses succeed one another, the terminal state conditions from one impulse become the initial state conditions of the next. In Figure 5.9, the arc from $(0, 0)$ to (x, y_1) represents the state conditions during the time interval t_1. The arc from (x_1, y_1) to (x_2, y_2) represents the state conditions during the time interval $t_2 - t_1$. The arc from (x_2, y_2) to (x_3, y_3) is for the time interval $t_3 - t_2$, and so on. In this

Fig. 5.8

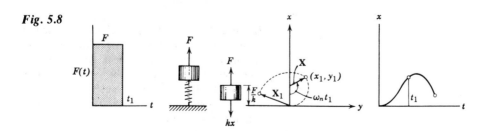

Fig. 5.9

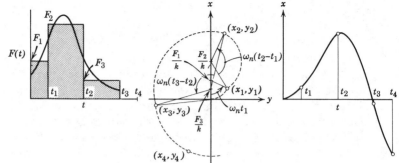

way, the response to any general forcing function can be found by approximating it as the response to a series of successive rectangular step impulses.

Figure 5.9 is a general forcing function, which can be approximated by the three step impulses. In Figure 5.10, the number of impulses has been doubled, giving a better approximation and a smoother state space curve. Note that the displacement time curve is little different from Figure 5.9. The velocity time curve is less exact, which could be expected. The use of the phase-plane is a very good way to find displacement, but it is a poor way to find velocity.

Fig. 5.10

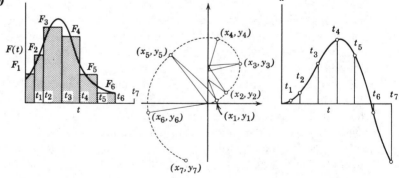

PROBLEM 5.17 The pendulum of Problem 4.22 has a length of 30.4 mm. Determine (a) the maximum excursion and (b) the double amplitude of steady state vibration of the pendulum after 1 s if the horizontal oscillation x_1 of the slider varies as shown on next page. Use the phase-plane for your solution.

Answer: (a) $+10$ mm, -11.62 mm; (b) 3.78 mm

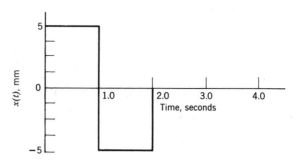

PROBLEM 5.18 The instrument package is mounted on commercial isolators within an aircraft. The isolators deflect 4.76 mm under static load. Using the phase-plane method, graphically determine the maximum amplitude of the instrument package if the aircraft fuselage moves vertically as shown.

Answer: 15.5 mm

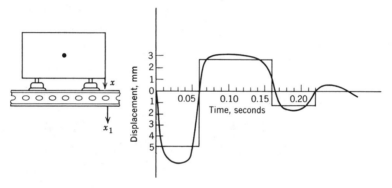

PROBLEM 5.19 Ground motion at a structure is determined to vary as shown in the displacement–time curve. Using the phase-plane method, graphically determine the lateral response of the structure with a period of 0.4 s, assuming that it is initially at rest. Determine the maximum excursion of motion and the double amplitude of steady state vibration at the end of 1 s.

Answer: 5.88 mm; 3.88 mm

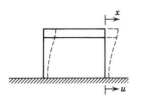

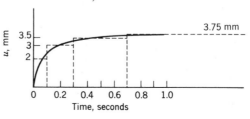

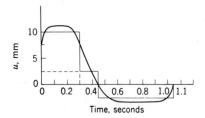

Time, seconds

PROBLEM 5.20   Ground motion at the structure of Problem 5.19 is determined to vary as shown in the displacement–time curve. Using the phase-plane method and assuming that the structure is initially at rest, graphically determine the maximum excursion of motion and the double amplitude of steady state vibration at the end of 1.05 s.

Answer: 29.6 mm; 8.8 mm

PROBLEM 5.21   A test specimen impacts against a sand-filled rubber bag, which has a linear constant of deformation of 25 N/mm for deformations up to 0.44 m. The table supporting the specimen is hydraulically driven as shown. Determine the maximum deflection and the steady state amplitude after 0.030 s. The natural period of oscillation of the system is 0.015 s.

Answer: 29.8 mm; 9.4 mm

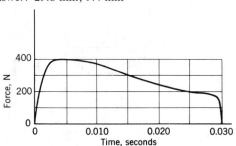

Time, seconds

PROBLEM 5.22   A static force of 500 N deflects the table 2 mm. The table receives a series of short hammer blows with a duration of 0.1 s and with 1 s between blows. What is the maximum horizontal displacement of the table, if the natural frequency of the table is 1 cps, and the table is initially at rest?

Answer: 4 mm

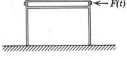

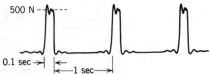

5.8. THE USE OF THE LAPLACE TRANSFORM METHOD

The Laplace transform is also used in the solution of vibration problems, primarily because the method is simple to use, has wide acceptance, and can be applied to simple differential equations without going through a derivation or integration. Tables of transforms are readily available, and the process of solving differential equations dissolves to judging the correct transform or the inverse transform.

By definition, the Laplace transform of $x(t)$ is

$$\mathscr{L}[f(t)] = \int_0^\infty f(t)e^{-st}\, dt = f(s) \tag{5.25}$$

Here, s is the transform parameter and is treated as a constant. For simple functions, the integration is easy.

$$\mathscr{L}(e^{-at}) = \int_0^\infty e^{-at}e^{-st}\, dt = \int_0^\infty e^{-(s+a)t}\, dt = \left[\frac{e^{-(s+a)t}}{-(s+a)}\right]_0^\infty = \frac{1}{s+a}$$

$$\mathscr{L}(\cos at) = \int_0^\infty (\cos at)e^{-st}\, dt = \left[\frac{e^{-st}}{s^2+a^2}(-s\cos at + a\sin at)\right]_0^\infty = \frac{s}{s^2+a^2}$$

$$\mathscr{L}(\sin at) = \int_0^\infty (\sin at)e^{-st}\, dt = \left[\frac{e^{-st}}{s^2+a^2}(-s\sin at - a\cos at)\right]_0^\infty = \frac{a}{s^2+a^2}$$

and for the step function,

$$\mathscr{L}(u(t)) = \int_0^\infty 1\cdot e^{-st}\, dt = \left[\frac{e^{-st}}{-s}\right]_0^\infty = \frac{1}{s}$$

But, these elementary functions are simple, and the power of the Laplace transformation is not needed to solve problems using exponential and trigonometric functions, e^{-at}, $\sin at$, and $\cos at$. It is convenient, but the method does require the equation of motion, and stating it is often the more difficult task.

In practice, the equation of motion is Laplace transformed, term by term. The result is an algebraic equation that can be solved for the transfer function and the characteristic values of the response function. The time response is found by inverting the Laplace transform. A short table of Laplace transforms (Table 5.1) is included, but for involved problems the student should refer to a larger collection and one of the many good reference books on transforms and their application to engineering problems.

Table 5.1 Useful Laplace Transforms

$$\mathcal{L}(u(t)) = \frac{1}{s}$$

$$\mathcal{L}(t) = \frac{1}{s^2}$$

$$\mathcal{L}(e^{-at}) = \frac{1}{s+a}$$

$$\mathcal{L}(\sin at) = \frac{a}{s^2+a^2}$$

$$\mathcal{L}(\cos at) = \frac{s}{s^2+a^2}$$

$$\mathcal{L}(e^{-bt}\sin at) = \frac{a}{[(s+b)^2+a^2]}$$

$$\mathcal{L}(e^{-bt}\cos at) = \frac{s+b}{[(s+b)^2+a^2]}$$

$$\mathcal{L}(te^{-at}) = \frac{1}{(s+a)^2}$$

$$\mathcal{L}(t\sin at) = \frac{2as}{(s^2+a^2)^2}$$

$$\mathcal{L}(t\cos at) = \frac{(s^2-a^2)}{(s^2+a^2)^2}$$

$$\mathcal{L}\left[\frac{d}{dt}(f(t))\right] = sF(s) - f(0^+)$$

$$\mathcal{L}\left[\frac{d^2}{dt^2}(f(t))\right] = s^2F(s) - sf(0^+) - \frac{df}{dt}(0^+)$$

$$\mathcal{L}\left[\frac{e^{-at}}{a^2+\omega^2} + \frac{1}{\omega\sqrt{a^2+\omega^2}}\sin(\omega t - \phi)\right] = \frac{1}{(s+a)(s^2+\omega^2)}$$

5.9. CONVOLUTION

The response of a simple single degree of freedom system also can be obtained to any general forcing function, $F(t)$, by using a method called convolution. Unfortunately, if the forcing function is mathematically complicated, this method is difficult to apply, since it does require integration. But, for simple expressions, it is useful and it is the basis of the more powerful transform methods.

The motion of an undamped, unforced single degree of freedom system is

$$x = x_0 \cos \omega_n t + \frac{v_0}{\omega_n}\sin \omega_n t \tag{5.26}$$

In this equation, the initial amplitude of motion is $x(0) = x_0$, and the initial velocity is $\dot{x}(0) = v_0$. This is a general expression which describes how the motion continues with time, once initiated. Both the velocity and the displacement are uniquely known at any time. All that is needed are the two initial conditions, the time interval since the beginning of motion and the system constant, ω_n.

Fig. 5.11

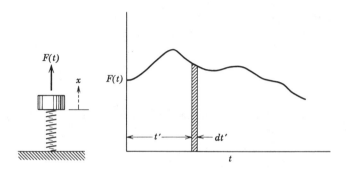

Let us now consider a general forcing function $F(t)$, as in Figure 5.11, which forces a simple spring and mass system. The general function $F(t)$ can be considered to be a series of infinitesimally short impulses, each affecting the motion of the system. If after a time t', the system is forced by an infinitesimal impulse $F(t')\,dt'$ from Newton's second law of motion, there will be a resultant infinitesimal change in momentum.

$$F(t')\,dt' = m\,d\dot{x} \tag{5.27}$$

It is important to recognize that only the momentum will be changed. If mass is not a variable, this means that only velocity will be changed.

$$d\dot{x} = \frac{F(t')\,dt'}{m} \tag{5.28}$$

The change in the displacement, x, at any time t, caused by a change in momentum at some other time t' will then depend on the time interval between t and t' or $t - t'$. Referring back to equation 5.26, by analogy, the change in the displacement due to the impulse $F(t')\,dt'$ will be

$$dx = \frac{d\dot{x}}{\omega_n}\sin \omega_n(t - t')$$

Compare this statement to the last term in equation 5.26, which is the effect of velocity on displacement. Instead of the initial velocity, v_0, each infinitesimal impulse initiates a change in the velocity $d\dot{x}$. After a time interval, $t - t'$, the change in the displacement will be dx.

Substituting for $d\dot{x}$,

$$dx = \frac{F(t')}{m\omega_n}\sin \omega_n(t-t') \, dt' \qquad (5.29)$$

This then is the change in displacement wrought by a change in momentum due to one short impulse $F(t') \, dt'$. If there is a series of short impulses, each impulse will also change the momentum and affect the displacement. The concept of convolution is that for a linear system, the response to the general forcing function can be found as the superposition of the responses to the sum of all the individual impulses. If $F(t')$ is integrable, and $x_0 = 0$,

$$x = \frac{1}{m\omega_n}\int_0^t F(t')\sin \omega_n(t-t') \, dt' \qquad (5.30)$$

Equation 5.30 is known as Duhamel's integral and is a special form of the convolution integral.

Mathematically, we normally express the convolution integral as

$$x(t) = \int_0^t F(t')G(t-t') \, dt' = F(t)*G(t) \qquad (5.31)$$

In equation 5.30, $\sin \omega_n(t-t')$ is a special form of the function $G(t-t')$, valid for an undamped system. The convolution integral is not that restrictive. The variable t' is merely a dummy variable in time.

Borel's theorem states that the convolution of two functions is the inverse of the product of their Laplace transforms. Thus,

$$x(t) = F(t)*G(t) = \mathcal{L}^{-1}[f(s) \cdot g(s)]$$

This is of extreme practical importance because it does allow us to use Laplace transforms to solve transient problems.

EXAMPLE PROBLEM 5.23

A force $F(t)$ is suddenly applied to a mass m, which is supported by a spring with a constant modulus k. After a short period of time τ, the force is suddenly removed. During the time the force is active, it is a constant, F. Determine the response of the system for $t > \tau$. The spring and mass are initially at rest before the force $F(t)$ is applied.

Solution:

For any time $t > \tau$, the equation of motion can be written as

$$m\ddot{x} + kx = F[u(t) - u(t-\tau)]$$

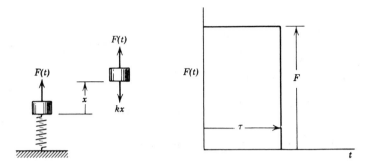

The designation of $u(t)$ and $u(t-\tau)$ are unit functions or the Heavyside step function,

$$u(t) = \begin{cases} 0 & t<0 \\ 1 & t>0 \end{cases}$$

$$u(t-\tau) = \begin{cases} 0 & t<\tau \\ 1 & t>\tau \end{cases}$$

The initial conditions are as before, $x(0)=0$ and $x(0)=0$. Transforming each side of the differential equation,

$$\mathcal{L}[x(t)] = x(s)$$

$$\mathcal{L}[\ddot{x}(t)] = s^2 x(s) - sx(0) - \dot{x}(0) = s^2 x(s)$$

$$\mathcal{L}[u(t)] = \frac{1}{s}$$

$$\mathcal{L}[u(t-\tau)] = \frac{e^{-s\tau}}{s}$$

Substituting the transforms for each of the functions,

$$ms^2 x(s) + kx(s) = F\left(\frac{1}{s} - \frac{e^{-s\tau}}{s}\right)$$

$$x(s) = \frac{F(1-e^{-s\tau})}{s(ms^2+k)}$$

$$x(s) = \frac{F}{m}\left[\frac{(1-e^{-s\tau})}{s(s^2+\omega_n^2)}\right]$$

From tables of Laplace transforms, the inverse,

$$\mathcal{L}^{-1}\left[\frac{1}{s(s^2+\omega_n^2)}\right] = \frac{1}{\omega_n^2}(1-\cos \omega_n t)\cdot u(t)$$

and

$$\mathcal{L}^{-1}\left[\frac{e^{-s\tau}}{s(s^2+\omega_n^2)}\right]=\frac{1}{\omega_n^2}[1-\cos\,\omega_n(t-\tau)]u(t-\tau)$$

For $0<t<\tau$, the solution is

$$x(t)=\frac{F}{m\omega_n^2}(1-\cos\,\omega_n t)$$

and for $t>\tau$, the complete solution is

$$x(t)=\frac{F}{m\omega_n^2}\{(1-\cos\,\omega_n t)-[1-\cos\,\omega_n(t-\tau)]\}$$

$$=\frac{F}{m\omega_n^2}[\cos\,\omega_n(t-\tau)-\cos\,\omega_n t]$$

This is a satisfactory expression for the displacement x as a function of time, but more can be learned by a simple trigonometric manipulation.
    Let,

$$t=\left(t-\frac{\tau}{2}\right)+\frac{\tau}{2}$$

and,

$$t-\tau=\left(t-\frac{\tau}{2}\right)-\frac{\tau}{2}$$

Substituting for t and $t-\tau$, and using the trigonometric identity

$$\cos(\alpha-\beta)-\cos(\alpha+\beta)=2\,\sin\,\alpha\,\sin\,\beta$$

$$x=\frac{F}{m\omega_n^2}\left[2\,\sin\,\omega_n\left(t-\frac{\tau}{2}\right)\sin\frac{\omega_n\tau}{2}\right]$$

This is also a valid expression for the displacement x as a function of time, but additionally, it is easy to see that $x(t)=0$ for $t>\tau$, if

$$\frac{\omega_n\tau}{2}=0,\,\pi,\,2\pi,\,3\pi,\,\ldots,\,n\pi$$

In each of these cases, the mass m will make one, two, three, or n complete oscillations, return to the initial position with zero velocity at which time $F(t)$ is removed, and the system will then remain at rest. No steady state oscillation is possible.

EXAMPLE PROBLEM 5.24

Repeat Example Problem 5.23, using the convolution integral.

Solution:

With rest conditions $x(0) = 0$; when $\dot{x}(0) = 0$, the convolution integral is

$$x = \frac{1}{m\omega_n} \int_0^t F(t')\sin \omega_n(t-t') \, dt'$$

This may be separated into intervals, 0 to τ when $F(t') = F$, and τ to t when $F(t') = 0$

$$x = \frac{1}{m\omega_n} \int_0^\tau F(t')\sin \omega_n(t-t') \, dt' + \frac{1}{m\omega_n} \int_\tau^t F(t')\sin \omega_n(t-t') \, dt'$$

Integrating,

$$x = \frac{F}{m\omega_n^2}[\cos \omega_n(t-t')]_0^\tau$$

$$x = \frac{F}{m\omega_n^2}[\cos \omega_n(t-\tau) - \cos \omega_n t]$$

which is the same result as Problem 5.23.

PROBLEM 5.25 A spring-mass system initially at rest is subjected to a triangular impulse as shown. Determine the displacement x of the mass at a time $t = \tau/2$ where τ = natural period in seconds.

Answer: 0.75 mm

$m = 50$ kg
$k = 96$ N/mm

PROBLEM 5.26 Using convolution, determine an expression for the transient response of the system of Problem 5.3 to the versed sine impulse for $t > \tau$, assuming the system to be initially at rest.

PROBLEM 5.27 Using the result of Problem 5.26, show that for $t > \tau$, the system will have no response if

$$\frac{\omega}{\omega_n} = 1, \frac{1}{2}, \frac{1}{3}, \frac{1}{4}, \frac{1}{5}, \cdots$$

PROBLEM 5.28 Discuss the transient response of a linear system with a single degree of freedom to a triangular impulse. In particular, what values of ω/ω_n will yield no response after $t = 2\pi/\omega$, and what value of ω/ω_n will have a minimum deflection at $t = \pi/\omega$?

Answer: $x = 0$, for $\dfrac{\omega_n}{\omega} = 2, 4, 6, \ldots$

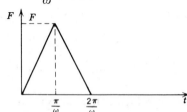

PROBLEM 5.29 Using convolution, determine the response of the system of Problem 5.28 to the sinusoidal pulse

$$F(t) = \begin{cases} F_0 \sin \omega t, & 0 \le t \le \pi/\omega \\ 0, & t > \dfrac{\pi}{\omega} \end{cases}$$

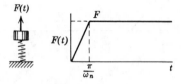

PROBLEM 5.30 Determine the response for $t > \pi/\omega_n$ of a simple mechanical oscillator to a forcing function that has a rise time of half a period, using the convolution integral. Compare the steady state amplitude of motion to that for the same system with a forcing function with zero rise time. Assume rest conditions, initially.

Answer: $x = \dfrac{F}{k}\left(1 - \dfrac{2}{\pi}\sin \omega_n t\right)$

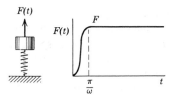

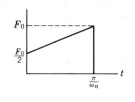

PROBLEM 5.31 Using convolution, determine the transient response for $t > \pi/\omega$ of a linear single degree of freedom system to the forcing function

$$F(t) = \frac{F}{2}(1 - \cos \omega t)$$

for $0 < t \le \pi/\omega$, $F(t) = F$ for $t > \pi/\omega$, assuming rest conditions initially.

Answer:

$$x = \frac{F}{2k\left(1 - \dfrac{\omega^2}{\omega_n^2}\right)}\left[2 - \frac{\omega^2}{\omega_n^2}\left(1 - \cos \pi\frac{\omega_n}{\omega}\right)\right]$$

$$+ \frac{F}{k}\left[1 - \cos \omega_n\left(t - \frac{\pi}{\omega}\right)\right]$$

PROBLEM 5.32 Using convolution, determine the steady state response for the system in Problem 5.30, if the rise time is $\pi/2\omega_n$.

Answer:

$$x = \frac{F}{k}\left[1 - \frac{2}{\pi}\left(\cos \omega_n t + \sin \omega_n t\right)\right]$$

PROBLEM 5.33 Using convolution, determine the steady state response for the system in Problem 5.30, if the rise time is $\pi/4\omega_n$.

PROBLEM 5.34 Using convolution, determine the response for $t \le \pi/\omega_n$ of a simple mechanical oscillator to the forcing function shown.

Answer:

$$x = \frac{F_0}{2m\omega_n^2}(1 - \cos \omega_n t) + \frac{F_0}{2m\omega_n^2}\left(\frac{t}{\tau} - \frac{\sin \omega_n t}{\omega_n \tau}\right)$$

PROBLEM 5.35 Using convolution, determine the response for $t \ge \pi/\omega_n$ for Problem 5.34.

Answer:

$$x = \frac{F_0}{m\omega_n^2}[2 \cos \omega_n(t - \tau) - \cos \omega_n t$$

$$+ \frac{1}{\omega_n \tau}(\sin \omega_n(t - \tau) - \sin \omega_n t)]$$

PROBLEM 5.36 A bridge truss can be simplified and represented as a deck, with a moment of inertia, I, and a supporting spring with a modulus k. The effect of an automobile moving across the bridge can be approximated as a moving concentrated load of weight W. The moment of the car increases linearly to a maximum at midspan and then decreases. Find an equation of motion using the Laplace transform as a function of time. *Hint:* Note that the applied moment is Wvt, where v is the velocity of the car.

Answer:

$$\theta = \frac{Wv}{I\omega_n^3}(\omega_n t - \sin \omega_n t)$$

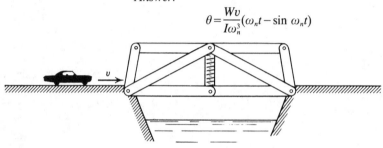

PROBLEM 5.37 Repeat Problem 5.10 using the Laplace transform.

PROBLEM 5.38 A spring with an elastic constant k and a mass m rests so that in the position shown the spring is unstretched. At $t=0$, the supports under the mass are suddenly released, permitting the mass to fall. Referring to the position at $t=0$ as the origin, $x(0)=0$, determine the differential equation of motion using the Laplace transform.

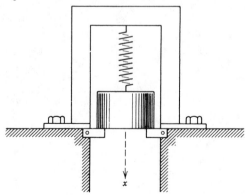

PROBLEM 5.39   Repeat Problem 5.26, using the Laplace transform.

PROBLEM 5.40   Repeat Problem 5.5, using the Laplace transform.

PROBLEM 5.41   Repeat Problem 5.6 using the Laplace transform.

CHAPTER SIX

DAMPING

6.1. INTRODUCTION

The process by which vibration steadily diminishes in amplitude is called damping. In damping, the energy of the vibrating system is dissipated as friction or heat, or transmitted as sound. The mechanism of damping can take any of several forms, and often more than one form may be present at a time.

Fluid damping may be either *viscous* or *turbulent*. In viscous damping, the damping force is proportional to velocity. In turbulent damping, the force is proportional to velocity squared. In *dry friction* or *coulomb damping* the damping force is constant. It is caused by kinetic friction between sliding dry surfaces. *Solid damping* or *hysteretic damping* is caused by the internal friction or hysteresis when a solid is deformed. Stress amplitude is a measure of solid damping.

The most commonly used damping mechanism is viscous damping, in which the damping force is proportional to velocity. Strictly, this is only valid for damping such as that caused by the laminar flow of a viscous fluid through a slot, as in a shock absorber, around a piston in a cylinder, or the oscillation of a journal in a bearing. In each of these cases the constant of proportionality is dependent on the absolute viscosity of the fluid, the surface area, and the fluid film thickness. All of the latter can be made constant for a given set of physical conditions. These examples are about the only practical examples of viscous damping. Other forms of damping do approximate viscous damping, however, if the dissipative forces are small. Lord Rayleigh made this approximation when he used viscous damping to approximate the combined effects of air damping and hysteresis in a tuning fork. The use of viscous damping has the advantage of linearizing the equation of motion. If the dissipative forces are not small, considerable error can be introduced by assuming damping to be viscous when it actually is not.

6.2. VISCOUS DAMPING

In Figure 6.1, a dash pot has been added to the simple single degree of freedom system which had consisted of only a spring and a mass. A dash pot is a loose fitting piston in a cylinder filled with a real fluid, such as water or oil. The piston is rigidly connected to the mass m, but ideally, the mass of the piston is negligible. It is loose fitting so that the fluid can flow around the piston through the clearance from one side to the other. This flow will be proportional to the pressure difference, the fluid viscosity, and the time rate of change in volume, which is the piston velocity. All of this is lumped into a damping constant of proportionality such that the damping force is

$$\mathbf{F} = -c\mathbf{v} \qquad (6.1)$$

The negative sign indicates that the damping force is opposite to the direction of velocity. The constant of proportionality, c, is the damping constant. The units are N·s/m, impulse per unit of displacement, or kilograms per second (kg/s).

The equation of motion for a single degree of freedom with spring, mass, and damper is

$$-kx - c\dot{x} = m\ddot{x}$$
$$m\ddot{x} + c\dot{x} + kx = 0 \qquad (6.2)$$

This is a familiar linear, second-order differential equation with constant coefficients. For the solution, the trial exponential function, Ce^{rt}, yields

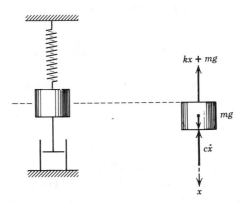

Fig. 6.1

a characteristic equation,

$$r^2 + \frac{c}{m}\,r + \frac{k}{m} = 0$$

the roots of which are complex.

$$r_{1,2} = -\frac{c}{2m} \pm \sqrt{\frac{c^2}{4m^2} - \frac{k}{m}} \tag{6.3}$$

The general solution for the displacement is then

$$x = e^{-(c/2m)t}[C_1 e^{+\sqrt{(c^2/4m^2)-(k/m)}\,t} + C_2 e^{-\sqrt{(c^2/4m^2)-(k/m)}\,t}] \tag{6.4}$$

C_1 and C_2 are again arbitrary constants that depend on the initial conditions of motion. There must be two arbitrary constants in the solution of a second-order differential equation.

Mathematically, this is an exponential equation, but physically, the displacement–time curve can be described as having three distinct forms, depending on whether the radical $\sqrt{(c^2/4m^2)-(k/m)}$ is real, zero, or imaginary.

Case I. $c^2/4m^2 > k/m$. In this case, the radical is real and the motion of the system is dominated by damping. On displacement and release, the system will approach equilibrium exponentially. No oscillation occurs and, theoretically, the system will never return to its original position.

Examples of heavily damped elastic systems for which this motion applies are recoil mechanisms, such as the common automatic door closer. This motion, expressed in equation 6.4, is shown in Figure 6.2a (p. 167).

Case II. $c^2/4m^2 = k/m$. In this case, Figure 6.2b, the radical is zero, and the system is said to be critically damped. That value of the damping constant for which the system is critically damped is called the *critical damping constant*. It is indicated by the damping constant with a subscript, c_{cr}. Its value is strictly a function of the system constants, m and k.

$$\begin{aligned} c_{cr}^2 &= 4mk \\ c_{cr} &= \sqrt{4mk} = 2m\omega_n \end{aligned} \tag{6.5}$$

The ratio of the actual damping constant to the critical damping constant is the *damping ratio*, ζ,

$$\frac{c}{c_{cr}} = \zeta \tag{6.6}$$

which is a dimensionless parameter.

In critically damped motion, the damped system is restored to equilibrium in a minimum of time and without oscillation. Instruments used to

Fig. 6.2

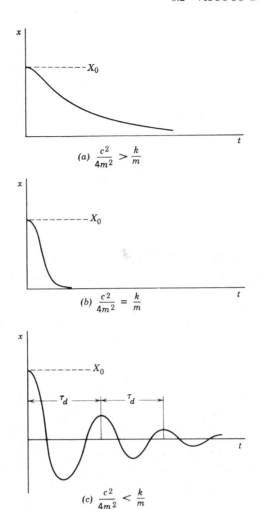

(a) $\dfrac{c^2}{4m^2} > \dfrac{k}{m}$

(b) $\dfrac{c^2}{4m^2} = \dfrac{k}{m}$

(c) $\dfrac{c^2}{4m^2} < \dfrac{k}{m}$

measure steady state values, such as a scale measuring dead weight, are usually critically damped. Mathematically, the two characteristic roots r_1 and r_2 of the equation of motion are identical. In that case the displacement would be

$$x = (C_1 + C_2 t)e^{-(c/2m)t} \tag{6.7}$$

Case III. $c^2/4m^2 < k/m$. This is the case of the damped harmonic in which oscillation about an equilibrium position occurs, with each successive amplitude diminished from the preceding amplitude. Rearranging equation 6.4 the displacement can be expressed as

$$x = e^{-(c/2m)t}[C_1 e^{i\sqrt{(k/m)-(c^2/4m^2)}\,t} + C_2 e^{-i\sqrt{(k/m)-(c^2/4m^2)}\,t}] \tag{6.8a}$$

Fig. 6.3

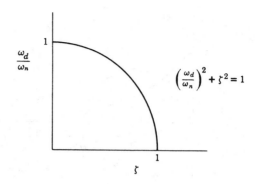

or in trigonometric form,

$$x = e^{-(c/2m)t}[A \cos \omega_d t + B \sin \omega_d t] \tag{6.8b}$$

where $\omega_d = \sqrt{(k/m) - (c^2/4m^2)}$ is the frequency in radians per second of the damped harmonic, and A, B, C_1, and C_2 are arbitrary constants. The motion is shown in Figure 6.2c.

The damped natural frequency and the undamped natural frequency are related to one another through the dimensionless damping ratio ζ, the viscous damped natural frequency always being less than the un-damped,

$$\omega_d = \omega_n \sqrt{1 - \zeta^2} \tag{6.9}$$

Figure 6.3 is a measure of the decrease in natural frequency with increased amounts of damping. It is equation 6.9, slightly transformed algebraically. It is obvious that the decrease in frequency will be small, unless the damping present is a large fraction of that necessary to critically dampen the system. In mechanical engineering systems, damping is normally quite a small fraction of critical damping. It is not negligible, but system damping above $\zeta = 0.2$ must be by design. The natural damping of real engineering materials or structures is quite small. Typical values of damping are shown in Table 6.1.

Table 6.1 Typical Values of Damping

	ζ
Automobile shock absorbers	0.1–0.5
Rubber	0.04
Riveted steel structures	0.03
Concrete	0.02
Wood	0.003
Cold rolled steel	0.0006
Cold rolled aluminum	0.0002
Phosphor bronze	0.00007

6.3. LOGARITHMIC DECREMENT

In a viscously damped harmonic, successive amplitudes have a simple logarithmic relation with one another. Referring again to Figure 6.2c, the maximum amplitude at time $t = t_0$ is X_0. One cycle later, the amplitude has diminished to an amplitude X_1, where $t = t_0 + \tau_d$. Two cycles later, the amplitude has diminished to X_2, with $t = t_0 + 2\tau_d$. The constant A is arbitrary. Its magnitude depends on the amplitude of the motion X_0 at $t = t_0$. The period for damped motion, which is the time between successive cycles, is $2\pi/\omega_d$.

$$X_0 = Ae^{-(c/2m)t_0}$$

$$X_1 = Ae^{-(c/2m)(t_0 + \tau_d)} = X_0 e^{-(c/2m)\tau_d}$$

$$X_2 = Ae^{-(c/2m)(t_0 + 2\tau_d)} = X_0 e^{-(c/2m)2\tau_d}$$

and after n cycles,

$$X_n = X_0 e^{-(c/2m)n\tau_d} \tag{6.10}$$

The quantity $(c/2m)\tau_d = \delta$ is a measure of the system damping called the *logarithmic decrement*,

$$X_n = X_0 e^{-n\delta}$$

or

$$\delta = \frac{1}{n} \ln \frac{X_0}{X_n} \tag{6.11}$$

It is dimensionless and is actually another form of the dimensionless damping ratio ζ

$$\delta = \frac{c}{2m}\left(\frac{2\pi}{\omega_d}\right) = \frac{2\pi\zeta}{\sqrt{1-\zeta^2}} \quad \text{if } \delta \text{ small} \tag{6.12}$$
$$\cdot \ \delta \approx 2\pi\zeta$$

The logarithmic decrement and the damping ratio are system constants since they do not have arbitrary values but are dependent on clearances, surface conditions, temperatures, size, shape, and other factors. As an example, $\delta = 4$ is a typical value for the logarithmic decrement of a shock absorber system in an automobile when new. Six months later, the logarithmic decrement will have decreased. A typical value of the logarithmic decrement for the shock absorber system of a used automobile would be $\delta = 2$.

The values of the logarithmic decrement for thin wires of various materials suspended in a torsional pendulum are shown in Figure 6.4. These data would have comparative value if the specimens were all of the same size, but the absolute number is not significant. As an example, for the wires tested, tungsten had a lower logarithmic decrement at 500°F than

(margin notes:) $\zeta = 0.33?$ $0.30?$ $\zeta = 0.15?$

Fig. 6.4

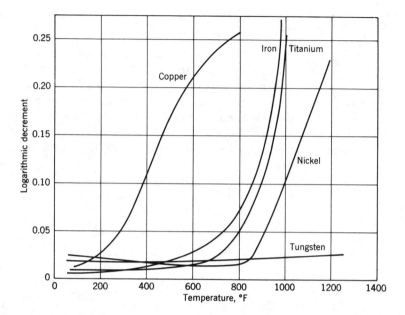

copper. This is an indication of comparative damping in wires. To extrapolate this knowledge to parts, where material properties, stresses, and weight would be significant, is difficult, since geometry and construction are so important. But, comparative knowledge is valuable to design engineers.

6.4. ENERGY DISSIPATED IN VISCOUS DAMPING

It is important at this stage to consider the energy dissipated per cycle in viscously damped harmonic motion. The rate of energy dissipation with time is

$$\frac{dU}{dt} = f\frac{dx}{dt} = c\left(\frac{dx}{dt}\right)^2 \tag{6.13}$$

For simple harmonic motion, $x = X \sin \omega t$, the energy dissipated over a complete cycle will be,

$$\Delta U = \int_0^{2\pi/\omega} c\left(\frac{dx}{dt}\right)^2 dt = \int_0^{2\pi} cX^2\omega \cos^2 \omega t \ d(\omega t)$$

$$\Delta U = \pi c\omega X^2 \tag{6.14}$$

Fig. 6.5

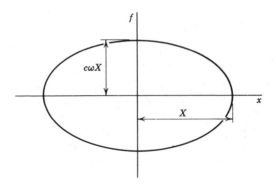

Alternatively, if we had plotted force as a function of displacement, where $x = X \sin \omega t$,

$$f = c\left(\frac{dx}{dt}\right) = c\omega X \cos \omega t$$

we would have an ellipse, shown in Figure 6.5

$$\left(\frac{x}{X}\right)^2 + \left(\frac{f}{c\omega X}\right)^2 = 1 \tag{6.15}$$

The area of the ellipse is the energy dissipated per cycle, $\Delta U = \pi c \omega X^2$.

The energy dissipated is proportional to the square of the amplitude of motion. It should be noted that it is not a constant value for a given amount of damping and amplitude, since the energy dissipated is also a direct function of frequency.

6.5. SPECIFIC DAMPING

The specific damping capacity is that fractional part of the total energy of the vibration system which is dissipated during each cycle of motion, $\Delta U / U$. For a simple system, with one generalized coordinate, the specific damping factor is directly related to the logarithmic decrement and the dimensionless damping ratio

$$\frac{\Delta U}{U} = \frac{\pi c \omega X^2}{\frac{1}{2} k X^2} \approx 2\delta \approx 4\pi\zeta \tag{6.16}$$

The total energy U can be expressed as the maximum potential energy, $\frac{1}{2} k X^2$.

Specific damping is not often used in mechanical vibrations, since it has a useful definition only for light damping, ($\delta < 0.01$). It is useful in comparing the damping capacity of engineering materials.

EXAMPLE PROBLEM 6.1

A simple viscometer consists of a thin steel disc, 200 mm in diameter and 2.5 mm thick, suspended on a steel wire as a torsional pendulum. In operation, the lower face of the disc just touches a 8-mm layer of oil in a shallow pan. When the pendulum is in contact with oil, the damped natural frequency, f_d, is lower than the undamped natural frequency, f_n, when the pendulum oscillates freely in air. Starting with Newton's law of viscosity, $F = \mu A (dv/dh)$, where μ is viscosity, A is the area of contact, and dv/dh is the velocity gradient in the oil, determine the expression for the viscosity of the oil in terms of the damped and undamped natural frequencies. Solve for the viscosity in the case where $f_d = 1.15$ Hz and $f_n = 1.20$ Hz. Discount any end effects on the disc.

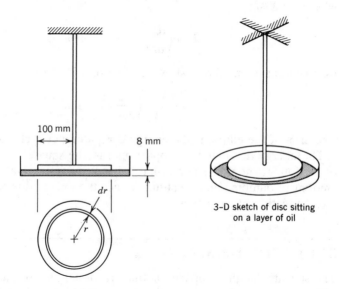

3-D sketch of disc sitting on a layer of oil

Solution:
The viscous drag on an element of area, which is a thin ring of radius r is

$$dF = \mu(2\pi r \, dr)\frac{dv}{dh}$$

A thin ring was selected as our element, since the velocity of all points at the same radius r will be the same, $v = r\dot{\theta}$. If we assume a linear velocity gradient in the oil

$$\frac{dv}{dh} = \frac{r}{h}\dot{\theta}$$

the moment of the drag dF is

$$dT = dFr = r\mu(2\pi r \ dr)\left(\frac{r}{h}\right)\dot{\theta} = \frac{2\pi\mu\dot{\theta}}{h}r^3 \ dr$$

Integrating over the face of the entire disc, the viscous torque which tends to dampen motion is

$$T = \int_0^R \frac{2\pi\mu}{h}\dot{\theta}r^3 \ dr = \frac{\mu\pi R^4}{2h}\dot{\theta}$$

This is a linear function of the angular velocity $\dot{\theta}$. The fraction $\mu\pi R^4/2h$ is the damping coefficient, with units of Newton-seconds per metre.

$$c = \frac{\mu\pi R^4}{2h}$$

Equation 6.9 states that the damped natural frequency is less than the undamped natural frequency by the factor $\sqrt{1-\zeta^2}$

$$\omega_d = \omega_n\sqrt{1-\zeta^2}$$

Solving for ζ, the dimensionless damping ratio

$$\zeta = \frac{c}{c_{cr}} = \sqrt{1-\left(\frac{\omega_d}{\omega_n}\right)^2} = \sqrt{1-\left(\frac{f_d}{f_n}\right)^2}$$

The critical damping coefficient is defined in terms of the natural frequency and the inertial mass,

$$c_{cr} = 2I\omega_n = 4\pi I f_n$$

Substituting for the damping coefficient, the viscosity of the oil can be found in terms of f_n and f_d and the physical constants I, h, and R.

$$\mu = 8\frac{Ihf_n}{R^4}\sqrt{1-\left(\frac{f_d}{f_n}\right)^2}$$

For this particular case,

$$f_d = 1.15 \ \text{Hz}$$
$$f_n = 1.20 \ \text{Hz}$$
$$R = 100 \ \text{mm}$$
$$I = \tfrac{1}{2}\pi(0.1)^2(0.0025)(7.85)(0.1)^2 = 3.08 \times 10^{-6}\text{kg·m}^2$$

and therefore,

$$\mu = 8\frac{(3.08 \times 10^{-6})(1.2)(0.008)}{(0.1)^4}\sqrt{1-\left(\frac{1.15}{1.20}\right)^2} = 1.93 \times 10^{-6}\text{Pa·s}$$

or, 790 Reyns.

PROBLEM 6.2 For a damping ratio of 0.2, what is the difference between the damped and undamped natural frequencies.

Answer: 2%

PROBLEM 6.3 Plot a curve of the double amplitude of motion as a function of the number of cycles elapsed for data shown. Plot on semi-logarithmic paper. Determine the logarithmic decrement and the damping factor.

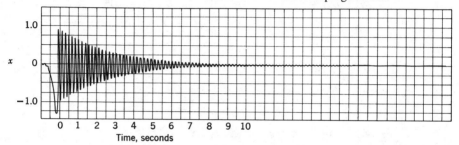

PROBLEM 6.4 A 2-kg piston supported on a helical spring vibrates freely with a natural frequency of 125 cpm. When oscillating within an oil-filled cylinder, the frequency of free oscillation is reduced to 120 cpm. Determine the damping constant c.

Answer: 14.33 kg/s

PROBLEM 6.5 What is the ratio of successive amplitudes of vibration for a simple mechanical system if the viscous damping ratio is known to be $\zeta = 0.5$.

Answer: 37.5:1

PROBLEM 6.6 What is the ratio of successive amplitudes of vibration for a simple mechanical system, if the viscous damping ratio is known to be $\zeta = 0.05$

Answer: 1.37:1

PROBLEM 6.7 Many devices have an adjustable viscous damping apparatus. In one such device the ratio of successive amplitudes is 10:1.

If the amount of damping is doubled, what will the ratio of successive amplitudes then be?

Answer: 380:1

PROBLEM 6.8 If the ratio of successive amplitudes of damped free vibration is 2:1, what is the logarithmic decrement and damping ratio ζ?

Answer: $\delta = 0.693$; $\zeta = 0.110$

PROBLEM 6.9 A shock absorber design must be limited to 10% overshoot, when displaced from equilibrium and released. Determine the necessary damping ratio ζ.

Answer: $\zeta = 0.591$

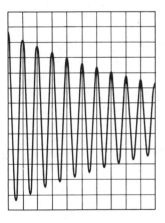

PROBLEM 6.10 Determine the logarithmic decrement and the damping ratio for the record of viscous damped free vibration shown here.

Answer: $\delta = 0.086$; $\zeta = 0.0137$

PROBLEM 6.11 For the free vibrations of a system, determine the viscous damping constant of the system, c. The following data are given:

Spring constant	8 kN/m
Mass	10 kg
Amplitude of first cycle	64 mm
Amplitude of second cycle	48 mm
Amplitude of third cycle	36 mm
Amplitude of fourth cycle	27 mm

Answer: $c = 0.82$ kg/s

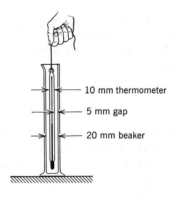

- 10 mm thermometer
- 5 mm gap
- 20 mm beaker

PROBLEM 6.12 A manometer consists of a tube 15 mm in diameter containing a column of SAE 5 oil, 0.25-m long overall, with a specific gravity of 0.8 at 15°C ($\mu = 35$ Cp or 35 MPa·s). Estimate the damping factor ζ for the manometer.

Hint: From Poiseuille's law for flow through a capillary tube, $v_{av} = \Delta P A / 8\pi \mu l$.

Answer: $\zeta = 0.351$

PROBLEM 6.13 A 0.25-m mercury in glass thermometer, which has a mass of 28 g in air, is suspended by a rubber band and is fully immersed in a slender beaker filled with a light oil. When displaced it oscillates at a frequency of 1 Hz and the motion is damped from an amplitude of 40 to 10 mm in two cycles.

(a) What is the logarithmic decrement?

(b) What is the dimensionless damping ratio, ζ?

(c) Estimate the viscosity, μ, of the oil.

Hint: Remember that Newton's equation for viscous motion is $F = \mu A (dv/dh)$; for a linear velocity gradient $F = \mu A (v/h)$.

PROBLEM 6.14 A 5-kg mass is suspended from a spring that has a stiffness of 2000 N/m. What damping constant added to the system would critically dampen it?

$$Hint: \frac{c}{c_{cr}} = 1, \quad \text{what is } c_{cr}?$$

Answer: 100 N·s/m

PROBLEM 6.15 A critically damped system consists of an elastic spring with a constant of 250 N/m supporting a 2-kg mass. The mass has an initial displacement of 100 mm and is given an initial velocity of 5 m/s in the direction opposite to the displacement. Construct a displacement–time diagram showing the time t after release when the mass passes the equilibrium position and the maximum overshoot.

PROBLEM 6.16 For the manometer of Problem 6.12, determine the diameter of the tube that will make the motion dead-beat, (critically damped).

Answer: 8.9 mm

PROBLEM 6.17 A testing device consists of a pneumatic cylinder that accelerates a piston and 5-kg assembly to a speed of 30 m/s. At that instant, the assembly is decelerated by engaging a spring and dash pot. The spring has a modulus of 50 N/mm and the dash pot is damped with a constant of 1000 N·s/m. Determine the maximum displacement of the piston after the spring and dash pot are engaged and the time of the engagement that this occurs.

Answer: 0.11 m

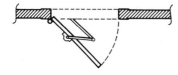

PROBLEM 6.18 A slab door, 2 m high, 0.75 m wide, 40 mm thick, and a mass of 36 kg, is fitted with an automatic door closer. The door opens against a torsion spring with a modulus of 10 N·m/rad. Determine the necessary damping to critically dampen the return swing of the door. If the door is opened 90° and released, how long will it take until the door is within 1° of closing?

Answer: $t = 5.36$ s

6.6. HYSTERETIC DAMPING

If we had considered the viscous damper to be in parallel with an elastic spring, as in Figure 6.6a, the total force $\mathbf{f}_t$ would be the sum of the damping force and the spring force,

$$\mathbf{f}_t = k\mathbf{x} + c\dot{\mathbf{x}} \tag{6.17}$$

If we again had used simple harmonic motion, $x = X \sin \omega t$, the component of the total force in the x direction would be

$$f_t = kX \sin \omega t + c\omega X \cos \omega t$$

Plotting force as a function of displacement in the x direction yields the skewed ellipse of Figure 6.6b

$$\left(\frac{x}{X}\right)^2 + \left(\frac{f_t - kx}{cX\omega}\right)^2 = 1$$

Fig. 6.6

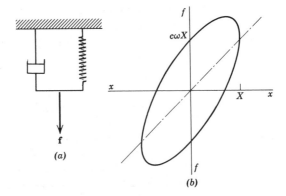

(a)

(b)

for which the force expressed in terms of the system constants and the maximum displacement is

$$f_t = kx \pm c\omega\sqrt{X^2 - x^2}$$

The energy dissipated for one cycle of motion is

$$\Delta U = \int_0^{2\pi/\omega} f_t\left(\frac{dx}{dt}\right) dt = \int_0^{2\pi/\omega} kx\left(\frac{dx}{dt}\right) dt + \int_0^{2\pi/\omega} c\left(\frac{dx}{dt}\right)^2 dt$$

$$= \int_0^{2\pi} kX^2\omega \sin \omega t \cos \omega t \, d(\omega t) + \int_0^{2\pi} cX^2\omega \cos^2 \omega t \, d(\omega t)$$

$$= \pi c\omega X^2$$

which is identical with equation 6.14. This is logical, since the spring force will do no net work over a complete cycle or some integral number of cycles. If damping were absent, the force–deflection curve would be a single line, $f = kx$, instead of a closed figure.

Using viscous damping to represent the internal damping of materials or the damping of a built-up structure incurs a serious error. The energy dissipated per cycle in viscous damping increases with frequency. *Solid damping, structural damping,* or *hysteretic damping* are all terms used to denote internal damping, and the energy dissipated per cycle in internal damping is invariably independent of frequency or decreases slightly with increasing frequency.

There are physical reasons why the damped natural frequency of a specimen will affect the apparent value of the decrement. One is the thermal effect of repeated elastic strain. A suddenly applied tensile stress causes a slight cooling of material as it expands. If one portion of a structure is stressed differently from another, as occurs in bending, there will be heat flow across grain boundaries dissipating energy and causing damping. At higher frequencies, there is no time for heat to flow and damping is less.

In the Introduction, we learned that if the dissipative forces are small, viscous damping is a good approximation of the actual mechanism of energy dissipation. The advantage of using a linearized equation of motion outweighs whatever compromises are necessary to make the approximation. We now have reached a conclusion that is contrary to our physical observations and viscous damping is not a satisfactory approximation as a damping mechanism.

In the early years of research leading to the design of high pressure steam turbines, damping capacity was one variable that was closely studied. High pressure steam jetting from nozzles meant large impulse forces on turbine blades. The first blades failed in a matter of hours from fatigue. Damping was one means of decreasing the number of blade oscillations. By changing the alloy composition of the blade material, the damping capacity was increased twentyfold. Aside, it was noted that the dimensionless damping ratio increased with temperature and decreased with frequency. The latter observation was in complete opposition to accepted viscous damping theory. Figure 6.7 shows the physical variation of the dimensionless damping factor with frequency, taken from the early literature. The numerical value of the damping parameters has been removed to avoid placing an absolute value on the data.

Fig. 6.7

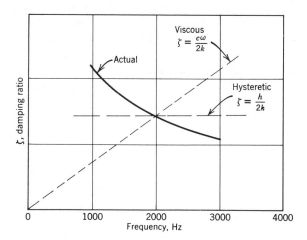

The simplest device that can be used to represent internal damping is to assume that the damping force is proportional to velocity and inversely proportional to frequency. This type of damping is called hysteretic damping, since it can be directly related to the hysteresis loop. A constant h is used in place of the product $c\omega$ and is called the *hysteretic damping*

constant. The units for h are the same as those of a spring constant, Newtons per millimeter or Newtons per metre.

$$\mathbf{f}_t = k\mathbf{x} + \frac{h}{\omega}\dot{\mathbf{x}} \qquad (6.18)$$

The usual symbol for hysteretic damping is a crossed box. Retracing our steps we can see that by replacing $c\omega$ by h, the energy dissipated per cycle is dependent only on the constant h and the amplitude of motion.

$$\Delta U = \pi h X^2 \qquad (6.19)$$

Referring to Figure 6.8, the hysteretic damping constant is a measure of the hysteresis loop and is a property of the material or structure. The intercept of the hysteresis loop on the force axis depends on the value of the hysteretic damping constant h and the maximum displacement X. If d is the intercept of the hysteresis loop on the displacement axis,

$$\frac{h}{k} = \frac{d}{\sqrt{X^2 - d^2}} \approx \frac{d}{X} \qquad (6.20)$$

For light damping, the value of d is small in comparison to X and it is sufficient to evaluate the hysteretic damping constant by the approximate ratio of equation 6.20. This is permissible, since hysteretic damping is not used where large plastic deformation is evident, for several reasons.

The free vibration response of a system with hysteretic damping is no different from the free vibration response of a system with viscous damping. The hysteretic logarithmic decrement can be defined in a similar manner as the viscous logarithmic decrement.

$$\delta = \frac{\pi h}{m\omega_d^2} \approx \pi \frac{h}{k} \qquad (6.21)$$

Fig. 6.8

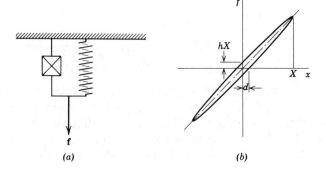

(a)   (b)

Calling ζ the ratio of the actual hysteretic damping to that amount necessary to provide critical damping

$$\zeta = \frac{\delta}{2\pi} \approx \frac{h}{2k} \qquad (6.22)$$

This is constant, which does not completely represent the data of Figure 6.7, but it does fit the data better than a function increasing linearly with frequency.

One other added note, in application to the design of turbine blades, for similar geometries, smaller blades have lower inherent damping because of their higher natural frequencies. This is another reason why it is meaningless to state absolute values for the logarithmic decrement of engineering materials. Logarithmic decrement is a function of geometry.

In some real materials, stress is not proportional to strain and the energy dissipated in each cycle is not proportional to the square of the amplitude of motion (stress). Examples are inelastic materials, such as rubber, and materials such as cast iron, where the damping mechanism is not thermoelastic. In cast iron it is due to the slipping of one grain on another with the aid of free carbon in the grain boundaries. An empirical relation for the energy dissipated in each cycle is

$$\Delta U = \pi h X^m \qquad (6.23)$$

m is an empirical exponent. If $m = 2$, the system response cannot be distinguished from viscous damping. If $m \neq 2$, then the specific damping capacity will itself be a function of amplitude.

$$\frac{\Delta U}{U} = \frac{\pi h X^m}{\frac{1}{2}kX^2} = 4\pi\zeta X^{m-2}$$

In most materials $m > 2$, which means that the specific damping capacity will increase with amplitude of motion.

6.7. COMPLEX STIFFNESS

Occasionally, the term complex stiffness is used for structural damping. It is a simple mathematical expedient, but awkward to defend physically. If motion is harmonic, and this is a necessary assumption, $x = Xe^{i(\omega t + \phi)}$ and $\dot{x} = i\omega Xe^{i(\omega t + \phi)} = i\omega x$. Substituting into equation 6.2, the equation of motion becomes

$$m\ddot{x} + k\left(1 + \frac{h}{k}i\right)x = 0 \qquad (6.24)$$

The complex quantity $[1 + (h/k)i]$ is the *complex stiffness* of the system. It represents both the elastic and the damping forces at the same time. There are advantages to the use of this terminology, if the student is familiar with complex notation, and it is particularly adaptable for vectors, but complex stiffness has no physical meaning in the same engineering sense as the elastic stiffness.

EXAMPLE PROBLEM 6.19

When a built-up structure is loaded and unloaded, recorded data yield the load deflection curve shown on p. 183. From these data, estimate the hysteretic damping coefficient, h, the logarithmic decrement, δ, and the dimensionless damping ratio, ζ.

Solution:

The area enclosed by the hysteresis curve is the energy dissipated during each full cycle. Counting squares, it is approximately 36 N·m

$$\Delta U = \pi h X^2 = 36 \text{ N·m}$$

For a maximum deflection of 20 mm, the hysteretic damping coefficient, h is

$$h = \frac{\Delta U}{\pi X^2} = \frac{36}{\pi (0.02)^2} = 28.65 \text{ N/mm}$$

The slope of the force–deflection curve is $k = 250$ N/mm, and the logarithmic decrement and dimensionless damping ratio are

$$\delta = \pi \frac{h}{k} = 0.36$$

$$\zeta = \frac{1}{2} \frac{h}{k} = 0.0575$$

All of these calculations are based on the assumption that damping is light.

If we had used the intercept to determine h, and symmetry to determine the spring constant,

$$hX = 500 \text{ N}$$
$$k = 225 \text{ N/mm}$$

Load, N	Deflection, mm
0	0
2000	8
3500	14
3850	16
4000	18
4000	20
3000	16
1000	8
−1000	0
−3000	−8
−4500	−14
−4850	−16
−5000	−18
−5000	−20
−4000	−16
−2000	−8

For $X = 20$ mm, $h = 25$ N/mm

$$\delta = \pi \frac{h}{k} = 0.349$$

$$\zeta = \frac{1}{2}\frac{h}{k} = 0.0555$$

which values are very close to those just obtained.

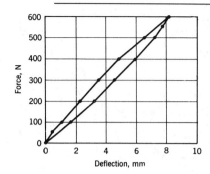

PROBLEM 6.20 Through experimental measurement, force–deflection data for a structure show the following hysteresis loop. Estimate the hysteretic damping coefficient, h.

Force, (N)	Deflection, (mm)
0	
50	0.5
100	1.1
200	2.3
300	3.7
400	5.1
500	6.6
600	8.2
550	7.7
500	7.1
400	5.9
300	4.6
200	3.3
100	1.7
0	0

Answer: $h = 7875$ N/m

PROBLEM 6.21 Data from a laboratory bending test on an intervertebral disc are given in the following figure. Estimate the hysteretic damping coefficient.

Answer: $h = 3.29$ N·m/rad

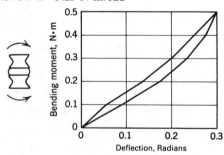

PROBLEM 6.22 A helical spring with a modulus of 50 N/m supports a mass of 1 kg. The mass is given an initial deflection of 20 mm and released. After 200 cycles of motion, it is noted that the amplitude has decayed to 10 mm. What

is the value of the hysteretic damping coefficient, h?

Answer: $h = 0.055$ N/m

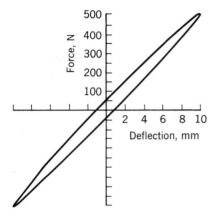

PROBLEM 6.23 When a simple elastic structure is loaded and unloaded, the force–deflection curve shows a hysteresis loop. Determine the damping ratio if the structure supports a 1-kg mass.

Answer: $\zeta = 0.05$

PROBLEM 6.24 The damped vibration curve of Problem 6.3 was taken from an experimental vibrating table consisting of a single large mass supported on four-leaf springs serving as legs. The tape speed was 10 mm/s. The table has a single degree of freedom and has a mass of 50 kg. Estimate the hysteretic damping constant, h.

PROBLEM 6.25 For the record of damped vibration given in Problem 6.10, determine the hysteretic damping constant, h, if $m = 5$ kg.

Answer: $h = 540.4$ N/m

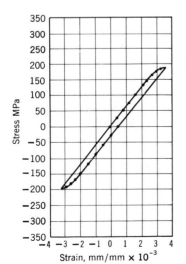

PROBLEM 6.26 Actual data of the stress strain characteristics for aluminum are shown. Determine the hysteretic damping constant from the given data.

Answer: $h = 4.2 \times 10^9$ N/m

PROBLEM 6.27 If a space frame deflects 2 mm for a 500-N load, applied so that the deflection approximates the first mode shape, for which the resonant frequency is 10 Hz, and $\zeta = 0.1$, determine the hysteretic damping coefficient.

Answer: $h = 50,000$ N/m

PROBLEM 6.28 A simplified diagram of the response of a steam turbine blade is shown as it passes through a turbine jet. This particular blade is a reaction blade 100 mm long, with a mass of 0.5 kg, with a damped natural frequency of 600 Hz. For the purposes of this problem the cross-section of the blade can be assumed to be uniform. Estimate the hysteretic damping coefficient h.

Answer: 21.6 N/mm

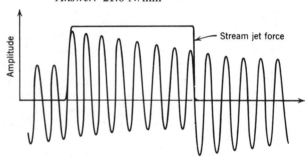

6.8. COULOMB DAMPING

The third important type of damping is Coulomb or dry friction damping. This is sometimes called constant damping, since the damping force is independent of displacement and its derivatives and only depends on the normal forces between sliding surfaces. The direction of the friction force does oppose motion however, and the sign of the friction force will change when the direction of motion changes. This necessitates two solutions to the equation of motion, one valid for one direction and the other valid when motion is reversed. Individually, the solutions are linear, but they are discontinuous after every half-cycle.

Referring to Figure 6.9 (p. 187), for the half-cycle moving from right to left, the equation of motion, $\Sigma F = m\ddot{x}$ becomes

$$-kx + \mu N = m\ddot{x}$$

Fig. 6.9

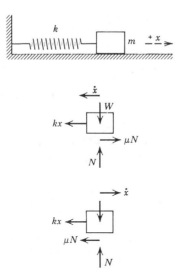

the solution is

$$x = A_1 \cos \omega_n t + B_1 \sin \omega_n t + \frac{\mu N}{k} \qquad (6.25a)$$

If the motion is from left to right, the equation of motion for that half-cycle becomes

$$-kx - \mu N = m\ddot{x}$$

for which the solution is

$$x = A_2 \cos \omega_n t + B_2 \sin \omega_n t - \frac{\mu N}{k} \qquad (6.25b)$$

A_1, B_1, A_2, and B_2 are arbitrary constants that depend on the initial conditions of each successive half-cycle. Different subscripts are used to show that the constants are not the same. The factor $\mu N/k$ is a constant, being the virtual displacement of the spring under the friction force μN, if it were applied as a static force.

For each half-cycle the motion is a harmonic, and the displacement–time curve is a pure half-sine curve with the equilibrium position changing from $+\mu N/k$ to $-\mu N/k$, each half-cycle.

As an example of motion under the influence of dry friction, let us study the motion of the simple system of Figure 6.9, starting with some given initial conditions and continuing until motion ceases.

At time $t = 0$, the mass m is displaced a distance $x(0) = X_0$ and released from rest, such that $\dot{x}(0) = 0$. A_1 and B_1 can now be evaluated explicitly from equation 6.25a

$$A_1 = X_0 - \frac{\mu N}{k} \quad \text{and} \quad B_1 = 0$$

$$x = \left(X_0 - \frac{\mu N}{k} \right) \cos \omega_n t + \frac{\mu N}{k}$$

(6.26a)

This is a cosine wave displaced in the positive direction by an amount $\mu N / k$. It is valid only for $0 \le t \le \pi / \omega_n$. At time $t = \pi / \omega_n$, $x = (2\mu N / k) - X_0$. For the second half-cycle, motion reverses and equation 6.25b must be used.

$$A_2 = X_0 - \frac{3\mu N}{k} \quad \text{and} \quad B_2 = 0$$

$$x = \left(X_0 - \frac{3\mu N}{k} \right) \cos \omega_n t - \frac{\mu N}{k}$$

(6.26b)

This is a cosine wave displaced by an amount $\mu N / k$, in the negative direction, and with a reduced amplitude of $X_0 - (3\mu N / k)$. It is valid only for $\pi / \omega_n \le t \le 2\pi / \omega_n$. At time $t = 2\pi / \omega_n$, $x = X_0 - (4\mu N / k)$. For the third half-cycle, $2\pi / \omega_n \le t \le 3\pi / \omega_n$, motion again reverses and equation 6.25a can be used again.

$$x = \left(X_0 - \frac{5\mu N}{k} \right) \cos \omega_n t + \frac{\mu N}{k}$$

(6.26c)

In each successive cycle, the amplitude is reduced by an amount equal to $4(\mu N / k)$. Motion stops at the end of the half-cycle for which the amplitude is less than $\mu N / k$. At that point, the spring force restoring equilibrium is less than the friction force and motion ceases, $kx < \mu N$. In Figure 6.10, this occurs at the end of the third cycle. The rest position is displaced from equilibrium and represents a permanent deformation in which the friction force is locked in. Shaking or tapping the system will usually jar it sufficiently to restore equilibrium.

The decrement for constant damping is not logarithmic but linear.

$$X_{n+1} = X_n - \frac{4\mu N}{k}$$

The displacement–time curve will lie within the envelope of a pair of straight lines that approach the equilibrium position with a slope of $2(\mu N \omega_n / \pi k)$ as shown in Figure 6.10.

Two interesting facets of coulomb damping or dry friction are significant. The first is its extent. All real damping must be partly due to coulomb damping, since only dry friction damping can stop motion. In viscous damping and hysteretic damping, motion theoretically continues forever, albeit at an infinitesimally small amplitude. This is an academic point,

Fig. 6.10

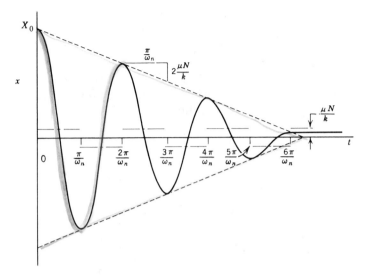

but it is basic to the understanding of damping mechanisms. The second is that the natural frequency is unaltered by dry friction damping.

6.9. EPILOGUE ON DAMPED FREE VIBRATION

Rarely does one form of damping occur exclusive of another. More often, all three kinds of damping occur together, perhaps with several other forms of damping, which have not been mentioned. In understanding a

Fig. 6.11

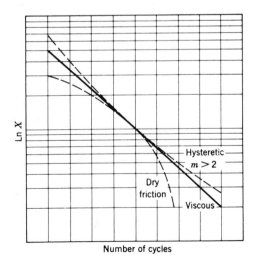

particular problem, if the amplitude is large, the damping may be considered as hysteretic, and viscous if the amplitude is small. A convenient indication of the damping present is the semilogarithmic plot of the logarithm of the maximum displacement versus the number of cycles of free vibration, such as that of Figure 6.11. If damping is viscous, or hysteretic with the energy dissipated per cycle being proportional to the square of the amplitude, the plot will be linear. The slope of the line is the loga-

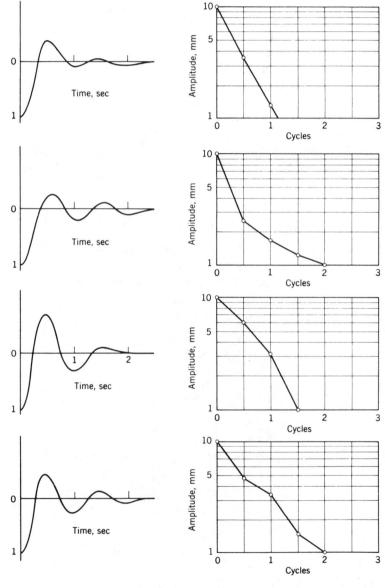

Fig. 6.12

rithmic decrement. If internal damping is present, and the energy dissipated per cycle is proportional to the amplitude raised to some power greater than two, the plot will be curved concave upwards. This will be particularly noticeable at higher amplitudes. If dry friction is present, the plot will be concave downwards.

Often $1\frac{1}{2}$ to 2 cycles are sufficient to indicate a change in damping. Such is the case in Figure 6.12. Here, four decrement curves are shown, all taken from the actual damped records of automobile motion. To obtain them, a recording pen was attached to either the front or rear bumper, while the car was standing still, and the body displaced. In the first example, damping is viscous due primarily to a well-functioning shock absorber system. In the second, the shock absorber action is poor and damping is primarily structural or chassis. In the third, the dry friction of rubbing structural parts is apparent. A fourth example, where the shock absorbers are ineffective, a loose fitting plunger makes the damping different in each direction. It is included as a curiosity. Regardless of their condition, most shock absorbers limit motion to two or three cycles and this limited motion is enough to provide a good indication of the spring and shock absorber system performance.

As a last remark, even for automobile shock absorber systems, damping is usually less than half critical, $\zeta<0.5$. This fact bears out one original statement: in mechanical engineering systems, damping is light.

EXAMPLE PROBLEM 6.29

A trailer is supported on a single axle and a pair of leaf springs. Dry friction between the spring leaves replaces a hydraulic shock absorber. When the trailer is fully loaded and then carefully emptied, the height of the trailer bed is measured to be 0.615 m from level ground. When the empty trailer is jacked up, so that all load is removed from the springs and axle, and then the trailer is carefully lowered, the trailer bed height is 0.635 m. Estimate the number of cycles of damped free vibration that

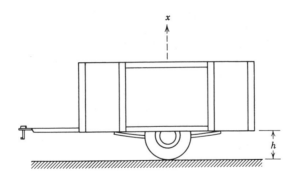

the trailer will experience, if it is depressed 65 mm and released from rest. What will be the rest position of the trailer bed?

Solution:

The equilibrium position of the trailer is at $h = 0.625$ m, midway between the two extremes. In the lower position, the interleaf friction restrains

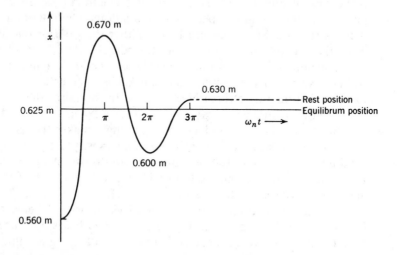

the springs from restoring the trailer to equilibrium, the same is true for the higher position, but the frictional forces are reversed

$$\frac{\mu N}{k} = 10 \text{ mm}$$

For a damped oscillation with an initial displacement of 65 mm, $4(\mu N/k) = 40$ mm of amplitude will be lost during each cycle. The trailer will come to rest after $1\frac{1}{2}$ cycles with the bed at a height of 0.630 m. At that position, the displacement from equilibrium is only 5 mm.

$$x = 5 \text{ mm} < \frac{\mu N}{k}$$

The spring force restoring the trailer to equilibrium is not sufficient to overcome the Coulomb frictional force and motion ceases. Needless to say, equilibrium can be reached by jarring or shaking the trailer.

PROBLEM 6.30   For a system with a natural frequency of 5 Hz, determine the equivalent viscous damping coefficient for the hysteresis loop of Problem 6.20.

Answer: 250 N·s/m

PROBLEM 6.31 A spring and a mass are constrained to move horizontally on a flat surface. A record of the motion is shown below. Determine the coefficient of friction.

Answer: $\mu = 0.1$

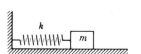

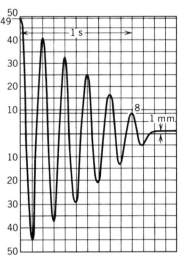

PROBLEM 6.32 The 15 kg mass of Problem 6.31, is given an initial displacement of 60 mm and released. The spring modulus is 180 N/m. Assuming that the kinetic coefficient of dry friction is constant, $\mu_k = 0.2$, determine the position at which the block comes to rest.

Answer: Block stops after second cycle, 5.4 mm to the left of equilibrium

PROBLEM 6.33 Consider the mass m of Problem 6.31. The spring is linear and friction between the mass and the horizontal surface can be described by the Coulomb coefficient of friction μ. The mass is initially displaced a small distance and released from rest. After one (1) complete cycle of motion, the mass comes to rest again. For what value(s) of μ is such a motion description possible?

Hint: Remember, the natural frequency, ω_n, is not affected by dry friction.

Answer: $0.258 < \mu < 0.429$

PROBLEM 6.34 An ordinary pocket watch was opened and the balance wheel was observed to oscillate at 5.2 Hz. The balance wheel is essentially a thin ring 12 mm in diameter and attached to a spindle which is 0.5 mm in diameter. The spindle runs in jewel bearings, which are conical pivot bearings. The friction moment can be shown to be $C \cdot \mu W d$ for conical pivot bearings, where μ is the coefficient of friction, d is the diameter of the spindle, and mg is the weight that the spindle supports. The constant C is 0.341 for 90° conical pivot bearings. When the watch is fully run down, if the balance wheel is displaced 90°, it will oscillate 40 times before stopping. Estimate the coefficient of dry kinetic friction, μ.

Answer: $\mu = 0.21$

PROBLEM 6.35 A small boy makes a swing in a tree. The coefficient of dry friction between the rope and the tree branch is μ. Determine an expression for the decrease in the angle of swing for each cycle caused by dry friction.

Answer: $\theta_\mu = \dfrac{2d}{l}\left(\dfrac{e^{\mu\pi}-1}{e^{\mu\pi}+1}\right)$

PROBLEM 6.36 A 3.10-kg connecting rod is suspended on a cylinder that fits loosely in the wrist pin bearing. It is displaced an angle of 6° and released. The coefficient of friction between the bearing and journal is $\mu = 0.05$. Determine the number of cycles until motion ceases and the angle of repose.

Answer: $5\frac{1}{2}$ cycles; 0.0014 rad

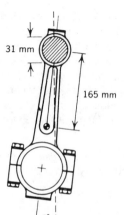

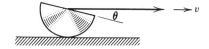

PROBLEM 6.37 When the solid semicylinder is pulled laterally at a constant velocity on the flat horizontal surface, the semicylinder will be inclined to the horizontal at an angle θ_1, which angle is a function of the kinetic coefficient of friction μ_k. Describe the motion of the semicylinder if it starts from rest and the static coefficient of friction μ_s exceeds μ_k. Determine the maximum amplitude of motion and the natural frequency.

Answer: $\theta_{max} = \dfrac{3\pi}{4}(\mu_s - \mu_k)$; $\omega_n = \sqrt{\dfrac{8g}{3\pi r}}$

CHAPTER SEVEN

DAMPED

FORCED

VIBRATIONS

7.1. INTRODUCTION

In real systems, damping is always present to some measurable degree. It may be difficult to predict its effects, particularly if the mechanism of damping is anything other than simple viscous damping, but it is always present. Many times, damping can be assumed to be viscous for engineering purposes, and other times, it can be wholly ignored. It is just as important to know when damping can be ignored as it is to know what damping mechanism is best to assume. As an example, there is little difference in the damped and undamped response of a system with a single degree of freedom, if the forcing frequency is many times greater than the natural frequency. In this case damping, regardless of the mechanism, can be ignored. The effect of damping at or near resonance is quite another matter.

Usually, we can categorize damping according to one or more of the three forms already discussed: viscous, hysteretic, or dry friction damping. Each is merely a model of the actual damping present, and at frequencies away from resonance, the differences are more or less academic. At resonance, the difference in the various forms of damping are pronounced. For example, the dry friction model does not limit resonant amplitudes, but hysteretic and viscous damping do. Viscous damping affects the frequency of the resonant peak, but hysteretic and dry friction damping do not.

7.2. FORCED DAMPED HARMONIC VIBRATION $F(t) = F_1 \sin \omega t$

Figure 7.1 shows the simple spring–mass system of Chapter 4 subjected to a harmonic forcing function $F_1 \sin \omega t$ with a viscous damper added.

Fig. 7.1

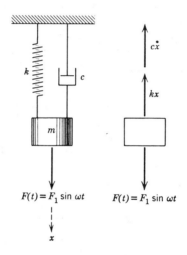

As we have learned, the viscous damping mechanism provides a damping force which is proportional to the velocity of the mass. From the free body diagram, the equation of motion is

$$-kx - c\dot{x} + F_1 \sin \omega t = m\ddot{x}$$
$$m\ddot{x} + c\dot{x} + kx = F_1 \sin \omega t \qquad (7.1)$$

This again is a linear second-order differential equation. It is a particular form of equation 6.2 and the integral must contain both the general integral, which is equation 6.4, and a particular integral that will be a solution of the equation of motion when the applied force is $F_1 \sin \omega t$. The solution is

$$x = e^{-(c/2m)t}[A \cos \omega_d t + B \sin \omega_d t] + C \cos \omega t + D \sin \omega t \qquad (7.2)$$

The first term is the homogeneous term and is a transient which damps out with time. The second and third terms are the particular terms and represent steady state vibration, which is present as long as the forcing function is active. In Chapter 4, we omitted the term involving damped free vibration, without explaining why this was permissible in considering the steady state. We can now see that it can be ignored if $ct/2m>5$.

Substituting the steady state solution in the equation of motion, we have

$$\left[-C\omega^2+D\frac{c\omega}{m}+\frac{k}{m}C\right]\cos\omega t$$

$$+\left[-D\omega^2-C\frac{c\omega}{m}+\frac{k}{m}D\right]\sin\omega t=\frac{F_1}{m}\sin\omega t \quad (7.3)$$

This equation simply states vectorially that for any given harmonic, the amplitude of the cosine component is zero and the amplitude of the sine component is F_1/m, for all time. These statements can be expressed in two simultaneous equations, which we can solve for the magnitudes of C and D. If we had chosen the cosine function instead of the sine function for our excitation, the equations would have been reversed. In matrix form,

$$\begin{bmatrix} (\omega_n^2-\omega^2) & +\dfrac{c\omega}{m} \\[2ex] -\dfrac{c\omega}{m} & (\omega_n^2-\omega^2) \end{bmatrix}\begin{bmatrix} C \\[2ex] D \end{bmatrix}=\begin{bmatrix} 0 \\[2ex] \dfrac{F_1}{m} \end{bmatrix} \quad (7.4)$$

The values of C and D are

$$C=-\frac{F_1 c\omega}{m^2\left[(\omega_n^2-\omega^2)^2+\left(\dfrac{c\omega}{m}\right)^2\right]}$$

$$D=\frac{F_1(\omega_n^2-\omega^2)}{m\left[(\omega_n^2-\omega^2)^2+\left(\dfrac{c\omega}{m}\right)^2\right]} \quad (7.5)$$

Fig. 7.2

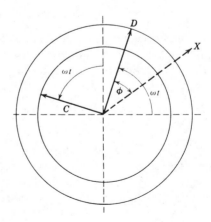

The negative value for the constant C merely indicates that the cosine component of the forced amplitude must lag the sine component instead of leading it. This can be shown in a vector diagram, Figure 7.2. Resolving the vectors C and D into the single vector X, at a phase angle to D, equation 7.2 becomes

$$x = e^{-(c/2m)t}[A \cos \omega_d t + B \sin \omega_d t] + X \sin(\omega t - \phi) \qquad (7.6)$$

The magnitude of X is

$$X = \sqrt{C^2 + D^2} = \frac{F_1}{m\sqrt{(\omega_n^2 - \omega^2)^2 + \left(\dfrac{c\omega}{m}\right)^2}}$$

$$= \frac{F_1}{k\sqrt{\left(1 - \dfrac{\omega^2}{\omega_n^2}\right)^2 + \left(2\zeta\dfrac{\omega}{\omega_n}\right)^2}} \qquad (7.7)$$

Recalling that $X/(F_1/k)$ is the *amplitude ratio*, the response of a viscous damped single degree of freedom system excited by the harmonic force $F_1 \sin \omega t$ is shown in Figure 7.3

$$\frac{X}{F_1/k} = \frac{1}{\sqrt{\left(1 - \dfrac{\omega^2}{\omega_n^2}\right)^2 + \left(2\zeta\dfrac{\omega}{\omega_n}\right)^2}} \qquad (7.8)$$

Two characteristics of the response are immediately evident. One is that damping decreases the amplitude ratio for all frequencies, somewhat in proportion to the amount of damping present. The reduction of the amplitude ratio in the presence of damping is most striking at or near resonance. The second observation is that with damping the maximum amplitude ratio occurs at a frequency lower than the resonant frequency $\omega = \omega_n$. One would expect the peak amplitude ratio to occur in resonance with the damped natural frequency, ω_d, but this does not happen. The maximum amplitude ratio occurs when

$$\frac{\omega}{\omega_n} = \sqrt{1 - 2\zeta^2} \qquad (7.9)$$

which is lower than the damped natural frequency, $\omega_d = \omega_n \sqrt{1 - \zeta^2}$, by a small amount.

This may seem to be trivial, but in resonance testing, it is sometimes convenient to obtain a measure of damping by determining the amplitude of vibration at resonance. Conversely, if the amount of damping is known, it is simple to make an estimate of the amplitude of vibration at resonance.

Fig. 7.3

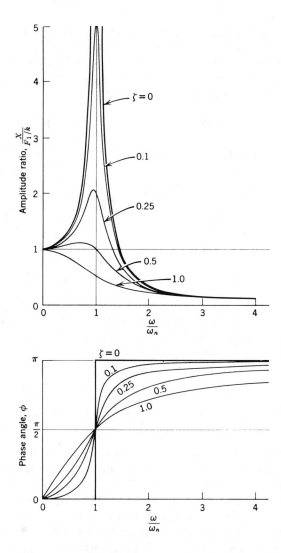

At $\omega = \omega_n$,

$$\left(\frac{X}{F_1/k}\right)_{\omega=\omega_n} = \frac{1}{2\zeta} = Q \tag{7.10}$$

This is one of the most important physical relations in vibration testing. If we were to use the *maximum amplitude* of vibration, instead of the amplitude at resonance, a small error would be introduced, since they are not the same, but the error would be negligible, for light damping. This is one of the academic points associated with viscous damping.

The value of the amplitude ratio at resonance is also called Q (for *Quality*) borrowing a term from electrical engineering circuitry that allows us to use a large number to designate an absence of damping.

The phase angle ϕ by which the response X lags the forcing function is defined as

$$\tan \phi = \frac{-C}{D} = \frac{c\omega}{m(\omega_n^2 - \omega^2)} = \frac{2\zeta \dfrac{\omega}{\omega_n}}{\left(1 - \dfrac{\omega^2}{\omega_n^2}\right)} \qquad (7.11)$$

The phase angle ϕ is very small for small values of ω/ω_n. For very large values of ω/ω_n, it approaches 180°, asymptotically. This means that the amplitude of vibration is in phase with the harmonic exciting force for $\omega/\omega_n \ll 1$ and out of phase for $\omega/\omega_n \gg 1$. At resonance, the phase angle is 90° for all values of viscous damping. Below resonance, the phase angle increases with increased damping. Above resonance, the phase angle decreases with increased damping. Figure 7.3 also shows the variation of phase angle with frequency and damping.

7.3. FORCED DAMPED VIBRATION CAUSED BY ROTATING UNBALANCED FORCES
$F(t) = m_0 \omega^2 e \sin \omega t$

In Figure 7.4, the force F_1 has been replaced by $m_0 \omega^2 e$, the magnitude of the force vector caused by the unbalanced mass m_0, rotating about the geometric axis O. The entire mass is m, which contains the eccentric rotor, and it is constrained to move only in the vertical direction. Lateral motion is ignored.

Fig. 7.4

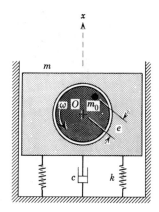

Replacing F_1 by $m_0\omega^2 e$ in equation 7.7,

$$X = \frac{m_0\omega^2 e}{k\sqrt{\left(1 - \frac{\omega^2}{\omega_n^2}\right)^2 + \left(2\zeta\frac{\omega}{\omega_n}\right)^2}}$$

Fig. 7.5

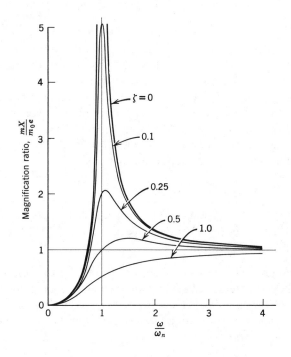

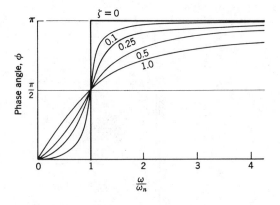

Since $\omega_n^2 = k/m$,

$$\frac{mX}{m_0 e} = \frac{\left(\dfrac{\omega}{\omega_n}\right)^2}{\sqrt{\left(1 - \dfrac{\omega^2}{\omega_n^2}\right)^2 + \left(2\zeta\dfrac{\omega}{\omega_n}\right)^2}} \tag{7.12}$$

This is the *magnification ratio* for forced damped vibration. Figure 7.5 is the nondimensional plot of $mX/m_0 e$ as a function of the frequency ratio ω/ω_n. It looks a lot like the amplitude ratio, except that it passes through the origin, and at very high frequencies, $\omega/\omega_n \gg 1$, the amplitude $X \to m_0 e/m$. The phase angle ϕ is identical with the phase angle for a harmonic force $F(t) = F_1 \sin \omega t$.

7.4. TRANSMITTED FORCES AND VIBRATION ISOLATION

Let us consider again the forces transmitted to the base or foundation of an elastic system and include the effect of damping. Referring to Figure 7.6, the mass m is subjected to the harmonic force, $F_1 \sin \omega t$. The resulting motion in the x direction will also be simple harmonic. The spring force and damping force are, respectively,

$$k = kX \sin(\omega t - \phi)$$
$$c\dot{x} = cX\omega \cos(\omega t - \phi)$$

Fig. 7.6

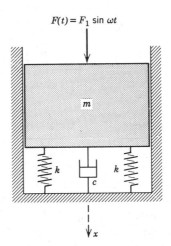

$F(t) = F_1 \sin \omega t$

These are orthogonal vectors and their sum represents the magnitude of the total transmitted force

$$|F_{TR}| = \sqrt{(kX)^2 + (c\omega X)^2} = X\sqrt{k^2 + c^2\omega^2}$$

This will include both the force transmitted through the damper as well as the force transmitted through the spring.

The *transmission ratio* has been defined as that fraction of the maximum impressed force, which is actually transmitted through to the foundation.

$$T.R. = \frac{X}{F_1}\sqrt{k^2 + c^2\omega^2}$$

Substituting for X from equation 7.8,

$$T.R. = \sqrt{\frac{1 + \left(2\zeta\dfrac{\omega}{\omega_n}\right)^2}{\left(1 - \dfrac{\omega^2}{\omega_n^2}\right)^2 + \left(2\zeta\dfrac{\omega}{\omega_n}\right)^2}} \qquad (7.13)$$

The transmission ratio for various amounts of damping is plotted in Figure 7.7.

Fig. 7.7

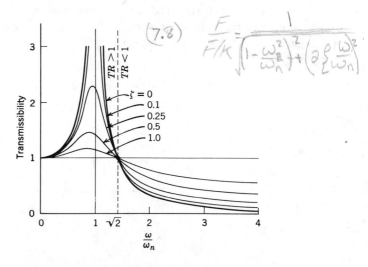

If we compare this result with equation 4.9, it is quite obvious that the forces transmitted cannot become infinite at resonance, unless damping is absent. With damping present, the denominator will never be zero. A careful examination of Figure 7.7 also shows several other interesting aspects. While damping decreases the amplitude of motion for all frequencies, damping decreases the maximum forces transmitted only if $\omega/\omega_n < \sqrt{2}$. Above that value, the addition of damping increases the forces transmitted. If the forcing frequency varies, the choice of elastic supports to minimize the transmission of force must be a compromise. They must have sufficient damping to limit the amplitude and forces transmitted while passing through resonance, and not enough to seriously add to the force transmitted at operating speeds. Luckily, natural rubber is a very satisfactory material and is often used for the isolation of vibration. For very delicate machines, requiring extreme isolation, only coiled springs can give the large static deflections that are needed for very low natural frequencies.

7.5. SEISMIC INSTRUMENTS

The measurement of vibration in all its aspects, amplitude, velocity, acceleration, and stress, just to mention several of the most desired quantities, is a field in itself. The instruments used for measurement and recording of vibration are highly developed, intricate, and expensive. At the core of these measurements, however, one usually finds some form of a seismic instrument. A seismic instrument, in its simplest form, is an elastically supported or seismic mass mounted inside a frame so that the relative motion between the mass and frame can be indicated or recorded. This relative motion is a direct measure of the vibration. Figure 7.8 is a schematic of a seismic vibration measuring device.

The mass is both elastically suspended from its supporting frame and viscously damped. Let us assume that the frame is moving with a motion corresponding to the displacement

$$u = b \sin \omega t$$

In this case, we are using the fundamental, but we could equally as well have chosen a harmonic of order n, $u = b \sin n\omega t$.

If the displacement of the seismic mass is x, the equation of motion is

$$-k(x-u) - c(\dot{x} - \dot{u}) = m\ddot{x} \tag{7.14}$$

We must be careful and note that the spring force is proportional to the relative displacement between the seismic mass and the frame and the

Fig. 7.8

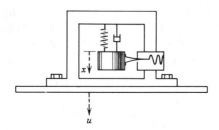

damping force is proportional to the relative velocity. The absolute motion of the seismic mass, x, is of secondary interest, since it is the relative motion that our instrument will measure, and it is the motion of the frame, u, that we seek to measure. Arranging the equation of motion in terms of the relative displacement and its derivatives, after first defining the relative displacement as the difference between x and u

relative displacement
$$z = x - u$$

The equation of motion becomes

$$m(\ddot{z} + \ddot{u}) + c\dot{z} + kz = 0$$
$$m\ddot{z} + c\dot{z} + kz = mb\omega^2 \sin \omega t \tag{7.15}$$

This is essentially the same as equation 7.1, with the force $F_1 \sin \omega t$ replaced by $mb\omega^2 \sin \omega t$. In the steady state, the relative motion would follow the motion of the frame according to the relation,

$$z = Z \sin(\omega t - \phi) \tag{7.16}$$

where Z/b is determined by the fraction,

$Z/b =$ (magnification ratio)

$$\frac{Z}{b} = \frac{\dfrac{\omega^2}{\omega_n^2}}{\sqrt{\left(1 - \dfrac{\omega^2}{\omega_n^2}\right)^2 + \left(2\zeta\dfrac{\omega}{\omega_n}\right)^2}} \tag{7.17}$$

and the phase angle is the same as equation 7.11.

If we wish to measure or indicate amplitude, the natural frequency of a seismic instrument must be several times less than the lowest harmonic of the motion to be measured. This means that the ratio ω/ω_n is large and the relative motion between the mass and frame approaches the absolute motion of the frame. There is a phase difference, which approaches 180°, but this does not affect the magnitude of the reading, which is usually all that matters. Essentially, the seismic mass does not respond to the forced motion and remains motionless in space. We can use the seismic mass as a reference for measuring motion. Damping has little effect on the relative amplitude of motion, and in general, amplitude-measuring seismic instruments are undamped.

If we wish to measure or indicate acceleration, the response of the instrument must match acceleration. The acceleration, $\ddot{u} = -b\omega^2 \sin \omega t$. The relative motion is an indication of the acceleration, if $\omega/\omega_n \ll 1$. Rephrasing equations 7.16 and 7.17,

$$z = \left[\frac{1}{\sqrt{\left(1 - \dfrac{\omega^2}{\omega_n^2}\right)^2 + \left(2\zeta\dfrac{\omega}{\omega_n}\right)^2}}\right] b\frac{\omega^2}{\omega_n^2} \sin(\omega t - \phi) \tag{7.18}$$

Fig. 7.9

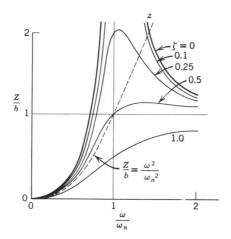

For $\omega/\omega_n \ll 1$, the entire bracket is approximately unity, and ω_n^2 is a constant of the system. In Figure 7.9, $b(\omega^2/\omega_n^2)$ is a parabola. It can be seen that the response of a seismic instrument fits this parabola very well for $\omega/\omega_n \lesssim 0.5$.

In the case of an accelerometer, the natural frequency must be higher than the highest harmonic of the motion to be measured. This is a considerable problem, since most motion that is experimentally measured is not sinusoidal and contains higher harmonics. If damping is added, $0.5 < \zeta < 0.7$, the range of the accelerometer can be extended up to $\omega/\omega_n \lesssim 0.75$, depending on the acceptable error. Seismic accelerometers with a frequency range to 10,000 Hz, are commercially available.

The equation of motion for the absolute displacement of the seismic mass, restating equation 7.14, would be

$$m\ddot{x} + c\dot{x} + kx = ku + c\dot{u}$$

or, since $u = b \sin \omega t$,

$$m\ddot{x} + c\dot{x} + kx = kb \sin \omega t + cb\omega \cos \omega t \tag{7.19}$$

This is the same mathematical expression as transmission ratio. With some reference to the derivation in Section 7.4, the maximum amplitude of the absolute motion of the seismic mass would be

$$\frac{X}{b} = \sqrt{\frac{1 + 2\zeta\dfrac{\omega^2}{\omega_u^2}}{\left(1 - \dfrac{\omega^2}{\omega_u^2}\right)^2 + \left(2\zeta\dfrac{\omega}{\omega_u}\right)^2}}$$

Again, this is only of academic interest, to show how all these expressions fit together. Seismic instruments indicate the relative displacement between the seismic mass and its frame, and not absolute displacement.

$F = Kx \doteq m\ddot{x}$

$\dfrac{K}{M} = \dfrac{9.81}{.005}$

EXAMPLE PROBLEM 7.1

A small motor driven paint compressor set has a mass of 27 kg and causes each of the four rubber isolators on which it is mounted to deflect 5 mm. The motor runs at a constant speed of 1750 rpm. The compressor piston has a 50-mm stroke. The piston and reciprocating parts have a mass of 0.5 kg, and for the purposes of this problem, the reciprocating motion of the piston can be assumed to be simple harmonic. Determine the amplitude of vertical motion at the operating speed. Assume the damping factor for rubber to be $\zeta = 0.2$.

Solution:
This is a clear application of equation 7.12, which states that the magnification ratio is

$$\frac{mX}{m_0 e} = \frac{\left(\dfrac{\omega}{\omega_n}\right)^2}{\sqrt{\left(1 - \dfrac{\omega^2}{\omega_n^2}\right)^2 + \left(2\zeta \dfrac{\omega}{\omega_n}\right)^2}}$$

In this case, the harmonic force varies at a frequency of 1750 rpm, or

$$\omega = \frac{2\pi}{60}(1750) = 183 \text{ s}^{-1}$$

The natural frequency is

$$\omega_n = \sqrt{\frac{9.806}{0.005}} = 44.3 \text{ s}^{-1} \qquad \dot{=} \quad \frac{K}{m} \doteq \frac{\ddot{x}}{x_n}$$

and the frequency ratio

$$\frac{\omega}{\omega_n} = \frac{183}{44.3} = 4.14$$

For a dimensionless damping factor $\zeta = 0.2$,

$$\frac{mX}{m_0 e} = \frac{(4.14)^2}{\sqrt{[1 - (4.14)^2]^2 + [2(0.2)(4.14)]^2}}$$

$$= \frac{17.12}{\sqrt{(-16.12)^2 + (1.655)^2}} = 1.056$$

The piston and reciprocating parts have a mass of 0.5 kg. The total mass of the compressor and engine is 27 kg. The eccentricity is 25 mm, half the stroke. This means that

$$X = \frac{0.5}{27}(25)(1.06) = 0.49 \text{ mm}$$

$$\underline{X} = \frac{m_0}{m} e \left(\frac{m\underline{X}_0}{m_0 e_0} \right)^{1.056}$$

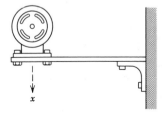

PROBLEM 7.2   An electric motor has a mass of 25 kg and is mounted on a beam cantilevered from a vertical wall. If the motor is displaced 16 mm, the vibration of the motor and beam is observed to dampen to less than 1 mm in 4 cycles. Estimate the value of the dimensionless parameter mX/m_0e for resonant forced vibration, if the armature were unbalanced.

Answer: 4.53

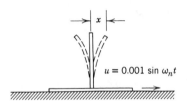

$u = 0.001 \sin \omega_n t$

PROBLEM 7.3   The free vibration of a cantilever beam is observed to decay from an amplitude of 20 mm to half that value in 10 cycles. Calculate the maximum amplitude of vibration at resonance, which is to be expected if the base is subjected to a harmonic vibration 1 mm in amplitude.

Answer: 46 mm

PROBLEM 7.4   A machine has a resonant frequency at 400 rpm when supported on four steel springs for which ζ can be neglected. At 1200 rpm the undamped amplitude of motion is 0.5 mm. What would be the amplitude if the steel springs are replaced by four rubber isolators, where $\zeta = 0.25$? The resonant frequency is unchanged.

Answer: 0.49 mm

PROBLEM 7.5   The enclosed graph shows the actual vertical displacement of a paper pulp screen mounted on rubber isolators as the speed varies from zero to 1500 rpm. Estimate the damping ratio ζ of the system.

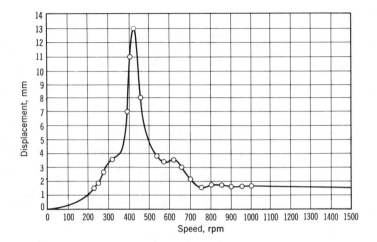

PROBLEM 7.6 When the electric motor of Problem 7.2 is displaced, the motion is damped so that the ratio of successive amplitudes of free vibration is 2:1. What is the transmission ratio for $\omega/\omega_n = 5$?

Answer: T.R. $= 0.062$

PROBLEM 7.7 The decay of free vibration for a complex structure that supports and houses a jet engine is shown. Estimate:

(a) The magnification ratio of vibration, mX/m_0e, at resonance

(b) The magnification ratio at the operating speed of 2200 rpm.

Answer: (a) 10.5; (b) 1.23

PROBLEM 7.8 An electric motor operates mechanical equipment at a speed of 1750 rpm. The system is supported on rubber pads, which have a static deflection of 5 mm. Determine the transmission ratio of force to the foundation, if the damping of the rubber pads is $\zeta = 0.25$.

Answer: 0.1414

PROBLEM 7.9 The modal response of a space frame when excited by a harmonic displacement of a constant magnitude is shown in the following figure. Estimate the damping of the system.

Answer: $\zeta = 0.1$

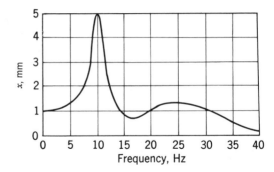

PROBLEM 7.10 In a resonance test under harmonic excitation, it was noted that the amplitude of motion at resonance was exactly twice the amplitude at an excitation frequency 20% greater than resonance. Determine the damping factor ζ of the system.

Answer: $\zeta = 0.138$

PROBLEM 7.11 For small amounts of damping, show that the damping ratio ζ can be approximated from the response curve as

$$\zeta = \frac{f_2 - f_1}{f_2 + f_1}$$

f_2 and f_1 being the frequencies on either side of resonance, where the amplitude is 0.707 of its maximum value.

PROBLEM 7.12 An electromechanical device is mounted on a set of rubber isolators. The system exhibits an amplitude resonance peak of 5:1 at a frequency of 500 cpm. Above what frequency would the transmission of force be reduced to one half?

Answer: 879 cpm

PROBLEM 7.13 The following resonance curve is the actual vertical motion of the floor of a factory measured near a punch press. Estimate the resonant speed and the damping ratio ζ. What is the transmission ratio at 1800 rpm?

Answer: $\zeta = 0.123$, T.R. $= 0.083$

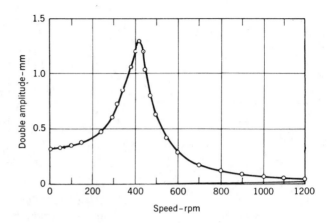

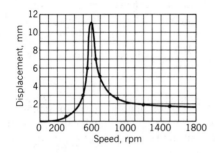

PROBLEM 7.14 The shown resonance curve is the actual vertical motion of the floor measured near a turbine driven water feed pump. Estimate the resonant speed and the damping ratio ζ. What is T.R. at 1800 rpm?

Answer: $f_n = 600$ rpm; $\zeta = 0.0625$
at 1800 rpm, T.R. $= 0.133$

PROBLEM 7.15 Show that the peak of a viscously damped resonance curve for $X/(F_1/k)$ occurs at $\omega = (\sqrt{(1 - 2\zeta^2)})\omega_n$.

PROBLEM 7.16 Show that the peak of the viscously damped resonance curve for (mX/me) occurs at $\omega_n = (\sqrt{1 - 2\zeta^2})\omega$. Note that this result is the inverse of the answer to Problem 7.15.

⁕ **PROBLEM 7.17** If a vibrometer is used to determine amplitudes of vibration at frequencies very much higher than its own natural frequency, what would be the optimum system damping ratio ζ for maximum accuracy?

Answer: $\zeta = 0.707$

PROBLEM 7.18 An electric motor that is used as a mechanical drive is mounted in the center of a light frame table. The mass of the motor added to the effective mass of the table is 40 kg. The armature and rotating parts have a mass of

10 kg and have eccentricity of 15 mm. The table is observed to deflect 3 mm when the motor is mounted. In free vibration, a displacement of 32 mm is damped to 1 mm in 1 s. The operating speed of the motor is 875 rpm. Calculate the amplitude of motion if the damping is assumed to be viscous.

Answer: 6.10 mm

PROBLEM 7.19 A vibrometer has a natural frequency of 5 Hz and is constructed with a damping factor of $\zeta = 0.6$. Determine the lowest frequency for which vibration can be measured with an accuracy of $\pm 2\%$.

Answer: 17.55 Hz

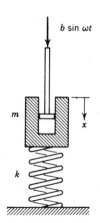

PROBLEM 7.20 A piston slides with viscous friction in a cylinder bored within a mass m. The piston has a sinusoidal motion $b \sin \omega t$. Determine the equation of motion of the mass m. Determine an expression for x as a function of ωt.

Answer:
$$x = \frac{2\zeta \dfrac{\omega}{\omega_n} b \sin(\omega t + \phi)}{\sqrt{\left(1 - \dfrac{\omega^2}{\omega_n^2}\right)^2 + \left(2\zeta \dfrac{\omega}{\omega_n}\right)^2}}$$

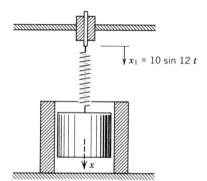

PROBLEM 7.21 A 4-kg piston slides with viscous friction inside a cylinder. The piston and cylinder are separated by an oil film, and viscous friction would be an exact model. The upper end of an elastic spring that supports the piston moves with a harmonic motion $x_1 = 10 \sin 12\, t$, in millimeters. The spring has a modulus of 400 N/m. The damping constant is 20 N·s/m or 20 kg/s. Calculate the maximum amplitude of the piston.

Answer: 13.3 mm

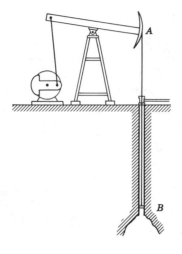

PROBLEM 7.22 An oil well pumping system consists of a motor-driven walking beam and sucker rods attached to a mechanical lift pump. The diameter of the pump barrel is 51 mm. The steel sucker rods are 19 mm in diameter and 1905 m long. The total mass of the sucker rods immersed in the column of oil is 3675 kg. The column of oil lifted at each stroke of the pump is 3175 kg. The speed of the pump motor is 18 rpm. Assume that the damping ratio ζ is 0.5. If the stroke of the sucker rods at the walking beam is 1.25 m, what is the stroke of the piston? Determine the output of the well in barrels per day.

Answer: 410 bbls/day

7.6. FORCED HARMONIC VIBRATION WITH HYSTERESIS DAMPING

For the system of Figure 7.10, a hysteresis damper replaces the viscous damper. The equation of motion becomes

$$-kx - (h/\omega)\dot{x} + F(t) = m\ddot{x}$$

$$m\ddot{x} + (h/\omega)\dot{x} + kx = F(t)$$

(7.20)

In this case, the damping force is not simply a function of velocity. It is also a function of the forcing frequency ω. This makes the equation of motion nonlinear, unless $\omega = \omega_n$.

The mathematical solution for motion with hysteresis damping is quite complex. We can simplify our task by considering only the steady state

Fig. 7.10

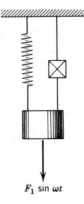

$F_1 \sin \omega t$

response to a sinusoidal forcing function $F(t) = F_1 \sin \omega t$ or $F(t) = F_1 e^{i\omega t}$
The motion will be the harmonic,

$$x = X \sin(\omega t - \phi)$$

where,

$$X = \frac{F_1}{m\sqrt{(\omega_n^2 - \omega^2)^2 + \left(\dfrac{h}{m}\right)^2}} \tag{7.21}$$

by dividing by ω_n^2, the amplitude ratio is

$$\frac{X}{F_1/k} = \frac{1}{\sqrt{\left(1 - \dfrac{\omega^2}{\omega_n^2}\right)^2 + \left(\dfrac{h}{k}\right)^2}} \tag{7.22}$$

and the phase angle becomes

$$\tan \phi = \frac{h}{m(\omega_n^2 - \omega^2)} = \frac{h}{k\left(1 - \dfrac{\omega^2}{\omega_n^2}\right)} \tag{7.23}$$

The steady state response of a single degree of freedom system with hysteresis damping is shown in Figure 7.11. Comparing Figure 7.11 with Figure 7.3, for viscous damping two differences are apparent. One is that the maximum amplitude ratio for hysteresis damping occurs at the resonant frequency, $\omega = \omega_n$, and not at a frequency below resonance. The second is that the phase angle has an intercept of $\phi = \tan^{-1}(h/k)$ at zero forcing frequency. Motion with hysteresis damping can never be in phase with the forcing frequency, unless we presume the trivial case when both the damping and the forcing frequency are zero. We have learned that the factor h/k is a measure of the hysteresis damping and it appears again here. Note that it takes the place of the dimensionless parameter $2\zeta(\omega/\omega_n)$.

For the spring and a hysteretic damper, the respective spring force and damping forces transmitted to the base or foundation are

$$kx = kX \sin(\omega t - \phi)$$
$$(h/\omega)\dot{x} = hX \cos(\omega t - \phi)$$

These also are orthogonal vectors, and the magnitude of the total transmitted force is

$$|F_{TR}| = X\sqrt{k^2 + h^2}$$

Fig. 7.11

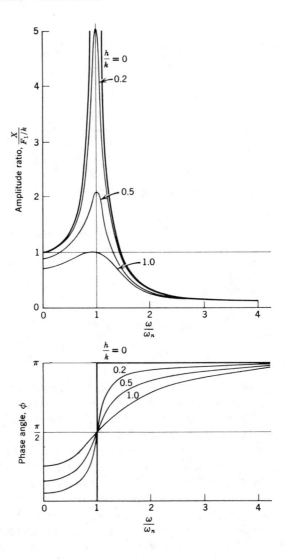

The transmission ratio will be

$$\text{T.R.} = \sqrt{\frac{1 + \left(\dfrac{h}{k}\right)^2}{\left(1 - \dfrac{\omega^2}{\omega_n^2}\right)^2 + \left(\dfrac{h}{k}\right)^2}} \tag{7.24}$$

The transmission ratio for hysteresis damping is plotted in Figure 7.12. Note that maximum force transmission occurs at resonance for hysteresis damping, but for viscous damping, it does not. The transmissibility is

Fig. 7.12

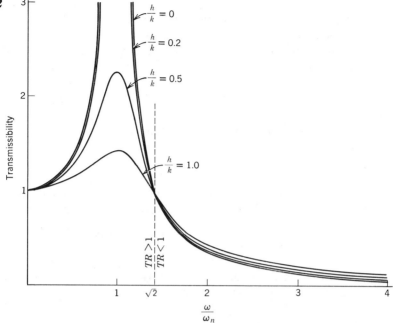

exactly the same for viscous damping as it is for hysteresis damping at $\omega = \omega_n$ and $\omega = \sqrt{2}\omega_n$. Below $\omega = \omega_n$ and above $\omega = \sqrt{2}\omega_n$, viscous damping gives greater force transmission. Between $\omega = \omega_n$ and $\omega = \sqrt{2}\omega_n$, hysteresis damping gives greater force transmission.

These differences between forced vibration with hysteresis damping and forced vibration with viscous damping are not significant, but they are the source of some difficulty in reconciling physical data. In most damped vibration, damping is not viscous, and to assume that it is without knowing its real characteristics is an assumption of some error. It is, however, usually a conservative assumption.

7.7. FORCED HARMONIC VIBRATION WITH DRY FRICTION DAMPING

The response of a single degree of freedom system subjected to a harmonic forcing function is very difficult to predict if dry friction is present. The equation of motion is also nonlinear.

$$m\ddot{x} + kx \pm \mu N = F_1 \sin \omega t \qquad (7.25)$$

The sign of the friction force changes with the direction of motion. An exact solution is known, due to the early work of J. P. Den Hartog.

If the dry friction force is small compared to the harmonic forcing function, an approximate solution is available that explains much of the phenomena associated with dry friction damping.

If the friction force is μN and the amplitude is X, the energy dissipated per quarter cycle is $\mu N X$. For a full cycle of motion, the energy dissipated is

$$\Delta U = 4\mu N X \tag{7.26}$$

The energy dissipated per cycle using an equivalent viscous damping constant is

$$\Delta U = \pi c_{eq} \omega X^2 \tag{7.27}$$

Equating the energy dissipated per cycle, we obtain an equivalent viscous damping factor that is measured in terms of dry friction

$$2\zeta_{eq}\frac{\omega}{\omega_n} = 2\frac{c_{eq}}{c_{cr}}\frac{\omega}{\omega_n} = \frac{4\mu N}{\pi k X} \tag{7.28}$$

Substituting this equivalent damping term into equation 7.8, we have an expression for the amplitude ratio

$$\frac{X}{F_1/k} = \frac{1}{\sqrt{\left(1-\frac{\omega^2}{\omega_n^2}\right)^2 + \left(\frac{4\mu N}{\pi k X}\right)^2}}$$

This contains X within the radical. Squaring and solving algebraically, the amplitude ratio is

$$\frac{X}{F_1/k} = \sqrt{\frac{1-\left(\frac{4\mu N}{\pi F_1}\right)^2}{\left(1-\frac{\omega^2}{\omega_n^2}\right)^2}} \tag{7.29}$$

This is a satisfactory expression for the amplitude ratio, if $\mu N/F_1 < \pi/4$. If $\mu N/F_1 > \pi/4$, the numerator of the fraction under the radical is negative, the radical is imaginary, and the approximate solution cannot be used. It is actually in serious error if $\mu N/F_1 > \frac{1}{2}$.

The exact solution as derived by Den Hartog is given in white in Figure 7.13 for reference. If the damping force is small enough to permit continuous motion, the approximate solution and the exact solution are close. If the dry friction force is large, discontinuous motion results. The latter problem is of academic interest and has some practical importance, but fortunately, the amplitudes of motion involved are usually small and rarely critical. In Figure 7.13, the area below the white dotted line represents discontinuous motion.

Fig. 7.13

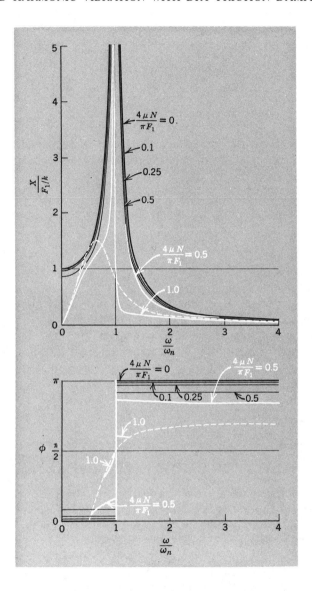

At resonance, the amplitude of motion is not limited by the dry friction damping that is present. The energy dissipated in dry friction is proportional to the friction force and the amplitude of motion. The work input of the harmonic forcing function is also proportional to the magnitude of the force and the amplitude of motion. Thus, as long as the friction force is less than the applied force, the amplitude will increase without limit.

The phase angle can be found in a similar way.

$$\tan \phi = \frac{\dfrac{4\mu N}{\pi k X}}{\left(1-\dfrac{\omega^2}{\omega_n^2}\right)} = \frac{\dfrac{4\mu N}{\pi F_1}}{\left(1-\dfrac{\omega^2}{\omega_n^2}\right)} \sqrt{\frac{\left(1-\dfrac{\omega^2}{\omega_n^2}\right)^2}{1-\left(\dfrac{4\mu N}{\pi F_1}\right)^2}}$$

For a given value of $\mu N/F_1$, the $\tan \phi$ is constant, but with a positive value of $\omega/\omega_n < 1$ and a negative value for $\omega/\omega_n > 1$,

$$\tan \phi = \pm \frac{\dfrac{4\mu N}{\pi F_1}}{\sqrt{1-\left(\dfrac{4\mu N}{\pi F_1}\right)^2}} \tag{7.30}$$

It follows that the phase angle is discontinuous at resonance for dry friction damping.

Experiments have shown what actually happens if μN is large. Motion stops and then starts again. That is, during the cycle, the mass will come to a complete stop, remain static for a brief time, and then start again. For low frequency ratios, the mass may start and stop more than once in each cycle. This is known as the "stick-slip" phenomenon or "chatter." It can be illustrated by moving a long piece of chalk perpendicular to a chalkboard.

7.8. EQUIVALENT VISCOUS DAMPING

As stated in Chapter 6 viscous damping may or may not be a good approximation of the actual damping in a real system. It is, however, the simplest form of damping to use since it does linearize the equations of motion. Often, complex forms of damping are lumped in terms of an equivalent viscous damping coefficient. This equivalence can be found in any one of several ways.

One definition of equivalent viscous damping has already been used. By equating the energy dissipated per cycle to the amount of energy which would be dissipated in viscous damping, equation 7.27 gives

$$c_{eq} = \frac{\Delta U}{\pi \omega X^2} \tag{7.31}$$

Another definition of equivalent viscous damping is that it is that amount of viscous damping which would limit the amplitude ratio at resonance to the value observed experimentally.

$$\zeta_{eq} = \frac{c_{eq}}{c_{cr}} = \frac{1}{2\left(\dfrac{X}{F_1/k}\right)_{\omega=\omega_n}}$$

It should be noted that this method can be extended for frequencies other than resonance. The energy input per cycle for the harmonic force $F_1 \sin \omega t$ is $F_1 X \pi \sin \phi$. The energy dissipated in damping per cycle is $c_{eq} X^2 \pi \omega$. Equating the energy input per cycle from the impressed force and the energy absorbed in damping

$$c_{eq} = \frac{F_1}{X\omega} \sin \phi \qquad (7.32)$$

This is identical with the previous result for $\omega = \omega_n$, except for reference to the phase angle. Measuring the phase angle of motion, however, can be difficult. At resonance, $\sin \phi = 1$, and the result of equation 7.32 becomes identical with equation 7.31.

A third method used to compute equivalent damping refers to the width of the response curve at resonance. This has application in resonance testing and is particularly useful when damping is light and only a portion of the frequency response spectrum has been tested. Use of the width of the response curve involves accurately determining the frequencies for which $\tan \phi = \pm 1$. Referring to Figure 7.14, one will occur at ω_a, slightly below the resonant frequency. The other will occur slightly above the resonant frequency, at ω_b.

From equation 7.11 for $\tan \phi = 1$,

$$\frac{c_{eq}\omega_a}{m(\omega_n^2 - \omega_a^2)} = 1$$

Fig. 7.14

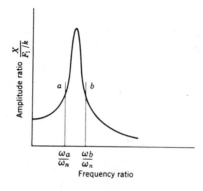

for tan $\phi = -1$,

$$\frac{c_{eq}\omega_b}{m(\omega_n^2 - \omega_b^2)} = -1$$

Adding $c_{eq}\omega_a$ to $c_{eq}\omega_b$

$$c_{eq}(\omega_a + \omega_b) = m(\omega_b^2 - \omega_a^2)$$

and

$$c_{eq} = m(\omega_b - \omega_a) \tag{7.33}$$

The difficulty in using this technique is that phase angle is very difficult to measure accurately. Resonance is avoided, however, and the equivalent viscous damping constant measured in this manner is much more an average than if it were taken from the single value at resonance. Using, $c_{cr} = 2m\omega_n$

$$\zeta_{eq} = \frac{(\omega_b - \omega_a)}{2\omega_n} \tag{7.34}$$

This method is used in measuring the damping present in the torsional vibration of internal combustion engines. It requires only a narrow portion of the resonance curve, which is usually all it is possible to obtain, since below a minimum speed an internal combustion engine will simply refuse to run. Also, compare this method with Problem 7.11.

EXAMPLE PROBLEM 7.23

For Example Problem 6.19, estimate the value of the dimensionless parameter $X/(F_1/k)$ at resonance under forced damped vibration, $F(t) = F_1 \sin \omega t$.

Solution:
From Example Problem 6.19, $h = 28.65$ N/mm and $k = 250$ N/mm. The maximum amplitude with hysteretic damping is

$$\frac{X}{F_1/k} = \frac{1}{\sqrt{\left(1 - \frac{\omega^2}{\omega_n^2}\right)^2 + \left(\frac{h}{k}\right)^2}}$$

At resonance, $\omega = \omega_n$, and

$$\left(\frac{X}{F_1/k}\right)_{\omega = \omega_n} = \frac{k}{h} = \frac{250}{28.65} = 8.73$$

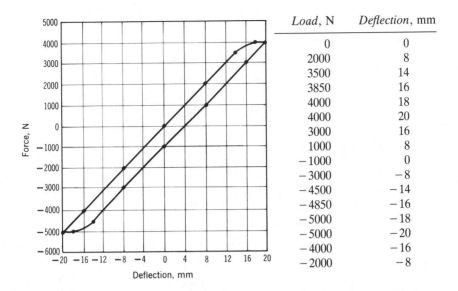

Load, N	Deflection, mm
0	0
2000	8
3500	14
3850	16
4000	18
4000	20
3000	16
1000	8
−1000	0
−3000	−8
−4500	−14
−4850	−16
−5000	−18
−5000	−20
−4000	−16
−2000	−8

EXAMPLE PROBLEM 7.24

The spring and mass of Problem 6.32 are shown here again. In this case, one quarter of the mass is contained in a rotor that rotates at an angular speed ω with an eccentricity of 2 mm. The kinetic coefficient of dry friction is 0.2. The mass is 15 kg and the spring modulus is 1800 N/m. What is the amplitude of motion at $\omega = 100$ s^{-1}? What happens when $\omega = 10$ s^{-1}. What does this physically mean?

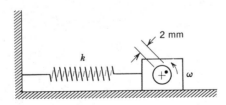

Solution:
The applied force, $F(t) = m_0\omega^2 e \sin \omega t$. With the given system constants, for $\omega = 10$ s^{-1},

$$F_1 = m_0\omega^2 e = (\tfrac{15}{4})0.002(10)^2 = 0.75 \text{ N}$$

and,

$$\frac{\mu N}{F_1} = \frac{\mu N}{m_0\omega^2 e} = \frac{0.2(15)(9.806)}{0.75} = 39.2$$

Since $39.2 > \pi/4$, motion is unlikely. The applied force is not sufficient to overcome static friction.

For $\omega = 100 \text{ s}^{-1}$, it is quite a different matter. In this case,

$$F_1 = m_0 \omega^2 e = \tfrac{15}{4}(0.002)(100)^2 = 75 \text{ N}$$

$$\frac{\mu N}{F_1} = \frac{\mu N}{m_0 \omega^2 e} = \frac{(0.2)(15)(9.806)}{75} = 0.392$$

Now, motion is possible and it is continuous, since $0.392 < \pi/4$. The amplitude can be found using equation 7.29.

$$\omega_n^2 = \frac{k}{m} = \frac{1800}{15} = 120 \text{ s}^{-2}$$

$$\frac{X}{F_1/k} = \sqrt{\frac{1 - \left[\dfrac{4\mu N}{\pi F_1}\right]^2}{\left[1 - \dfrac{\omega^2}{\omega_n^2}\right]^2}}$$

$$= \sqrt{\frac{1 - \left[\dfrac{4}{\pi}(0.392)\right]^2}{\left[1 - \dfrac{100^2}{120}\right]^2}} = 0.0118$$

$$X = \frac{75}{1800}(0.0105) = 0.4 \text{ mm}$$

PROBLEM 7.25 Repeat Example Problem 7.23, using hysteretic damping where successive amplitudes of free vibration are halved.

Answer: $\dfrac{k}{h} = 4.52$

PROBLEM 7.26 Repeat Example Problem 7.23, using four isolators made from the helical springs of Problem 6.22.

Answer: $\dfrac{k}{h} = 9.1$

PROBLEM 7.27 The 200-kg induction motor of Problem 4.12 is mounted on 16 rubber isolators, each of which exhibits the shown force deflection characteristics

(a) What is the value of the dimensionless parameter mX/m_0e for resonant forced vibration using these isolators?

(b) What is the transmission ratio for $\omega = 1800$ rpm.

Answer: (a) 10; (b) 0.047

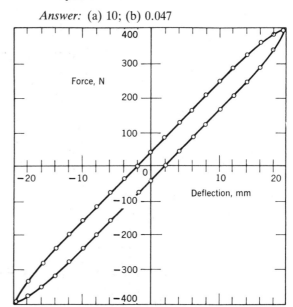

PROBLEM 7.28 Repeat Problem 7.18 with hysteretic damping.

Answer: 6.14 mm

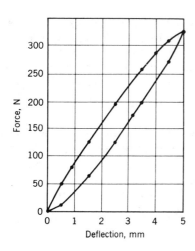

PROBLEM 7.29 A small 20-kg electronic instrument is mounted in an aircraft panel on four rubber isolators that deflect 3 mm under load. The instrument must be isolated from a dominant excitation of 1200 cpm. The load–deflection curve for the rubber isolators is shown here. What is the transmission ratio of the vibration transmitted to the mounted instrument?

Answer: 0.266

PROBLEM 7.30 An electromechanical device is mounted on a set of rubber isolators. The system exhibits a resonance amplitude ratio of 5. Above what frequency ratio would the transmission of force be reduced to one quarter, if

(a) damping is ignored.

(b) viscous damping is assumed,

(c) hysteretic damping is assumed.

PROBLEM 7.31 A light frame structure shows the following amplitude–time trace for damped-free vibration. The frame deflects 2 mm under a force of 500 N. What would be the approximate amplitude of motion, if the force F is replaced by the harmonic force $F = 500 \sin \omega t$, where $\omega = 960$ rpm?

Answer: 0.316 mm

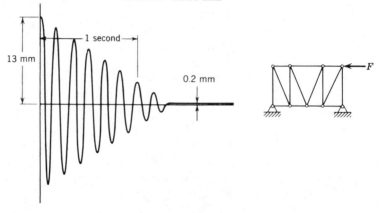

PROBLEM 7.32 The connecting rod of Problem 2.48 and Problem 6.36 is subjected to a small oscillating couple, $M(t) = 0.1 \sin 20\, t$, where the couple is measured in N·m. What is the approximate amplitude of steady-state motion?

Answer: 0.022 rad (1.27°)

PROBLEM 7.33 For Problem 6.31, the spring has a constant of 12,500 N/m. Determine the amplitude of steady-state motion if a force $F(t) = 10 \sin \omega t$ is applied to the mass m. $F(t)$ oscillates with a harmonic frequency of 1200 rpm.

Answer: 0.1 mm

CHAPTER EIGHT

NONLINEAR VIBRATION

8.1. LINEARITY AND NONLINEARITY

For the most part, in the preceding sections, we have treated only linear systems. That is, the differential equation that describes the motion of the system is a linear differential equation. A differential equation is said to be linear when each term of the equation contains only the first power of the dependent variable or any of its derivatives. All other equations are called nonlinear. At times, some approximations had to be made to make a differential equation linear, but these were simple approximations, and if we considered only small amplitudes of motion, they were readily accepted. With the consideration of damping, however, linearity could not be so quickly accepted. In fact, real damped systems are more often nonlinear than linear, which makes it important to recognize the differences between linearity and nonlinearity in vibration.

First, mass can be nonlinear, but in mechanical engineering there are few problems in which the mass of a system cannot be assumed to be constant for engineering purposes. A significant exception is a reciprocating engine or a reciprocating compressor. In Figure 8.1, the inertia of the piston that moves up and down within the cylinder adds to the moment of inertia of the rotating crank and crankshaft when the crank angle θ is at an angle that is nearly a right angle to the axis of piston travel. The actual angle depends on the length of the connecting rod and the throw of the crank. At crank angles of $\theta = 0°$ and $\theta = \pi$, the piston passes through top and bottom dead center, respectively, and the inertia of the piston

Fig. 8.1

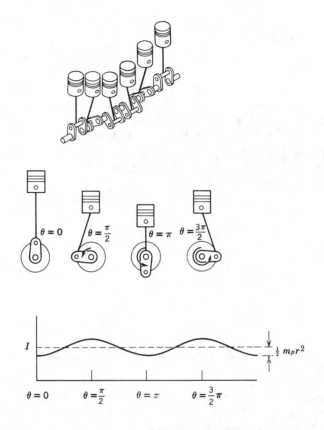

does not add to the rotating inertia. Thus, the rotating inertia of a recip-rocating engine varies with crank displacement, reaching a maximum and minimum twice in each revolution of the crankshaft. For the effective rotating inertia I_e, the moment of inertia of the crankshaft I_c, the recip-rocating mass m_p, and crank throw r and crank angle θ

$$I_e = I_c + \tfrac{1}{2}m_p r^2 - \tfrac{1}{2}m_p r^2 \cos 2\theta$$

The mass moment of inertia, I_e, is a constant plus a quantity that varies with crank displacement, the coordinate θ. It is a case of nonlinear mass. As stated, this is still only an approximation.

For multicylinder engines the reciprocating inertia of one cylinder is balanced by the inertia of another. This is one of the reasons that in-line engines have offset cranks on the crankshaft. As a result, in multicylinder engines, the variation of mass moment of inertia with time is also ignored. The equivalent mass moment of inertia is simply calculated to be

$$I_e = I_c + \tfrac{1}{2}m_p r^2 \tag{8.1}$$

Second, of the three most prevalent forms of damping, two are nonlinear. The entire subject of damping, both linear and nonlinear, is treated extensively in Chapter 6.

By far the most common case of nonlinearity in mechanical vibration is that of the elastic member. In a built-up structure, the effective spring constant often increases with displacement. For small displacements, the structure is quite flexible, but one can imagine a structure stiffening as the clearances within it are overcome. For large displacements, it is quite possible to have a situation where the effective spring constant decreases with displacement. The moduli of most engineering materials decrease with increased strain. In some cases, the decreased load-carrying ability with increased strain is so pronounced that the strain mechanism is assumed to be plastic. Buckling also may considerably lessen the load-carrying ability of any member or structure. Either a stiffening spring or a softening spring may be created by design or by accident. Figure 8.2 shows two cases of *nonlinear elasticity*, one increasing with displacement, the other decreasing.

As stated, a system is said to be linear if the differential equation of motion of the system is a linear differential equation. Consider the equation

$$m\ddot{x} + c\dot{x} + kx = F(t) \tag{8.2}$$

For this equation to be linear, the coefficients of $\ddot{x}$, $\dot{x}$, and x cannot be functions of the dependent variable x, or any of its time derivatives, and the function x and its time derivatives, if they appear, must appear to the first power, only. Here, mass is a constant coefficient of acceleration,

Fig. 8.2

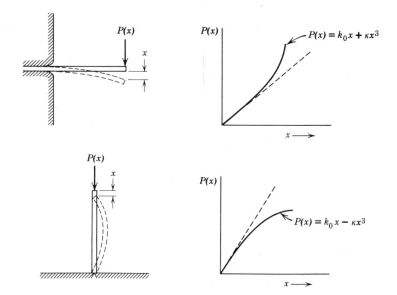

damping is a constant coefficient of velocity, and the elastic constant is a constant coefficient of displacement. This equation is described as a linear, second-order differential equation with constant coefficients.

The fundamental advantage that linearity has is that the principle of superposition may be used. This principle states that any linear combination of solutions to a linear differential equation is also a solution to that equation. It does not hold for nonlinear differential equations. Nonlinear systems do not lend themselves to general solutions, and in most cases, the solutions to nonlinear equations do not exist. If solutions do exist, and some solutions of nonlinear equations are known, these solutions are very specific. Some examples will follow.

When faced with nonlinearity in mechanical vibration, all is not hopeless. There is always the strong possibility that the equation of motion can be *linearized*, that is, some linear approximation to the nonlinear differential equation can be found and accepted. We did just that in accepting viscous damping as an approximate model of real damping. There is also the possibility that some important information can be learned from a nonlinear equation of motion, regardless of the nonlinearity. The period, if the motion is periodic, and the maximum amplitude, if the vibration is forced, are good examples. There are ways to find the frequency and amplitude of a nonlinear system. The closed form solution of the nonlinear differential equation may be wholly unnecessary for an engineering treatment of the problem.

8.2. DISCONTINUOUS LINEARITY

A simple, but very common, form of nonlinearity occurs when the linearity of the system is discontinuous. In Figure 8.3, the elastic springs are linear, but the motion of the mass m is not a continuous function. It is actually discontinuous twice in each cycle. Systems such as this are easy to analyze, taking them step by step. Discontinuous linearity can be used to approximate nonlinearity through linear segments. Free vibration with Coulomb or dry friction damping is a case of nonlinearity that is really discontinuous linearity.

In Figure 8.3, the mass m is not in contact with the springs, except at the extreme excursions of each cycle, and then, it is in contact with only one spring at a time. While in contact with the right spring, the motion is harmonic,

$$x = B \sin \omega_n(t - t_1) + x_1 \tag{8.3}$$

where x_1 is half the distance the mass must travel between springs, and t_1 is the time when contact is first made with the right spring. This equation is valid for $t_1 \leqslant t \leqslant t_2$.

Fig. 8.3

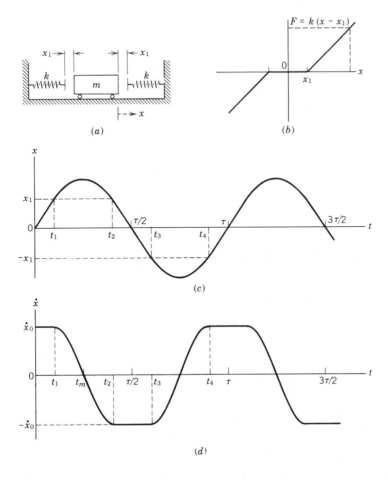

(a)

(b)

(c)

(d)

The natural circular frequency of the harmonic motion is ω_n

$$\omega_n = \sqrt{\frac{k}{m}}$$

t_2 is the time when contact is lost with the right spring and the total time in contact with the spring is $t_2 - t_1$,

$$t_2 - t_1 = \pi \sqrt{\frac{m}{k}} \tag{8.4}$$

The velocity would be

$$\dot{x} = B\omega_n \cos \omega_n (t - t_1) \tag{8.5}$$

The maximum velocity occurs at the time the mass makes contact with the spring, at $t = t_1$.

$$\dot{x}_1 = B \omega_n$$

From this, it can be deduced that the arbitrary coefficient B is the maximum deflection of the spring. This yields the time required to cross the clearance x_1.

$$t_1 = \frac{x_1}{\dot{x}_1} \tag{8.6}$$

While in contact with the left spring, the motion is again harmonic, but negative.

$$x = -B \sin \omega_n(t - t_3) - x_1 \tag{8.7}$$

Counting all the pieces needed to make one complete cycle, the period for this periodic motion would be

$$\tau = \frac{4x_1}{\dot{x}_1} + 2\pi \sqrt{\frac{m}{k}} \tag{8.8}$$

The frequency of the periodic motion would be its reciprocal.

8.3. LINEARIZATION BY APPROXIMATION

A second thought in the solution of a nonlinear equation of motion would be to attempt to replace the nonlinear terms with linear terms. How nonlinear is the problem? If a nonlinear term cannot be replaced with a linear term, can it be replaced with a linear and a nonlinear term, separating the linearity from the nonlinearity. When this can be done, the system is quasilinear. Very few mechanical systems are strictly linear. We have linearized quasilinear systems before.

Example Problem 8.1 repeats the compound pendulum, which can be quasilinear or nonlinear, depending on the amplitude of motion.

EXAMPLE PROBLEM 8.1   THE COMPOUND PENDULUM

Repeat Problem 2.29, (also Problem 3.2)

Solution:
The differential equation of motion for rotation is

$$\sum M_0 = I_0 \ddot{\theta}$$

$$-mg\frac{l}{2} \sin \theta = I_0 \ddot{\theta}$$

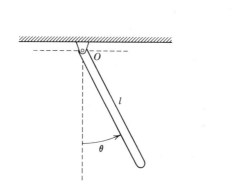

This is a nonlinear equation, since the displacement term is not a linear function of θ.

A first integral can be obtained by multiplying by $\dot{\theta}$

$$-mg\frac{l}{2}(\sin\,\theta)\dot{\theta} = I_0\ddot{\theta}\cdot\dot{\theta}$$

Assuming the initial conditions, at $t=0$, $\theta=0$, $\dot{\theta}=\dot{\theta}_0$, and integrating with respect to θ

$$mgl(1-\cos\,\theta) = I_0\dot{\theta}_0^2 - I_0\dot{\theta}^2$$

$$\dot{\theta} = \sqrt{\dot{\theta}_0^2 - \frac{mgl}{I_0}(1-\cos\,\theta)}$$

Integrating again, and using $I_0 = \frac{1}{3}ml^2$,

$$t = \int_0^\theta \frac{d\theta}{\sqrt{\dot{\theta}_0^2 - \frac{3g}{l}(1-\cos\,\theta)}}$$

$$t = \frac{1}{\dot{\theta}_0}\int_0^\theta \frac{d\theta}{\sqrt{1 - \frac{3g}{l\dot{\theta}_0^2}(1-\cos\,\theta)}}$$

$$t = \frac{1}{\dot{\theta}_0}\int_0^\theta \frac{d\theta}{\sqrt{1 - \frac{6g}{l\dot{\theta}_0^2}\sin^2\frac{\theta}{2}}}$$

Integrals such as this cannot be expressed in terms of simple functions. They are known as elliptic integrals of the first kind. When $6g/l\dot{\theta}_0^2<1$, the pendulum will oscillate, but the motion will not be simple harmonic. One important observation is that the period will depend on the maximum amplitude, a characteristic of nonlinear oscillations.

Linearization by approximation. Now use, as we did in Example Problem 2.29, the Maclaurin expansion for sin θ

$$\sin \theta = \theta - \frac{\theta^3}{3!} + \frac{\theta^5}{5!} - \frac{\theta^7}{7!} + \cdots$$

Replacing sin θ with the first two terms of its expansion, we have an approximation of the exact differential equation of motion, which contains a linear and a cubic function of θ

$$\frac{1}{3}ml^2\ddot{\theta} + \frac{mgl}{2}\left(\theta - \frac{\theta^3}{6}\right) = 0$$

$$\ddot{\theta} + \frac{3}{2}\frac{g}{l}\left(\theta - \frac{\theta^3}{6}\right) = 0$$

This is a form of Duffing's equation, about which much is known. It is named after G. Duffing who first presented and discussed the problem. Its general solution is unknown, however.

This approximation is valid up to amplitudes of about 60°. Replacing sin θ with only the first term *linearizes* the differential equation of motion. The result is similar to equation 2.2, and the solution is simple harmonic.

$$\ddot{\theta} + \frac{3}{2}\frac{g}{l}\theta = 0 \tag{2.2}$$

EXAMPLE PROBLEM 8.2.   SPRING LOADED PENDULUM

A helical spring is fixed at its upper end, is stretched and attached at its lower end to the mass m or bob of a simple pendulum. For simplicity, the unstretched length of the spring is equal to the length of the pendulum. The rod supporting the pendulum bob must be rigid and inelastic, but again, for the sake of simplicity, assume that it has no mass. Determine the equation(s) of motion of the pendulum.

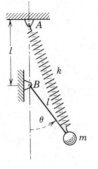
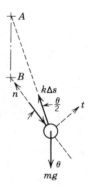

Solution:

Note that in phrasing the problem, we already have one simplifying assumption, which is that the supporting rod is rigid and inelastic, but it is assumed to have no mass. This is an approximation. It is only a good approximation if the mass of the pendulum is very much greater than the mass of the rod, but using the approximation does make the problem much easier to solve.

Summing forces in the t direction,

$$\sum F_t = ma_t$$

$$-mg \sin \theta + k(\Delta s) \sin \frac{\theta}{2} = m(l\ddot{\theta})$$

where Δs is the stretch of the spring from its free length.

$$\Delta s = \sqrt{l^2 + l^2 - 2l^2 \cos(\pi - \theta)} - l$$

$$\Delta s = \sqrt{2l^2 + 2l^2 \cos \theta} - l$$

$$\Delta s = 2l \cos \frac{\theta}{2} - l$$

Substituting back, the equation of motion becomes

$$m(l\ddot{\theta}) + mg \sin \theta - kl\left(2 \cos \frac{\theta}{2} - 1\right)\left(\sin \frac{\theta}{2}\right) = 0$$

This is the exact differential equation of motion. It is nonlinear, and no attempt will be made to solve it.

Now let us use Maclaurin expansions for $\sin \theta$, $\sin \theta/2$, and $\cos \theta/2$.

$$\sin \theta = \theta - \frac{\theta^3}{3!} + \frac{\theta^5}{5!} - \frac{\theta^7}{7!} + \cdots$$

$$\sin \frac{\theta}{2} \doteq \frac{\theta}{2} - \frac{\theta^3}{(2^3)3!} + \frac{\theta^5}{(2^5)5!} - \frac{\theta^7}{(2^7)7!} + \cdots$$

$$\cos \frac{\theta}{2} = 1 - \frac{\theta^2}{4(4!)} + \frac{\theta^4}{(2^4)(4!)} - \frac{\theta^6}{(2^6)(6!)} + \cdots$$

Substituting these expansions, dividing by ml and simplifying,

$$\ddot{\theta} + \frac{g}{l}\left(\theta - \frac{\theta^3}{6} + \frac{\theta^5}{120} - \cdots\right) - \frac{k}{m}\left(1 - \frac{\theta^2}{4} + \frac{\theta^4}{48} - \cdots\right)\left(\frac{\theta}{2} - \frac{\theta^3}{48} + \frac{\theta^5}{3840} - \cdots\right) = 0$$

or,

$$\ddot{\theta} + \frac{g}{l}\left(\theta - \frac{\theta^3}{6} + \frac{\theta^5}{120} - \cdots\right) - \frac{k}{2m}\left(\theta - \frac{\theta^3}{6} + \frac{11\theta^5}{1920} - \cdots\right) = 0$$

Considering only two terms in each expansion, one linear and one non-linear, yields an approximate, but very good, differential equation. In this form, it is a form of Duffing's equation.

$$\ddot{\theta} + \left[\frac{g}{l} - \frac{k}{2m}\right]\left[\theta - \frac{\theta^3}{6}\right] = 0$$

If only one term of each expansion is used, a *linear* differential equation results.

$$\ddot{\theta} + \left[\frac{g}{l} - \frac{k}{2m}\right]\theta = 0$$

This is valid for small displacements. We also have the means at our resource to evaluate the difference between assuming linearity and non-linearity.

EXAMPLE PROBLEM 8.3

The 5-kg mass is firmly connected to *two* springs, each with a constant of $k_1 = 1000$ N/m. There are clearances of 10 mm on either side of the mass. If the mass travels further than 10 mm, it comes in contact with either of two additional springs, one on each side, each with a constant of $k_2 = 2500$ N/m. Determine the natural frequency of free vibration of the system, if (a) the amplitude of motion is 10 mm and (b) the amplitude of motion is 15 mm.

Solution:

For (a), only the two 1000-N/m springs and the 5-kg mass need to be considered. The natural circular frequency is

$$\omega_1 = \sqrt{\frac{2k_1}{m}} = \sqrt{\frac{2(1000)}{5}} = 20 \text{ s}^{-1}$$

$$f_n = \frac{1}{2\pi}\sqrt{\frac{2(1000)}{5}} = 3.183 \text{ Hz}$$

For (b) the mass is in contact with the two 1000-N/m springs for 10 mm on either side of equilibrium. For deflections of 10 to 15 mm, on either side, it is in contact with these two springs and one or the other of the 2500-N/m springs. The force-deflection diagram is shown.

Let us consider the motion of the mass in steps, each step being linear and harmonic. At time $t = 0$, the mass moves with a velocity $\dot{x}_0$ through its position of equilibrium. The equation of motion is

$$x = \frac{\dot{x}_0}{\omega_1}\sin\omega_1 t = \frac{\dot{x}_0}{20}\sin 20t$$

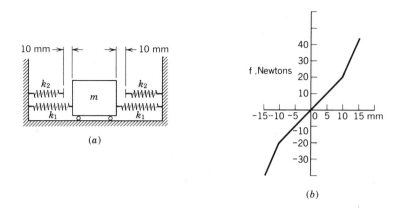

(a)

(b)

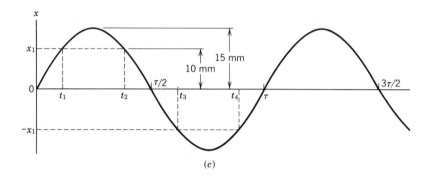

(c)

At time $t = t_1$, when the mass makes contact with the second spring, the amplitude is 10 mm. With the amplitude expressed in millimeters and the velocity in millimeters per second, the time t_1 is

$$t_1 = \frac{1}{20} \sin \frac{20(10)}{\dot{x}_0} \tag{1}$$

Both t_1 and $\dot{x}_0$ are unknown.

Between time $t = t_1$ to time $t = t_2$, the mass is in contact with the two 1000-N/m springs and one additional 2500-N/m spring. The natural circular frequency is

$$\omega_2 = \sqrt{\frac{2k_1 + k_2}{m}} = \sqrt{\frac{2(1000) + 2500}{5}} = 30 \text{ s}^{-1}$$

The equation of motion for linear, harmonic motion for this segment is

$$x = C \cos 30(t - t_1) + D \sin 30(t - t_1)$$

The constants C and D are determined from the first linear segment, since the velocity and displacement must be continuous at time $t = t_1$.

$$C = 10 \text{ mm} \quad \text{and} \quad D = \frac{\dot{x}_1}{30} = \frac{\dot{x}_0}{30} \cos 20t_1$$

Substituting for C and D,

$$x = 10[\cos 30(t - t_1)] + \frac{\dot{x}_0}{30} \cos 20t \, [\sin 30(t - t_1)]$$

$$x = \left(\sqrt{(10)^2 + \left(\frac{\dot{x}_0}{30}\right)^2 \cos^2 20t_1} \right) \left(\sin[30(t - t_1) + \alpha] \right)$$

At the maximum amplitude of 15 mm, $\sin[30(t - t_1) + \alpha] = 1$, and,

$$15 \text{ mm} = \sqrt{(10)^2 + \left(\frac{\dot{x}_0}{30}\right)^2 \cos^2 20t_1} \tag{2}$$

Solving equations (1) and (2) simultaneously for x_0 and t_1.

$$t_1 = 0.02688 \text{ s} \quad \text{and} \quad \dot{x}_0 = 390.5 \text{ mm/s}$$

With a little extra work,

$$\tan \alpha = \frac{C}{D} = \frac{10(30)}{390.5}; \quad \alpha = 0.6551 \text{ rad}$$

The maximum deflection of 15 mm is found to occur at

$$\sin[30(t - t_1) + \alpha] = 1$$

or at $t = 0.05741$ s. This means that the period of the vibration at 15 mm would be 0.2296 s, and $f_n = 4.35$ Hz. Note that this is considerably higher than (a), where $f_n = 3.18$ Hz. Bringing the oscillating mass into contact with springs with larger moduli could be expected to change the frequency, and it does.

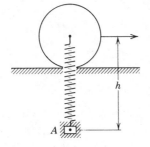

PROBLEM 8.4 A homogeneous solid cylinder rolls without slipping on a horizontal surface. The cylinder is restrained only by a linear spring fixed at point A. For simplicity, assume that the distance h is the stretched length of the spring.

(a) Determine an approximate nonlinear equation of motion.

(b) Determine an approximate linear equation of motion.

Answer: (a) $\ddot{x} + \dfrac{2}{3}\dfrac{k}{m} \times \left[1 + \left(\dfrac{x}{2h} \right)^2 \right] = 0$

(b) $\ddot{x} + \dfrac{2}{3}\dfrac{k}{m} x = 0$

PROBLEM 8.5 For a simple pendulum of mass m and length l, write the differential equation of motion, using the first two terms of the power series expansion for $\sin \theta$.

Answer: $\ddot{\theta} + \dfrac{g}{l}\left[\theta - \dfrac{\theta^3}{6} \right] = 0$

PROBLEM 8.6 Assuming harmonic motion, $\theta(t) = \Theta \sin \omega_n t$, determine the period of oscillation of the simple pendulum of Problem 8.5 when the length of the pendulum is $l = 300$ mm and the amplitude of oscillation is

(a) $\Theta = \dfrac{\pi}{36}$

(b) $\Theta = \dfrac{\pi}{6}$

Compare these results with the solution for small amplitude

$$\tau = 2\pi \sqrt{\dfrac{l}{g}}$$

PROBLEM 8.7 A simple pendulum of mass m and length b is altered by attaching two elastic springs to the mass. For simplicity, assume that the distance a is the free length of the spring. Using the first two terms of the power series expansion for $\sin \theta$, write the equation of motion for free vibration of the system in the $y - z$ ane. Note that the nonlinear terms compensate ach other.

Answer: $\ddot{\theta} + \dfrac{g}{b}\theta + \left[\dfrac{b^2}{a^2}\dfrac{\prime}{r} - \dfrac{g}{6b} \right]\theta^3 = 0$

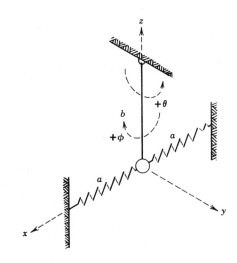

PROBLEM 8.8 For the simple pendulum of Problem 8.7, determine

(a) The equation of motion for free vibration of the system in the $x - z$ plane
(b) An approximate nonlinear equation of motion
(c) An approximate linear equation of motion.

PROBLEM 8.9 Determine the nonlinear equation of motion for large oscillations of the inverted pendulum of Problem 2.39.

Answer: $\ddot{\theta} - \dfrac{3g}{2l} \sin \theta + \dfrac{3k}{m} \sin 2\theta = 0$

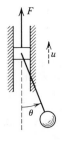

PROBLEM 8.10 A simple pendulum has a mass m and a length l. The pivot of the pendulum is given a harmonic displacement $u = b \sin \omega t$ by an applied force F. This is an example of a time-dependent nonlinearity. Determine the differential equation of motion.

Answer: $\theta + \left(\dfrac{g}{l} - \dfrac{b\omega^2}{l} \sin \omega t \right) \theta = 0$, which is Mathieu's equation. It is linear, but nonautonomous.

PROBLEM 8.11 The linear springs have a modulus of k and are in series, with a clearance c between the mass and each spring. The natural frequency of the springs and mass, where the clearance is zero, is 30 Hz. What is the reduced natural frequency for a total amplitude of 1 mm if the clearance c equals 0.5 mm?

Answer: 18.3 Hz

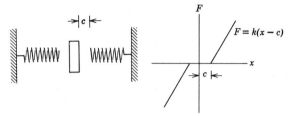

PROBLEM 8.12   A 1-kg mass is attached as a pendulum to a cord that is 1 m long. It is released from rest when $\theta = 5°$. Determine the amplitude at the other end of the swing of the pendulum, and the period of the pendulum.

Answer: 1.713 s

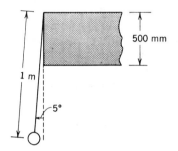

PROBLEM 8.13   The two springs of the system shown are not attached to the 5-kg mass. In its equilibrium position, which is the position shown, each spring just touches the mass and each is at its free length, neither extended or compressed. If the mass is displaced 20 mm to the right and released, what is the frequency of the vibration?

Answer: 3 Hz

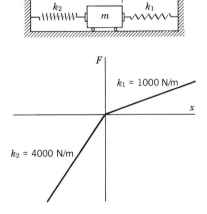

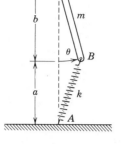

PROBLEM 8.14 The slender rod of Example Problem 2.29 has one end of a spring attached to its lower end. The other end of the spring is fixed directly below the pivot of the rod, and it is fixed so that the free length of the spring is the length a. Determine the nonlinear equation of motion of the rod and spring.

Answer: $\ddot{\theta} + \dfrac{3g}{2b}\sin\theta +$

$$\dfrac{3k}{m}\left[1 - \dfrac{a}{\sqrt{a^2 + 2b(b+a)(1-\cos\theta)}}\right]\sin\theta\cos\theta = 0$$

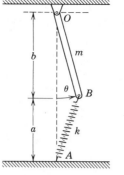

PROBLEM 8.15 Determine the nonlinear equation of motion for the system shown. In the rest position, which is the position shown, neither spring is extended or compressed.

Answer: $\ddot{x} + \dfrac{k}{m}\left[2 - \dfrac{a}{\sqrt{a^2 + x^2}}\right]x = 0$

8.4. FORCED VIBRATION OF A NONLINEAR SYSTEM

The forced vibration of a system containing a nonlinear spring presents several unique characteristics that are important, even in an elementary study of mechanical vibration. The equation of motion for forced vibration is, $\Sigma F = m\ddot{x}$, or

$$m\ddot{x} + P(x) = F(t) \tag{8.9}$$

The term, $P(x)$ is used to denote a nonlinear spring force. Various functions may be used for $P(x)$, but one of the simplest is the polynomial

$$P(x) = k_0 x + \kappa x^3 \tag{8.10}$$

shown in Figure 8.2. Additional terms could be used, but the cubic equation gives an approximate solution, which is usually satisfactory. For symmetry, only odd powers of x should be used. If even powers were used, say $P(x) = k_0 x + \kappa x^2$, the force $P(x)$ would stiffen with positive x and soften with negative x, which would be abnormal. k_0 is the initial elastic constant, with the units being Newtons per metre. The units for κ are Newtons per metre cubed (N/m^3).

Substituting the cubic expression in the equation of motion, we have

$$m\ddot{x} + k_0 x + \kappa x^3 = F(t) \tag{8.11}$$

Let us now excite this nonlinear system with the simple harmonic forcing function $F(t) = F_1 \sin \omega t$. The result is a nonlinear differential equation,

$$m\ddot{x} + k_0 x + \kappa x^3 = F_1 \sin \omega t \qquad (8.12)$$

An approximation of the response for this simple system can be found by assuming the displacement to be simple harmonic. This would be the true motion if $\kappa = 0$. Let us assume that $P(x)$ is almost linear and the resulting motion is near enough to being simple harmonic to let

$$x = X \sin \omega t$$

This assumption satisfies the equation of motion at $\omega t = 0, \pi, 2\pi, 3\pi, \ldots,$ $n\pi$, since the displacement and forcing function are identically equal to zero at these times. It does not satisfy the equation of motion at any other time, however, and it cannot be an exact solution. We can determine the value of the displacement X, which will make the trial solution further satisfy the equation of motion at $\omega t = \pi/2, 3\pi/2, 5\pi/2, \ldots, [(2n+1)/2]\pi$, by equating the magnitude of the vector amplitudes on each side of equation 8.12, since

$$\sin \frac{2n+1}{2}\pi = 1 \qquad \text{for} \qquad n = 1, 3, 5, \ldots$$

$$-m\omega^2 X + k_0 X + \kappa X^3 = F_1 \qquad (8.13)$$

Rearranging terms,

$$F_1 + m\omega^2 X = k_0 X + \kappa X^3 = P(x)$$

Graphically, these two equations are mutually satisfied where they intersect. If the elasticity is linear, there is a unique intersection, but if the elasticity is nonlinear, more than one intersection may be possible. In Figure 8.4, each amplitude X is plotted as a function of frequency, ω. At $\omega = 0$, the amplitude is X_0. If $\omega = \omega_1$, the amplitude increases to X_1, where $F_1 + m\omega_1^2 X_1 = k_0 X_1 + \kappa X_1^3$. At a frequency $\omega = \omega_2$, there will be two amplitudes that satisfy the equality $F_1 + m\omega_2^2 X_2 = k_0 X_2 + \kappa X_2^3$, one at negative amplitude, where the straightline and the curve are tangent. For $\omega = \omega_3$, and any higher frequencies, there will be three amplitudes, two negative and one positive. The result is a distorted response curve, which is typical of nonlinear forced vibration. The curve has two branches, one where the exciting force and the amplitude have the same sign and one where they have opposite signs. Compare this curve to Figure 4.2.

The reason for this distortion is that the natural frequency of the nonlinear system is itself a function of displacement. This is not the case in linear vibration. If we define the frequency ω_n as the natural frequency of the system at zero amplitude X_0, and if we call ω_n' the natural frequency

Fig. 8.4

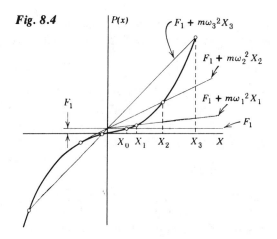 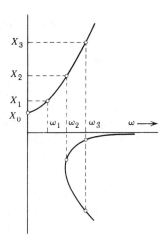

that is amplitude dependent,

$$\omega_n' = \omega_n \sqrt{1 \pm \frac{3}{4}\left(\frac{\kappa}{k_0}\right)^2 X^2} \qquad (8.14)$$

The + or − is used to designate a stiffening or softening spring.

The distorted response curve has one very pronounced effect on forced vibration. This is the so-called "jump" phenomenon, which is explained dramatically in Figure 8.5. At high frequencies, three amplitude ratios are possible. If $\omega_C < \omega_B$ the amplitudes from C to B are not attainable. If the frequency is decreased from some value above resonance, the amplitude will increase along line DC. At C the amplitude will "jump" to C' with no change in frequency. If the frequency increases from some frequency below resonance, the amplitude will increase along the higher curve AB. Without damping, the amplitude continues to increase with an increase in frequency until a catastrophic failure occurs. Fortunately, some form of damping is always present. With damping, the amplitude of resonance is limited and the amplitude "jumps" to B' with no change in frequency. Even with the amplitude-limiting effect of damping, the degree of damping may not be sufficient to prevent catastrophic failure.

One other phase of nonlinear response deserves being mentioned. If we go back to equation 8.12 and substitute our trial solution for x but not for $\ddot{x}$, we have,

$$m\ddot{x} + k_0 X \sin \omega t + \kappa X^3 \sin^3 \omega t = F_1 \sin \omega t$$

Substituting the trigonometric identity

$$\sin^3 \omega t = \frac{1}{4}(3 \sin \omega t - \sin 3\omega t)$$

Fig. 8.5

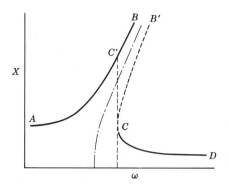

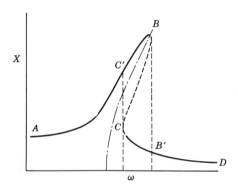

we have

$$\ddot{x} = -\left(\frac{k_0}{m}X + \frac{3\kappa X^3}{4\ m} - \frac{F_1}{m}\right)\sin\omega t - \frac{\kappa X^3}{4}\sin 3\omega t$$

Integrating for x, the displacement will be an equation of the type

$$x = B_1 \sin \omega t + B_3 \sin 3\omega t \tag{8.15}$$

containing a first and a third harmonic. This is a closer approximation to the actual displacement than simple harmonic motion, which had been assumed, but in most practical cases the amplitude of the third harmonic is so small that it can be neglected. The rare exception is the case when the forcing function *contains* a third harmonic. If damping is light, the third harmonic present in the amplitude will match the third harmonic in the forcing function and energy will be added to the system over an integral number of cycles, in which case the amplitude of motion will build. The third harmonic cannot build separately, without the amplitude of the fundamental also increasing. If resonance occurs, the observed vibration will be at a frequency of one third that of the natural frequency. This very rare phenomenon is *subharmonic resonance*. It is caused by the nonlinearity of the system and never occurs in linear systems.

EXAMPLE PROBLEM 8.16

In the nonlinear system shown, the mass m is attached to two springs. When the mass is in the middle position, the springs have zero initial tension. Determine the nonlinear differential equation of motion, if an harmonic exciting force $F_1 \sin \omega t$ is applied to m in the x direction. The two springs and the mass m are all in the horizontal plane.

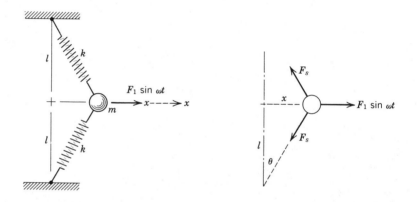

Solution:

A free body diagram of the mass m shows three active forces, the two spring forces and the harmonic exciting force $F_1 \sin \omega t$. The weight force is vertical.

From geometry, the spring deflections are $\sqrt{l^2+x^2}-l$, the spring forces are

$$F_s = k(\sqrt{l^2+x^2}-l)$$

and

$$\sin \theta = \frac{x}{\sqrt{l^2+x^2}}; \qquad \cos \theta = \frac{l}{\sqrt{l^2+x^2}}$$

Using these terms, the equation of motion for the mass m is

$$\Sigma F_x = m\ddot{x}$$

$$F_1 \sin \omega t - 2F_s \sin \theta = m\ddot{x}$$

$$m\ddot{x} + 2k\left[1 - \frac{l}{\sqrt{l^2+x^2}}\right]x = F_1 \sin \omega t$$

and, it is nonlinear.

We can approximate this equation of motion by using the first two terms of the binomial expansion

$$(l^2+x^2)^{1/2}=(l^2)^{1/2}+\frac{1}{2}(l^2)^{-1/2}\cdot x^2-\frac{1}{8}(l^2)^{-3/2}x^4+\cdots$$

$$=l+\frac{x^2}{2l}-\frac{x^4}{8l^3}+\frac{x^6}{16l^5}-\cdots$$

This is valid for small x. For example, for $x=0.2l$, the error in using just two terms is 0.02%. For $x=0.5l$, the error would be 0.7%, and for $x=l$, the error would be 8.6%. Clearly, truncating the expansion at two terms would be satisfactory for $x<0.2l$, but not for large x.

At the same time, let us make the additional approximation

$$\sin\theta\sim\frac{x}{l}$$

which has an error of 0.4% for $x=0.2l$.

The result is another nonlinear differential equation, but this equation is similar to equation 8.12.

$$m\ddot{x}+\frac{kx^3}{l^2}=F_1\sin\omega t$$

If we now assume that the resulting motion is near enough to being simple harmonic to let

$$x=X\sin\omega t$$

then

$$-m\omega^2X\sin\omega t+\frac{kX^3}{l^2}\sin^3\omega t=F_1\sin\omega t$$

This equation can be solved for the maximum amplitude X, at $\omega t=\pi/2$, $3\pi/2$, $5\pi/2\ldots$, knowing the other system constants.

$$-m\omega^2X+\frac{kX^3}{l^2}=F_1$$

$$X^3-\frac{m\omega^2l^2}{k}X-\frac{F_1l^2}{k}=0$$

PROBLEM 8.17 In Problem 8.16, assume that an exciting force $F_1\sin\omega t$ is applied to m in the x direction. If $F_1=50$ N, $k=22.5$ N/mm, $m=1.5$ kg, $l=50$ mm, and $X_0=25$ mm, determine the approximate value of the steady state amplitude of forced vibrations for $\omega=200$ rad/s.

Answer: 0.83 mm

PROBLEM 8.18 A patented rubber spring for heavy duty vehicles consists of a spherically shaped elastomer permanently bonded between two metal plates. The spring stiffens as the vehicle load is increased, softens as the load decreases. The result is a more constant natural frequency, regardless of the load being carried. The deflection of the vehicle suspension system as a function of the gross weight is shown. The gross mass of the vehicle is 2000-kg empty and 15,000-kg loaded. Determine the motion of the empty vehicle traveling at 100 km/h over a road that has tar strips that project 7 mm above the road surface every 5 m, for both a linear suspension and the nonlinear suspension.

Answer: linear, 17.9 mm; nonlinear, 4.9 mm

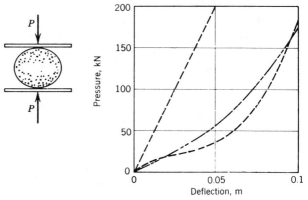

PROBLEM 8.19 An electric motor has a mass of 12.5 kg and is mounted on a structure cantilevered from a wall. For an additional load $P(x)$ applied at the motor mounting, the load deflection curve of the structure is shown here. If the armature of the motor has a mass of 5 kg and is known to be 0.1 mm off center, determine the vertical amplitude of motion when the motor is running at 600 rpm.

Answer: 0.042 mm

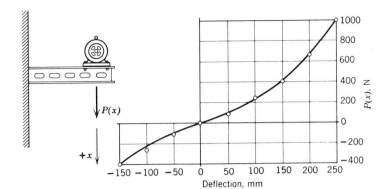

CHAPTER NINE

TWO
DEGREES
OF
FREEDOM

9.1. FREE VIBRATION AND THE FREQUENCY EQUATION

While it is possible to simplify the dynamics of complex systems and use one degree of freedom to approximate many, there are times when this simply cannot be done. A single coordinate may not be enough to describe motion. Two or more geometric coordinates may be needed. The system may also have more than one mass. These aspects of dynamics cannot be approximated, without affecting some of the very results we want to preserve.

These systems have more than one degree of freedom and are called multidegree of freedom systems. Instead of one resonant condition, there are several. Each resonant condition has its own characteristic mode shape. The study of multidegree of freedom systems is elaborate, but the general principles involved for many degrees of freedom can be set forth with only two. If you understand what happens in a system with two degrees of freedom, it is easy to extend your understanding to many degrees of freedom. The principal differences between one and two degrees of freedom are more than the differences between two and ten degrees of freedom.

Fig. 9.1

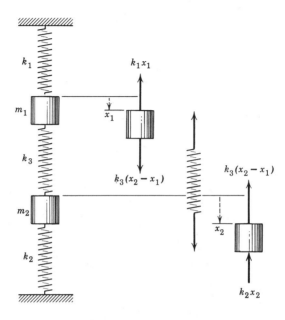

As an example, with the two mass system of Figure 9.1 constrained to move in one direction only, the system has two degrees of freedom. If these two masses were free to move in three-dimensional space, each would have three degrees of freedom, and six coordinates would be needed to describe their motion. As it is, the generalized coordinates x_1 and x_2 are both needed to describe the motion of masses m_1 and m_2. Consider the first mass m_1 to be displaced a distance x_1 from its equilibrium position and the mass m_2 to be displaced a distance x_2 from its equilibrium position. No external forces act on either mass, which means that the system vibrates freely. Isolating each mass, for $x_2 > x_1$, the equations of motion are

$$\sum \mathbf{F} = m\ddot{\mathbf{x}}$$
$$-k_1 x_1 + k_3(x_2 - x_1) = m_1 \ddot{x}_1 \tag{9.1}$$
$$-k_2 x_2 - k_3(x_2 - x_1) = m_2 \ddot{x}_2$$

These two equations can be rewritten in matrix form.

$$\begin{bmatrix} m_1 & 0 \\ 0 & m_2 \end{bmatrix} \begin{Bmatrix} \ddot{x}_1 \\ \ddot{x}_2 \end{Bmatrix} + \begin{bmatrix} (k_1 + k_3) & -k_3 \\ -k_3 & (k_2 + k_3) \end{bmatrix} \begin{Bmatrix} x_1 \\ x_2 \end{Bmatrix} = \begin{Bmatrix} 0 \\ 0 \end{Bmatrix} \tag{9.2}$$

The use of matrices is a shorthand way of expressing sets of simultaneous equations with convenience. The matrix

$$\begin{bmatrix} m_1 & 0 \\ 0 & m_2 \end{bmatrix}$$

is the *mass* matrix.

The matrix

$$\begin{bmatrix} (k_1+k_3) & -k_3 \\ -k_3 & (k_2+k_3) \end{bmatrix}$$

is the *stiffness* matrix. Each is a scalar that modifies a vector. The matrices

$$\begin{Bmatrix} \ddot{x}_1 \\ \ddot{x}_2 \end{Bmatrix} \quad \text{and} \quad \begin{Bmatrix} x_1 \\ x_2 \end{Bmatrix}$$

are column matrices representing vector quantities. They are also referred to as column vectors. Later, we will use matrix algebra to solve sets of many equations of motion simultaneously, but at this point, it is sufficient to establish that the differential equations 9.1, and 9.2 are identical expressions of the equations of motion. They are linear second-order differential equations and are the basis for the analysis of the vibration of the two degree of freedom system.

If motion is assumed to be in a principal mode, both generalized coordinates will have harmonic motion of the same frequency, ω.

$$x_1 = X_1 \sin \omega t$$

$$x_2 = X_2 \sin \omega t$$

Substituting these harmonic expressions in the equations of motion, will give us two algebraic equations involving X_1, X_2, and ω^2 as unknowns.

$$X_1(k_1 + k_3 - m_1\omega^2) - k_3 X_2 = 0$$

$$X_2(k_2 + k_3 - m_2\omega^2) - k_3 X_1 = 0$$

In each equation, the fraction X_2/X_1 can be determined in terms of the system constants and ω^2. Equating, X_2 and X_1 can be eliminated.

$$\chi = \frac{X_2}{X_1} = \frac{k_1 + k_3 - m_1\omega^2}{k_3} = \frac{k_3}{k_2 + k_3 - m_2\omega^2} \tag{9.3}$$

This fraction is called the *modal fraction*, since its value determines the mode of motion.

Cross-multiplying will give a quadratic in ω^2.

$$(k_1 + k_3 - m_1\omega^2)(k_2 + k_3 - m_2\omega^2) = k_3^2$$

$$m_1 m_2 \omega^4 - \omega^2[m_1(k_2 + k_3) + m_2(k_1 + k_3)] + k_1 k_2 + k_3(k_1 + k_2) = 0 \tag{9.4}$$

Equation 9.4 is called the *frequency equation*, and the roots of the frequency equation are the characteristic values or *eigenvalues* of the system. Since these frequencies satisfy the original assumption, that all parts of the system have the same harmonic frequency, ω, without an application of external force, they are natural frequencies. In the case of two degrees of freedom there are two natural frequencies, each corresponding to one of the two principal modes.

Fig. 9.2

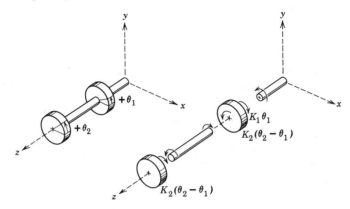

As another example, the torsional system of Figure 9.2 consists of two discs, each with mass moment of inertia, joined by one elastic torsional spring, and fixed to a rigid wall by another. If the discs are displaced about the z-axis and released, the system will vibrate torsionally about the z-axis. θ_1 and θ_2 are the generalized coordinates that describe the motion of the discs and the distortion of the springs. The equations of motion for angular displacements are,

$$\sum \mathbf{M} = I\ddot{\boldsymbol{\theta}}$$
$$-K_1\theta_1 + K_2(\theta_2 - \theta_1) = I_1\ddot{\theta}_1$$
$$-K_2(\theta_2 - \theta_1) = I_2\ddot{\theta}_2$$

(9.5)

For vibration in a principal mode.

$$\theta_2 = \Theta_2 \sin \omega t$$
$$\theta_1 = \Theta_1 \sin \omega t$$

The modal fraction Θ_2/Θ_1 is

$$\chi = \frac{\Theta_2}{\Theta_1} = \frac{K_1 + K_2 - I_1\omega^2}{K_2} = \frac{K_2}{K_2 - I_2\omega^2}$$

(9.6)

and this yields the frequency equation

$$I_1I_2\omega^4 - \omega^2[I_1K_2 + I_2(K_1 + K_2)] + K_1K_2 = 0$$

(9.7)

It again is a quadratic in ω^2, with two roots.

The existence of a *frequency equation* involving the physical constants of the system, the roots of which are the characteristic values or natural frequencies, is typical of multiple degrees of freedom. For two degrees of freedom, it is easy to establish a frequency equation and seek its characteristic values algebraically. This becomes increasingly difficult for multiple number of degrees of freedom, since the order of the frequency equation increases with the number of degrees of freedom. For three degrees of freedom, the frequency equation will have three roots, for four degrees of freedom, the frequency equation will be a quartic in ω^2, and so on. For any larger number, we must seek a means to find the characteristic values of the frequency equation without knowing the frequency equation itself. To determine the frequency equation explicitly becomes too arduous a task, and there are other ways of finding the roots of algebraic equations, numerically. It is important to recognize, however, that when we do use numerical means to determine the characteristic values of a system of many degrees of freedom, these characteristic values are the roots of a frequency equation that we have not determined, but which does exist.

9.2. MODES AND MODAL FRACTIONS

A mode is a description of motion. There are various kinds of modes, many with a modifying phrase, such as the first mode, the second mode, a principal mode or a coupled mode—all describing a particular manner of motion.

At a natural frequency, a vibrating system moves in a *principal mode*. This mode is also called a natural mode. If the amplitude of one mass is one unit of displacement, the mode is said to be normalized or is simply called a normal mode. All these descriptions mean the same thing, that all parts of the system have the same harmonic motion, with maximum displacements at identical times and maximum velocities at still other identical times. The number of principal modes that exist will correspond to the number of degrees of freedom.

The coordinates used to describe motion also describe the mode. These coordinates are not stated in absolute quantities, but as numerical ratios. That is, the value of one coordinate relative to all others is fixed for any given mode, and the absolute value of any one coordinate determines the value of all the other coordinates.

As an example, the value of X_2/X_1 and Θ_2/Θ_1 establishes the modes of motion for the two systems we have considered. It makes no difference what the absolute values of Θ_2, Θ_1, X_2, or X_1 really are, since the relative value of one with respect to the other sets the mode. To find the mode for a specific value of ω^2, simply substitute that value in the equation of motion, or in the modal fraction, equation 9.3.

When we refer to the modal fraction for a given characteristic value, a superscript in parentheses is conventional. This should not be confused as being an exponent, which it is not. A superscript (1) is used for the first mode, a superscript (2) for the second mode, a superscript (i) for the ith mode, and so on.

$$\chi^{(1)} = \frac{X_2^{(1)}}{X_1^{(1)}} = \frac{k_1 + k_3 - m_1 \omega_1^2}{k_3}$$

$$\chi^{(2)} = \frac{X_2^{(2)}}{X_1^{(2)}} = \frac{k_1 + k_3 - m_1 \omega_2^2}{k_3}$$

(9.8)

If one of the modes has a displacement of one unit, then the modes are *normalized* and the modal fraction has the same numerical value as the other coordinate. Thus for $X_1^{(1)} = 1$, $\chi^{(1)} = X_2^{(1)}$, and for $X_1^{(2)} = 1$, $\chi^{(2)} = X_2^{(2)}$. Normal modes are very useful, as we will later see. It is usual to assign unit displacement to the first coordinate, but which coordinate is X_1 and which is X_2 is arbitrary.

An array of all coordinates for any one mode is called a *modal column matrix*.

$$\begin{Bmatrix} X_1^{(1)} \\ X_2^{(1)} \\ \cdots \\ X_n^{(1)} \end{Bmatrix} \begin{Bmatrix} X_1^{(2)} \\ X_2^{(2)} \\ \cdots \\ X_n^{(2)} \end{Bmatrix} \begin{Bmatrix} X_1^{(i)} \\ X_2^{(i)} \\ \cdots \\ X_n^{(i)} \end{Bmatrix}$$

(9.9)

The first is the modal column for the first mode for n coordinates; the second is for the second mode, and the third is for the ith mode. For two degrees of freedom, there will only be two coordinates, two modes, and two modal columns.

EXAMPLE PROBLEM 9.1

Let us assume a two mass system consisting of two identical masses and three identical springs, with each mass constrained to move only in the vertical direction. Determine the frequency equation and solve for the two natural frequencies and the mode shapes corresponding to those two frequencies.

Solution:

Referring to equation 9.1, the equations of motion are

$$-2k x_1 + k x_2 = m \ddot{x}_1$$

$$-2k x_2 + k x_1 = m \ddot{x}_2$$

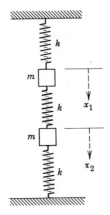

For harmonic motion at a frequency, ω

$$x_1 = X_1 \sin \omega t$$

$$x_2 = X_2 \sin \omega t$$

Substituting for x_1 and x_2 in the equations of motion,

$$\frac{X_2}{X_1} = \frac{2k - m\omega^2}{k} = \frac{k}{2k - m\omega^2}$$

Eliminating X_1 and X_2, frequency equation is

$$\omega^4 - 4\frac{k}{m}\omega^2 + \frac{3k^2}{m^2} = 0$$

Solving the frequency equation for the natural frequencies or characteristic values,

$$\omega_{1,2}^2 = \frac{2k}{m} \pm \frac{k}{m}$$

and,

$$\omega_1^2 = \frac{k}{m}$$

$$\omega_2^2 = \frac{3k}{m}$$

There are two natural frequencies ω_1 and ω_2, as expected for two degrees of freedom. Substituting the characteristic value of $\omega_1^2 = k/m$ in the first equation of motion, the modal fraction for the first mode is

$$\chi^{(1)} = \frac{X_2^{(1)}}{X_1^{(1)}} = \frac{2k - m\omega_1^2}{k} = +1$$

Substituting the second characteristic value, $\omega_2^2 = 3k/m$, the modal fraction for the second mode is

$$\chi^{(2)} = \frac{X_2^{(2)}}{X_1^{(2)}} = \frac{2k - m\omega_2^2}{k} = -1$$

Physically, this means that x_2 and x_1 are in phase and equal for the first mode, whatever their absolute value. In the second mode, they are equal but out of phase.

The results would have been the same, if we had used the second equation of motion to determine the modal fraction instead of the first equation.

EXAMPLE PROBLEM 9.2

Considering only small oscillations, determine the two natural frequencies of motion for a double pendulum. All motion is in one plane.

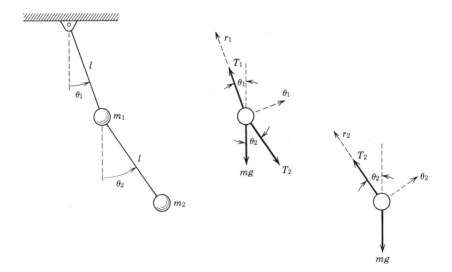

Solution:

θ_1 and θ_2 are the generalized coordinates that recognize constraint on the double pendulum.

The equations of motion are, for mass m_1, in the θ_1 direction,

$$-mg \sin \theta_1 + T_2 \sin (\theta_2 - \theta_1) = ml\ddot{\theta}_1$$

$$m_1 = m_2$$

and for mass m_2, in the θ_2 direction,

$$-mg \sin \theta_2 = m(l\ddot{\theta}_1 + l\ddot{\theta}_2)$$

In the direction normal to the path of motion, and neglecting any normal acceleration,

$$mg \cos \theta_2 - T_2 = 0$$

Considering θ_1 and θ_2 as small displacements, we can replace $\sin \theta$ and $\cos \theta$.

$$\sin \theta \sim \theta, \qquad \cos \theta \sim 1, \qquad T_2 \sim mg$$

These approximations make the equations of motion linear.
   For harmonic motion at a frequency, ω,

$$\theta_1 = \Theta_1 \sin \omega t$$

$$\theta_2 = \Theta_2 \sin \omega t$$

Substituting for θ_1 and θ_2 in the equations of motion,

$$ml\omega^2\Theta_1 + 2mg\Theta_1 - mg\Theta_2 = 0$$

$$m(l\omega^2\Theta_1 + l\omega^2\Theta_2) - mg\Theta_1 = 0$$

Eliminating Θ_1 and Θ_2, the frequency equation is

$$\omega^4 - \frac{4g}{l}\omega^2 + \frac{2g^2}{l^2} = 0$$

Solving the frequency equation for the two natural frequencies

$$\omega_1^2 = (2 - \sqrt{2})\frac{g}{l}$$

$$\omega_2^2 = (2 + \sqrt{2})\frac{g}{l}$$

Substituting the characteristic value of $\omega_1^2 = (2 - \sqrt{2})g/l$ into the first equation of motion, the modal fraction for the first mode is

$$\chi_1 = \frac{\Theta_2^{(1)}}{\Theta_1^{(1)}} = \frac{2mg - ml\omega_1^2}{mg} = +\sqrt{2}$$

To visualize this mode, assume an arbitrary value for say $\Theta_1 = 6°$. Then, $\Theta_2 = 6\sqrt{2}°$, or 8.5°. At the maximum displacements, the mode would look like the illustration (*a*) on the next page.

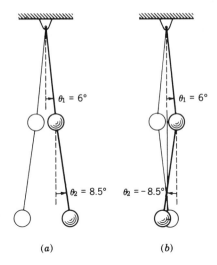

(a)   (b)

Substituting the characteristic value of $\omega_2^2 = (2 + \sqrt{2})g/l$ in the first equation of motion, the modal fraction for the second mode is

$$\chi_2 = \frac{\Theta_2^{(2)}}{\Theta_1^{(2)}} = \frac{2mg - ml\omega_2^2}{mg} = -\sqrt{2}$$

For $\Theta_1 = 6°$, $\Theta_2 = -8.5°$. The second mode looks like (b) above.

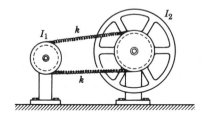

PROBLEM 9.3 A movie projector reel drive consists of a tightly wound helical spring, stretched around two grooved pulleys. The pulleys have radii r_1 and r_2 and centroidal moments of inertia I_1 and I_2, respectively. Derive an expression for the natural frequency or frequencies in cycles per second.

Answer: $\omega_1^2 = 0$; $\omega_2^2 = 2k\left(\dfrac{r_2^2 I_1 + r_1^2 I_2}{I_2 I_1}\right)$

PROBLEM 9.4 An automobile m_1 uses a bumper hitch to pull a loaded trailer m_2. The bumper acts as an elastic spring, k. The trailer with a 3800-kg mass is hitched to a 1750-kg automobile. The flexibility of the hitch is 175 N/mm. What are the frequencies or what is the frequency of free oscillation of the automobile and trailer. Illustration on next page.

Answer: only one, 1.92 cps

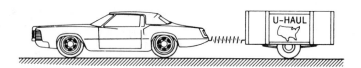

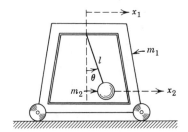

PROBLEM 9.5   Determine the two natural frequencies for the frame and pendulum. The pendulum swings free in the same plane in which the frame moves.

Answer: $\omega_1^2 = 0$; $\omega_2^2 = \dfrac{m_1 + m_2}{m_1}\left(\dfrac{g}{l}\right)$

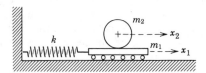

PROBLEM 9.6   Two identical solid circular cylinders, each with a mass m, are connected by a single spring with a modulus k. Determine the natural frequency(es) of small oscillations, if the cylinders are displaced from their equilibrium positions and roll without slipping on the horizontal surface.

Answer: $\omega_1^2 = 0$; $\omega_2^2 = \dfrac{4}{3}\dfrac{k}{m}$

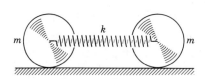

PROBLEM 9.7   A platform supports a circular cylinder and is elastically suspended from the wall by a spring with a modulus of k. Each has a mass m. Determine the natural frequencies and mode shapes of the system, if the cylinder rolls without slipping.

Answer: $f_n = \dfrac{1}{2\pi}\sqrt{\dfrac{3k}{4m}}$; $3x_2 = x_1$

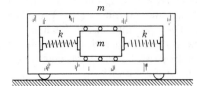

PROBLEM 9.8   The seismic mass m is mounted in a frame between two springs, each with a modulus of k. The frame has identical mass of m. Determine the natural frequencies of free vibration. There is no friction.

Answer: $\omega_1^2 = 0$; $\omega_2^2 = \dfrac{4k}{m}$

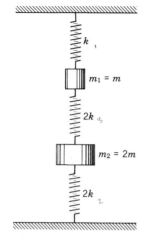

PROBLEM 9.9 Determine the two natural frequencies and mode shapes of the two-mass systems shown.

Answer: $\omega_1^2 = \dfrac{k}{m}$; $\chi^{(1)} = +1$

$\omega_2^2 = 4\dfrac{k}{m}$; $\chi^{(2)} = -\tfrac{1}{2}$

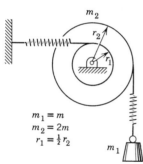

PROBLEM 9.10 Determine the natural frequencies for the system given here. The pulley can be considered as a solid circular cylinder.

Answer: $\omega_1^2 = 0.117\dfrac{k}{m}$;

$\omega_2^2 = 2.133\dfrac{k}{m}$

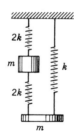

PROBLEM 9.11 Determine the natural frequencies and mode shapes for the two-mass system. Both masses move only vertically. Do not consider rotation of the lower mass.

Answer: $\omega_1^2 = 1.439\dfrac{k}{m}$; $\chi_1 = 1.281$

$\omega_2^2 = 5.562\dfrac{k}{m}$; $\chi_2 = -0.781$

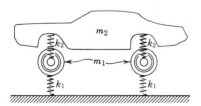

PROBLEM 9.12 A large automobile manufacturer analyzed the problem of the automobile by taking an entire automobile apart. By weighing each section, the following values of equivalent masses were found.

m_1	axle mass	180 kg
m_2	body mass	670 kg
k_2	springs	45.5 N/mm
k_1	tires	538 N/mm

Discount the natural frequencies such as pitch, roll, and yaw. Determine the two natural frequencies of motion.

Answer: 75.5 cpm, 544 cpm

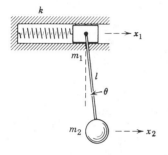

PROBLEM 9.13   A mass m is constrained to move in a smooth horizontal guide and is elastically supported by the spring with constant k. The sliding mass is also the point of support of the simple pendulum. The mass of the piston and the mass of the pendulum have identical masses of 5 kg. The spring has a constant of 400 N/m and the pendulum is 0.25 m long. Determine the two natural frequencies of the system. Assume that oscillations will be small.

Answer: $f_1 = 0.767$ Hz; $f_2 = 1.85$ Hz

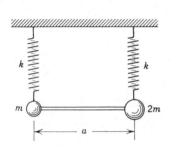

PROBLEM 9.14   Determine the two natural frequencies of the system shown here. Please think about coordinates, free bodies, and equations of motion before doing this problem. There are several choices of coordinates. The rod is rigid and massless.

Answer: $\omega_1^2 = \dfrac{1}{2}\dfrac{k}{m}$; $\omega_2^2 = \dfrac{k}{m}$

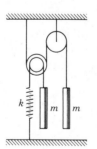

PROBLEM 9.15   Both masses are identical. The concentric pulleys are free to turn independently of each other. What are the natural frequencies and mode shapes? Neglect the mass and inertia of the pulleys.

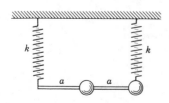

PROBLEM 9.16   Two identical springs support a rigid rod and two identical masses. Determine the two natural frequencies of motion for the two-mass system.

Answer: $\omega_1^2 = (3 - \sqrt{5})\dfrac{k}{m}$;

$\omega_2^2 = (3 + \sqrt{5})\dfrac{k}{m}$

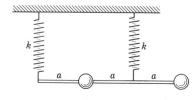

PROBLEM 9.17 Two identical springs support a rigid rod and two identical masses. Determine the two natural frequencies of motion for the two-mass system.

$$Answer:\ \omega_1^2 = \frac{(3-\sqrt{5})}{2}\frac{k}{m};$$

$$\omega_2^2 = \frac{(3+\sqrt{5})}{2}\frac{k}{m}$$

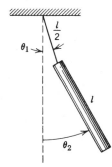

PROBLEM 9.18 Determine the two natural frequencies and mode shapes for a thin rod suspended as a pendulum.

$$Answer:\ \omega_1^2 = 0.917\frac{g}{l};\ \omega_2^2 = 13.083\frac{g}{l};$$

$$\chi_1 = 1.18,\ \chi_2 = -0.85$$

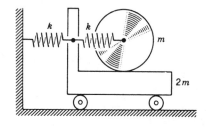

PROBLEM 9.19 A platform of mass $2m$ supports a solid cylindrical mass m, which is elastically separated from the platform by a spring. Determine the two natural frequencies.

$$Answer:\ \omega_1^2 = \frac{2}{7}\frac{k}{m};\ \omega_2^2 = \frac{k}{m}$$

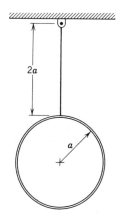

PROBLEM 9.20 A thin, circular ring of mass m is suspended by a string and oscillates in the plane of the paper. Write the equations of motion for small amplitude vibration. Determine the natural frequencies and the mode shapes for a small amplitude oscillation about equilibrium.

$$Answer:\ \omega_1^2 = \left(1 - \frac{1}{\sqrt{2}}\right)\frac{g}{a};\ \chi_1 = +\sqrt{2}$$

$$\omega_2^2 = \left(1 + \frac{1}{\sqrt{2}}\right)\frac{g}{a};\ \chi_2 = -\sqrt{2}$$

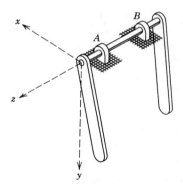

PROBLEM 9.21 Two slender uniform bars are supported at either end of a shaft with a torsional modulus of K. The shaft is free to rotate without frictions in journals at A and B. Determine the frequencies of free oscillation.

Answer: $\omega_1^2 = \dfrac{3}{2}\dfrac{g}{l}$; $\omega_2^2 = \dfrac{6K}{ml^2} + \dfrac{3}{2}\dfrac{g}{l}$

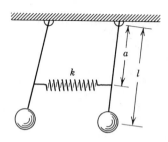

PROBLEM 9.22 Determine the natural frequencies of the system. The spring may be considered to be massless.

Answer: $\omega_1^2 = \dfrac{g}{l}$; $\omega_2^2 = \dfrac{g}{l} + 2\dfrac{k}{m}\dfrac{a^2}{l^2}$

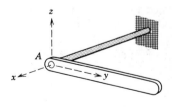

PROBLEM 9.23 A 2-kg slender uniform bar, 300 mm long, is fixed at one end to a circular shaft. In equilibrium, the end A of the shaft is deflected 12 mm in the z-direction and torsionally deflected 20° in the y–z plane, from its free position. Determine the two natural frequencies of motion in the y–z plane.

Answer: $f_1 = 1.77$ Hz; $f_2 = 9.69$ Hz

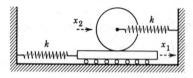

PROBLEM 9.24 At all times the cylinder rolls without slipping and the two masses are equal. The flat plate moves without friction on the horizontal surface. Determine the two natural frequencies of motion and the mode shapes.

Answer: $\omega_1^2 = \dfrac{1}{2}\dfrac{k}{m}$, $\chi_1 = -1$

$\omega_2^2 = \dfrac{k}{m}$, $\chi_2 = +1$

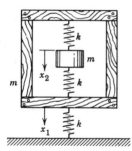

PROBLEM 9.25 Tension spring packaging is a means of supporting extremely fragile packages for shipping. Basically, the fragile package is supported inside the shipping container by springs in tension. To simplify the problem, idealize it as two masses and three springs. What are the two natural frequencies and mode shapes?

Answer: $\omega_1^2 = 0.438 \dfrac{k}{m}$; $\omega_2^2 = 4.56 \dfrac{k}{m}$

$$\begin{Bmatrix} X_1 \\ X_2 \end{Bmatrix} = \begin{bmatrix} 1.000 \\ 1.281 \end{bmatrix}; \begin{Bmatrix} X_1 \\ X_2 \end{Bmatrix} = \begin{bmatrix} 1.000 \\ -0.781 \end{bmatrix}$$

PROBLEM 9.26 Determine the two natural frequencies for the long slender bar supported by two springs, each having a stiffness, k. One spring is attached at the end of the bar, and the other is attached at a point that is one third of the length of the bar from the other end.

Answer: $\omega_1^2 = \dfrac{4}{3} \dfrac{k}{m}$; $\omega_2^2 = 4 \dfrac{k}{m}$

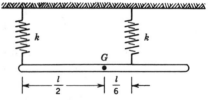

PROBLEM 9.27 Determine the two natural frequencies for the long slender bar supported by four springs, two having a stiffness k and two having a stiffness $\frac{1}{2}k$. Consider that the ends of the bar only move vertically, that is, the bar does not move laterally.

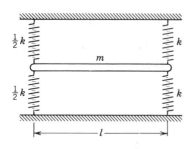

Answer: $\omega_1^2 = 2.536 \dfrac{k}{m}$;

$\omega_2^2 = 9.464 \dfrac{k}{m}$

PROBLEM 9.28 An electric motor and pump have a mass of 50 kg combined and are supported on a welded steel structure that has a mass of 100 kg separately. When the motor and pump are placed on the bed, it is noted that the support structure deflects 1.5 mm from its own static

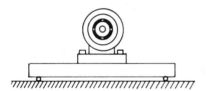

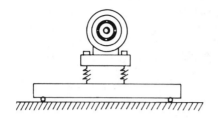

position. The resonant frequency is observed at 580 rpm, and the running speed is 600 rpm. To avoid resonance, vibration isolators between the motor and the support are suggested, specifically isolators that have a rated deflection of 4.5 mm. With isolators, this becomes a two degree of freedom system. What are the new natural frequencies?

(*Hint*: First, find the effective mass of the structure.)

Answer: $\omega_1^2 = 1548 \text{ s}^{-2}$; $\omega_2^2 = 11920 \text{ s}^{-2}$

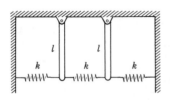

PROBLEM 9.29 Determine the two natural frequencies and mode shapes for the two coupled pendula.

Answer: $\omega_1^2 = \dfrac{3g}{2l} + \dfrac{3k}{m}$; $\theta_2 = +\theta_1$

$\omega_2^2 = \dfrac{3g}{2l} + \dfrac{9k}{m}$; $\theta_2 = -\theta_1$

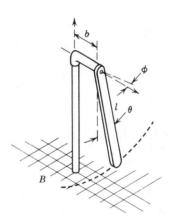

PROBLEM 9.30 The slender rod is pivoted at its upper end and swings freely from an *L*-shaped rod and bracket, which is fixed in the base at *B*. For the purpose of the problem, neglect the mass of the supporting rod and bracket. The rod and bracket have a torsional spring constant, *K*, and the mass of the slender rod is *m*. Determine the frequency equation for the natural frequencies of free vibration.

Answer: $\omega^4 - \omega^2 \left(\dfrac{4K}{mb^2} + \dfrac{6g}{l} \right) + \dfrac{6gK}{mb^2} = 0$

9.3. PRINCIPAL COORDINATES FOR TWO DEGREES OF FREEDOM

In a principal mode, if the motion of all parts of the system can be described by a single coordinate without reference to any other, that coordinate is a *principal coordinate*. To define the motion of *n*-degrees of freedom with a single coordinate seems impossible, but only because a principal coordinate is more a mathematical parameter than it is a geometric coordinate by which position is directly measured. In a three degree of freedom system, it is simple to express motion in terms of two or three coordinates, say the orthogonal coordinates x, y, and z, but it is difficult to physically accept one principal coordinate that expresses all motion for a three-mass system, but it can be done with equal ease.

Fig. 9.3

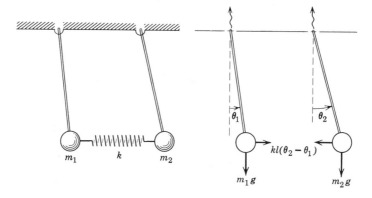

In Figure 9.3, two simple pendula are connected by a linear spring. The supporting rods are massless. The system has two degrees of freedom, since each pendulum can move independently of the other. θ_1 and θ_2 are generalized coordinates that physically describe the position of each mass and the distortion of the elastic spring. The equations of motion are

$$\sum \mathbf{M} = I\ddot{\boldsymbol{\theta}}$$

$$-k(\theta_2-\theta_1)l^2 \cos \theta_2 - m_2 gl \sin \theta_2 = m_2 l^2 \ddot{\theta}_2$$
$$+k(\theta_2-\theta_1)l^2 \cos \theta_1 - m_1 gl \sin \theta_1 = m_1 l^2 \ddot{\theta}_1 \qquad (9.10)$$

For small angles of displacement and for equal masses, the equations of motion can be simplified.

$$-kl^2(\theta_2-\theta_1) - mgl\theta_2 = ml^2\ddot{\theta}_2$$
$$+kl^2(\theta_2-\theta_1) - mgl\theta_1 = ml^2\ddot{\theta}_1 \qquad (9.11)$$

θ_1 and θ_2 are geometric descriptions of position, but they are not principal coordinates. The two principal modes both involve θ_1 and θ_2, but mathematically, two principal coordinates do exist, one corresponding to each principal mode. Adding and then subtracting these equations yield two alternate equations of motion,

$$ml^2(\ddot{\theta}_1+\ddot{\theta}_1) + mgl(\theta_2+\theta_1) = 0$$
$$ml^2(\ddot{\theta}_2-\ddot{\theta}_1) + (mgl+2kl^2)(\theta_2-\theta_1) = 0 \qquad (9.12)$$

Both of these equations are of the very familiar form,

$$\ddot{p} + \omega_n^2 p = 0$$

Each equation is independent of the other, and we can write by inspection,

$$\omega_1^2 = \frac{g}{l}$$

$$\omega_2^2 = \frac{g}{l} + \frac{2k}{m}$$

These are the two natural frequencies or eigenvalues for the system, and equations 9.12 are the two principal equations of motion. It should be noted that we have found the eigenvalues without having found the frequency equation. This is a very important concept that we will use again.

Substituting ω_1^2 and ω_2^2 in equations 9.11, the modal fractions are

$$\chi^{(1)} = \frac{\Theta_2^{(1)}}{\Theta_1^{(1)}} = +1$$

$$\chi^{(2)} = \frac{\Theta_2^{(2)}}{\Theta_1^{(2)}} = -1$$

(9.13)

Evidently, $(\theta_2 + \theta_1)$ and $(\theta_2 - \theta_1)$ have a direct relationship to the principal coordinates. The symbol p is used for a principal coordinate. It is simple to write principal equations of motion such as these, if the system has symmetry, that is, the springs are of equal moduli and the masses are equal in size. It is more complicated if unequal spring constants or unequal masses require weighting factors, but it can be done.

In linear vibration, principal coordinates must be harmonic functions of time, just as the geometric coordinates. Principal coordinates have all the same properties as the geometric coordinates θ_1 and θ_2. As examples, let us accept the two principal coordinates to be $p_1(t)$ and $p_2(t)$

$$p_1(t) = C_1 \sin(\omega_1 t + \phi_1)$$
$$p_2(t) = C_2 \sin(\omega_2 t + \phi_2)$$

(9.14)

The first is the principal coordinate for the first mode corresponding to the natural frequency ω_1. The second is the principal coordinate corresponding to the second natural frequency ω_2. Each has two arbitrary constants determined by initial conditions. If the displacements θ_1 and θ_2 involve both modes simultaneously, the θ_1 and θ_2 can be expressed as linear functions of p_1 and p_2. Furthermore, by superposition, these functions are additive.

$$\theta_1 = A_1 p_1 + B_1 p_2$$
$$\theta_2 = A_2 p_1 + B_2 p_2$$

(9.15)

In matrix form, this statement is clearer.

$$\begin{Bmatrix} \theta_1 \\ \theta_2 \end{Bmatrix} = \begin{bmatrix} A_1 & B_1 \\ A_2 & B_2 \end{bmatrix} \begin{Bmatrix} p_1 \\ p_2 \end{Bmatrix}$$

The linear relation between the generalized coordinates and the principal coordinates is the matrix

$$\begin{bmatrix} A_1 & B_1 \\ A_2 & B_2 \end{bmatrix}$$

which is nothing more than the modal matrix where each column represents one mode shape. The column $\begin{Bmatrix} A_1 \\ A_2 \end{Bmatrix}$ is the modal column matrix for the first mode and $\begin{Bmatrix} B_1 \\ B_2 \end{Bmatrix}$ is the modal column matrix for the second mode.

For the two pendula and a spring,

$$\begin{Bmatrix} \theta_1 \\ \theta_2 \end{Bmatrix} = \begin{bmatrix} \Theta_1^{(1)} & \Theta_1^{(2)} \\ \Theta_2^{(1)} & \Theta_2^{(2)} \end{bmatrix} \begin{Bmatrix} p_1 \\ p_2 \end{Bmatrix} = \begin{bmatrix} 1 & 1 \\ 1 & -1 \end{bmatrix} \begin{Bmatrix} p_1 \\ p_2 \end{Bmatrix} \tag{9.16}$$

These equations can be transformed to state the principal coordinates in terms of the generalized coordinates

$$\begin{aligned} p_1 &= \frac{\theta_1}{2} + \frac{\theta_2}{2} \\ p_2 &= \frac{\theta_1}{2} - \frac{\theta_2}{2} \end{aligned} \tag{9.17}$$

In this simple example, it is easy to verify that these are the principal coordinates corresponding to the principal modes, the lowest with the two masses moving together and the highest with the two masses moving in opposition. The numerical factor of 2 is something we could not deduce from adding or subtracting equations of motion. It would soon be apparent in any numerical evaluation of physical motion.

9.4. COUPLED MODES AND COUPLED COORDINATES

Coupling is the term used in mechanical vibration to indicate a connection between equations of motion. Its use is varied: there are coupled coordinates, coupled velocities, dynamic coupling, static coupling, coupled modes, and coupled masses.

If principal coordinates are used, at the natural frequencies, the equations of motion are completely separate. Each principal mode corresponds to one natural frequency or eigenvalue of the frequency equation. There are the same number of principal coordinates as there are equations of motion and principal modes, so each equation of motion is separate from every other. Only one coordinate, or any of its derivatives, with respect to time appear in each equation of motion. If an equation of motion contains cross-products of coordinates, or if the potential energy contains cross-products of coordinates, that equation of motion is *statically coupled*. If an equation of motion contains cross-products of velocity, or if the kinetic energy contains cross-products of velocity, that equation of motion is *dynamically coupled*. In matrix algebra, statically coupled equations of motion will cause off-diagonal terms in the stiffness matrix. Similarly, dynamically coupled equations of motion will cause off-diagonal terms to be in the mass matrix.

We must remember that a set of principal coordinates exist for every linear vibration system, but they may defy your geometric interpretation. It is within our ability to *decouple* any set of equations of motion by using principal coordinates. It is often preferable, however, to use a coupled system with coordinates that can be visualized than a decoupled system with principal coordinates that cannot. Our selection of coordinates is for our own convenience, and the selection will have no effect on eigenvalues or mode shapes.

Fig. 9.4

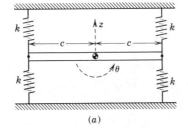

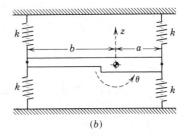

(a) (b)

In Figure 9.4a, a rigid body is supported on four equal and linear springs, located equidistant from the center of gravity at G. The plane of support contains the center of gravity.

In the z and θ directions, the equations of motion are

$$m\ddot{z}+4kz = 0$$
$$I_y\ddot{\theta}+4kc^2\theta = 0 \tag{9.18}$$

Stated in matrix algebra, there are no off-diagonal terms and

$$\begin{bmatrix} m & 0 \\ 0 & I_y \end{bmatrix}\begin{bmatrix} \ddot{z} \\ \ddot{\theta} \end{bmatrix}+\begin{bmatrix} 4k & 0 \\ 0 & 4kc^2 \end{bmatrix}\begin{bmatrix} z \\ \theta \end{bmatrix}=0 \tag{9.19}$$

Each mode is independent of all others. The characteristic values are

$$\omega_z^2 = \frac{4k}{m}$$
$$\omega_\theta^2 = \frac{4kc^2}{I_y}$$

(9.20)

If the center of gravity is not located midway between the supports, the two modes will be coupled. As an example, in Figure 9.4b, the center of gravity is located on the centerline, but closer to the right-hand support. Translation in the z-direction and rotation about the y-axis are coupled. We have the option of analyzing the coupled modes using the generalized coordinates of z and θ, or to seek the principal coordinates that would uncouple them. As explained in the last section, these two principal coordinates will be a linear transformation of θ and z.

Fig. 9.5

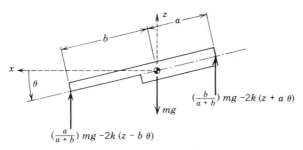

Figure 9.5 is a plane view considering only the two coordinates, z and θ. From the free body diagram, the equations of motion are

$$-2k(z-b\theta)-2k(z+a\theta) = m\ddot{z}$$
$$+2kb(z-b\theta)-2ka(z+a\theta) = I_y\ddot{\theta}$$

(9.21)

In matrix form, there are off-diagonal terms in the stiffness matrix.

$$\begin{bmatrix} m & 0 \\ 0 & I_y \end{bmatrix}\begin{bmatrix} \ddot{z} \\ \ddot{\theta} \end{bmatrix}+\begin{bmatrix} 4k & (2ka-2kb) \\ (2ka-2kb) & (2ka^2+2kb^2) \end{bmatrix}\begin{bmatrix} z \\ \theta \end{bmatrix}=0 \quad (9.22)$$

These equations are statically coupled. For normal modes where $z=Z \sin \omega t$ and $\theta=\Theta \sin \omega t$,

$$(4k-m\omega^2)Z+(2ka-2kb)\Theta = 0$$
$$(2ka-2kb)Z+(2ka^2+2kb^2-I_y\omega^2)\Theta = 0$$

Setting the determinant equal to zero yields the frequency equation,

$$\omega^4-\omega^2\left[\frac{4k}{m}+2\left(\frac{ka^2+kb^2}{I_y}\right)\right]+\left(\frac{2ka+2kb}{I_ym}\right)^2=0 \quad (9.23)$$

In the middle term, set $ka^2 + kb^2 = k(a+b)^2 - 4kab$. The eigenvalues for the uncoupled modes are $\omega_z^2 = 4k/m$ and $\omega_\theta^2 = k(a+b)^2/I_y = 4kc^2/I_y$. Substituting

$$\omega^4 - \omega^2 \left[\omega_z^2 + 2\omega_\theta^2 - \frac{4kab}{I_y} \right] + \omega_z^2\omega_\theta^2 = 0 \qquad (9.24)$$

Fig. 9.6

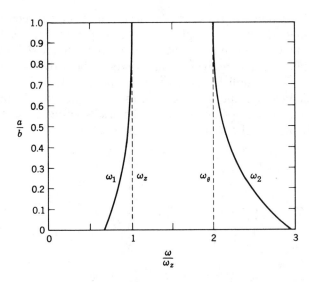

To show the effect of coupling, let us assume that the uncoupled natural frequency $\omega_\theta = 2\omega_z$. Figure 9.6 then shows the eigenvalues of the frequency equation, in terms of ω_θ and ω_z for different locations of the

Fig. 9.7

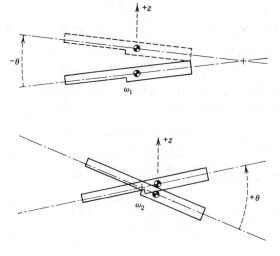

center of gravity. There still are two natural frequencies, ω_1 and ω_2, with $\omega_1 < \omega_z$ and $\omega_2 > \omega_\theta$.

The mode shapes for ω_1 and ω_2 are shown in Figure 9.7. For ω_1 the translation dominates, as would be expected, but there is a slight rotation. For ω_2, rotation dominates. The actual signs ($+$ or $-$) of the displacements in the modal fractions depend on the sign convention used for the equations of motion.

EXAMPLE PROBLEM 9.31

The two-mass system of Example Problem 9.1 is repeated. This time, the first mass m_1 is displaced 10 mm and released at time $t = 0$. The second mass, m_2, is also released at time $t = 0$, but is not displaced from the equilibrium position before being released. Determine the displacement of each mass as a function of time.

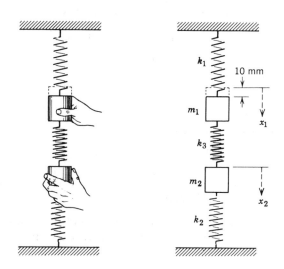

Solution:
The normal modes for this system have been determined to be

$$\left\{ \begin{matrix} X_1^{(1)} \\ X_2^{(1)} \end{matrix} \right\} = \left\{ \begin{matrix} 1 \\ 1 \end{matrix} \right\} \qquad \left\{ \begin{matrix} X_1^{(2)} \\ X_2^{(2)} \end{matrix} \right\} = \left\{ \begin{matrix} 1 \\ -1 \end{matrix} \right\}$$

The displacements $x_1(t)$ and $x_2(t)$ will exhibit both principal coordinates $p_1(t)$ and $p_2(t)$.

$$\left\{ \begin{matrix} x_1 \\ x_2 \end{matrix} \right\} = \left\{ \begin{matrix} 1 \\ 1 \end{matrix} \right\} [A_1 \cos \omega_1 t + B_1 \sin \omega_1 t] + \left\{ \begin{matrix} 1 \\ -1 \end{matrix} \right\} [A_2 \cos \omega_2 t + B_2 \sin \omega_2 t]$$

and the velocities are

$$\begin{Bmatrix} \dot{x}_1 \\ \dot{x}_2 \end{Bmatrix} = \begin{Bmatrix} 1 \\ 1 \end{Bmatrix} [B_1\omega_1 \cos \omega_1 t - A_1\omega_1 \sin \omega_1 t]$$

$$+ \begin{Bmatrix} 1 \\ -1 \end{Bmatrix} [B_2\omega_2 \cos \omega_2 t - A_2\omega_2 \sin \omega_2 t]$$

For the initial conditions, $\dot{x}_1(0) = 0$ and $\dot{x}_2(0) = 0$, which means that

$$0 = B_1\omega_1 + B_2\omega_2$$
$$0 = B_1\omega_1 - B_2\omega_2$$

If the solution is not trivial, $B_1 = B_2 = 0$.
For the initial conditions, $x_1(0) = 10$ mm and $x_2(0) = 0$,

$$10 = A_1 + A_2$$
$$0 = A_1 - A_2$$

Solving these two equations simultaneously,

$$A_1 = A_2 = 5 \text{ mm}$$

and,

$$x_1 = 5 \cos \omega_1 t + 5 \cos \omega_2 t$$
$$x_2 = 5 \cos \omega_1 t - 5 \cos \omega_2 t$$

Both principal modes will be observed in the motion of the geometric coordinates x_1 and x_2, as expected.

EXAMPLE PROBLEM 9.32

A rigid block with a mass m and a centroidal moment of inertia I is supported by eight equal springs, four vertical and four lateral. Determine

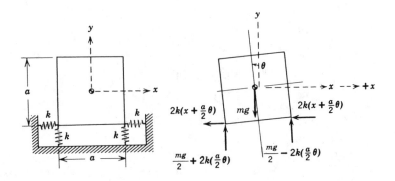

the coupled natural frequencies and modal fractions for translation in the
x direction and rotation in the x–y plane. The natural frequency in the
y direction is $\omega_y^2 = (4k/m)$ and is uncoupled, provided that the center of
gravity is located midway between the vertical supporting springs.

Solution:
In the horizontal x direction,

$$\sum \mathbf{F}_x = m\ddot{\mathbf{x}}$$

$$-4k\left(x + \frac{a}{2}\theta\right) = m\ddot{x}$$

and in the θ direction,

$$\sum \mathbf{M}_G = I_G\ddot{\boldsymbol{\theta}}$$

$$-4k\left(x + \frac{a}{2}\theta\right)\frac{a}{2} - \left[\frac{mg}{2} + 2k\left(\frac{a}{2}\theta\right)\right]\frac{a}{2} + \left[\frac{mg}{2} - 2k\left(\frac{a}{2}\theta\right)\right]\frac{a}{2} = I\ddot{\theta}$$

These are the two equations of motion for motion described by the θ
and x coordinates. These are coupled equations of motion, since θ and
x appear in each equation. For harmonic motion, $x = X \sin \omega t$ and $\theta = \Theta$
$\sin \omega t$. Substituting,

$$\begin{bmatrix} (-m\omega^2 + 4k) & +2ka \\ +2ka & (-I\omega^2 + 2ka^2) \end{bmatrix}\begin{bmatrix} X \\ \Theta \end{bmatrix} = 0$$

Setting the determinant equal to zero, the frequency equation is

$$(-m\omega^2 + 4k)(-I\omega^2 + 2ka^2) - 4k^2a^2 = 0$$

$$\omega^4 - \omega^2\left(\frac{4k}{m} + \frac{2k}{I}a^2\right) + \frac{4k^2a^2}{Im} = 0$$

The characteristic values are

$$\omega_1^2 = \frac{2k}{m} + \frac{ka^2}{I} - \sqrt{\frac{4k^2}{m^2} + \frac{k^2a^4}{I^2}}$$

$$\omega_2^2 = \frac{2k}{m} + \frac{ka^2}{I} + \sqrt{\frac{4k^2}{m^2} + \frac{k^2a^4}{I^2}}$$

If the rigid body is homogeneous, $I = \frac{1}{6}ma^2$, and

$$\omega_1^2 = 1.675\frac{k}{m} \quad \text{and} \quad \omega_2^2 = 14.32\frac{k}{m}$$

Substituting ω_1^2 and ω_2^2 in one of the equations of motion, the modal fractions are, for ω_1^2,

$$\left[-m\left(1.675\frac{k}{m} \right) + 4k \right] X + 2\,ka\Theta = 0$$

$$\frac{a\Theta}{X} = -1.162$$

and for ω_2^2

$$\left[-m\left(14.32\frac{k}{m} \right) + 4k \right] X + 2\,ka\Theta = 0$$

$$\frac{a\Theta}{X} = 5.16$$

In order to visualize these modes, let us put some values on the displacements X and Θ. For the first mode, let $X = 0.1a$. Since the mode shape fixes the relation between coordinates, the angular displacement $\Theta = 0.1162$-rad, clockwise, about 6.65°. Our coordinate convention designated counterclockwise rotation as positive. The first mode is seen to be predominantly a side to side motion.

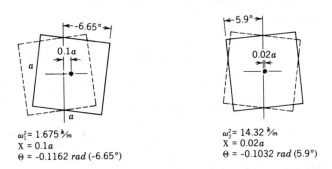

$\omega_1^2 = 1.675\ ^k\!/_m$
X = 0.1a
Θ = -0.1162 rad (-6.65°)

$\omega_2^2 = 14.32\ ^k\!/_m$
X = 0.02a
Θ = -0.1032 rad (5.9°)

For the second mode, again, let $X = 0.1a$. Then $\Theta = 0.516$ rad, or 29.6°, counterclockwise. The second mode is predominantly rotation. To better visualize it, let X be something smaller, say $X = 0.02a$. Remember, X and Θ are arbitrary, but when you fix one, you fix the other. For $X = 0.02a$, $\Theta = 0.1032$ rad or 5.9°. This is shown.

PROBLEM 9.33

(a) Can you determine the normal modes for Problem 9.21 by inspection? What are they?

(b) Can you determine the principal coordinates by inspection? What are they?

PROBLEM 9.34 Determine the principal coordinates of Problem 9.9.

Answer: $p_1 = \dfrac{x_1}{3} + \dfrac{2x_2}{3}$; $p_2 = \dfrac{2}{3}(x_1 - x_2)$

PROBLEM 9.35 Determine the principal coordinates of Problem 9.11.

Answer: $p_1 = 0.379x_1 + 0.485x_2$;
$p_2 = 0.621x_1 - 0.485x_2$

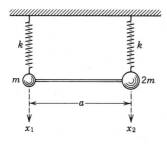

PROBLEM 9.36 Do Problem 9.14, using the coordinates x_1 and x_2. Sketch the normal modes for Problem 9.14. In terms of x_1 and x_2, sketch the mode shapes.

PROBLEM 9.37 Determine the normal modes for Problem 9.16. Sketch them.

PROBLEM 9.38 Determine the normal modes for Problem 9.18. Sketch them.

PROBLEM 9.39 Determine the normal modes for Problem 9.23. Sketch them.

PROBLEM 9.40 Determine the normal modes for Problem 9.26. Sketch them.

PROBLEM 9.41 Determine the normal modes for Problem 9.27. Sketch them.

PROBLEM 9.42 In Problem 9.5, the bob of the pendulum has a mass of 1 kg and the frame has a mass of 5 kg. The pendulum is 0.25 m long. Determine the amplitudes of free vibration for the bob and the frame if the pendulum is displaced 15° from the vertical and released from rest.

Answer: $x_1 = 0.436(1 - \cos \omega_2 t)$;
$\qquad x_2 = 0.436(1 + 5 \cos \omega_2 t)$

PROBLEM 9.43 The uniform bars of Problem 9.21 are released with one coincident with the vertical ($\Theta_1 = 0$), and the other inclined at an angle of 15° to the vertical ($\Theta_2 = \pi/12$). Determine the equation of motion for θ_1 and θ_2.

Answer: $\theta_1 = \dfrac{\pi}{24} \cos \omega_1 t - \dfrac{\pi}{24} \cos \omega_2 t;$

$\quad\quad \theta_2 = \dfrac{\pi}{24} \cos \omega_1 t + \dfrac{\pi}{24} \cos \omega_2 t$

PROBLEM 9.44 The entire two-mass system of Example Problem 9.31 and the frame within which it is suspended, is given an initial velocity of 2 m/s, with no initial displacement. (It could be dropped 200 mm.) Determine the equations of motion. Why is there only one mode present? For the purposes of the problem, $\omega_1 = 30$ s^{-1} and $\omega_2 = 52$ s^{-1}.

Answer: $X_1 = X_2 = \dfrac{1}{15} \sin 30t$

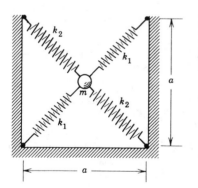

PROBLEM 9.45 The mass m moves in the plane of the paper. Describe the motion, if it is displaced a distance of 1 mm to the right and released, ($m = 2$ kg, $k_1 = 5$ N/mm, $k_2 = 2$ N/mm).

Answer: $x = \dfrac{1}{\sqrt{2}} \cos \omega_1 t; \ y = -\dfrac{1}{\sqrt{2}} \cos \omega_2 t$

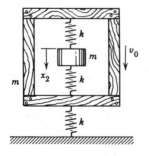

PROBLEM 9.46 The tension spring package of Problem 9.25 is repeated. Determine the maximum deflection between the box and the inner mass if the box is dropped such that $v_0 = \dot{x}_1(0) = \dot{x}_2(0) = 3.961$ m/s $k = 1930$ N/m, $m = 2.5$ kg

$$\omega_1^2 = 0.439 \frac{k}{m}; \ \omega_2^2 = 4.56 \frac{k}{m}$$

$$\begin{Bmatrix} X_1 \\ X_2 \end{Bmatrix} = \begin{bmatrix} 1.000 \\ 1.281 \end{bmatrix} \quad \begin{Bmatrix} X_1 \\ X_2 \end{Bmatrix} = \begin{bmatrix} 1.000 \\ -0.781 \end{bmatrix}$$

Answer: 68 mm

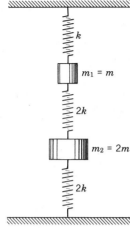

PROBLEM 9.47 The two degree of freedom system of Problem 9.9 has the following system constants:

$$[m] = \begin{bmatrix} m & 0 \\ 0 & 2m \end{bmatrix}; \quad [k] = \begin{bmatrix} 3k & -2k \\ 2k & 4k \end{bmatrix}$$

$$\omega_1 = \sqrt{\frac{k}{m}}; \qquad \chi_1 = \times 1$$

$$\omega_2 = 2\sqrt{\frac{k}{m}}; \qquad \chi_2 = -\frac{1}{2}$$

If the lower mass is displaced 10 mm from its equilibrium position, but the upper mass is not displaced, and then both are released simultaneously, what are the equations for x_1 and x_2?

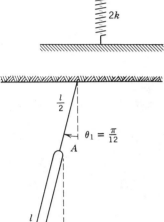

PROBLEM 9.48 The two natural frequencies and mode shapes of Problem 9.18 have previously been determined to be

$$\omega_1^2 = 0.917 \frac{g}{l}; \qquad \Theta_2^{(1)}/\Theta_1^{(1)} = 1.178$$

$$\omega_2^2 = 13.083 \frac{g}{l}; \qquad \Theta_2^{(1)}/\Theta_1^{(2)} = -0.847$$

The rod is now displaced 15° from the vertical and released from rest

$$\theta_1(0) = \frac{\pi}{12}; \qquad \theta_2(0) = \frac{\pi}{12}.$$

What are the equations of motion for the generalized coordinates $\theta_1(t)$ and $\theta_2(t)$?

Answer: $\theta_1 = 0.2388 \cos \omega_1 t + 0.0231 \cos \omega_2 t$
$\theta_2 = 0.2813 \cos \omega_1 t - 0.0196 \cos \omega_2 t$

9.5. RAYLEIGH'S PRINCIPLE

Rayleigh's energy method can be extended to find the two natural frequencies and their mode shapes for a two degree of freedom system, or the fundamental or lowest natural frequency of a vibrating system having more than two degrees of freedom. This extension can also be used to find other natural frequencies, but the use of this concept is deferred until a more detailed discussion can be made of multiple degrees of freedom.

As an example, consider a two-mass system such as that of Figure 9.8. If the motion of m_1 is x_1 and the motion of m_2 is x_2, the changes in kinetic

Fig. 9.8

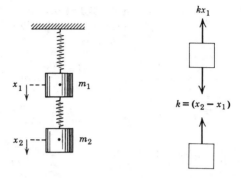

and potential energies can be written directly.

$$\Delta T = \tfrac{1}{2} m_1 \dot{x}_1^2 + \tfrac{1}{2} m_2 \dot{x}_2^2$$

$$\Delta V = \tfrac{1}{2} k x_1^2 + \tfrac{1}{2} k (x_2 - x_1)^2$$

The coordinates x_1 and x_2 are generalized coordinates but they are not principal coordinates.

For a two degree of freedom system, there are two natural frequencies, but at either natural frequency, both x_1 and x_2 will move with the same frequency. They will pass through maximum and minimum values, at the same time, and retain the ratio of x_2 to x_1 as a characteristic value of that particular natural frequency.

If we assume the motion to be simple harmonic, which we have found to be appropriate for Rayleigh's energy method,

$$x_1 = X_1 \sin \omega t \qquad \text{and} \qquad x_2 = X_2 \sin \omega t$$

$$\dot{x}_1 = X_1 \omega \cos \omega t \qquad \text{and} \qquad \dot{x}_2 = X_2 \omega \cos \omega t$$

Equating the maximum potential energy to the maximum kinetic energy, and solving for ω^2,

$$\Delta T_{\text{max}} = \Delta V_{\text{max}}$$

$$\tfrac{1}{2} m_1 \omega^2 X_1^2 + \tfrac{1}{2} m_2 \omega^2 X_2^2 = \tfrac{1}{2} k X_1^2 + \tfrac{1}{2} k (X_2 - X_1)^2 \qquad (9.26)$$

$$\omega^2 = \frac{k X_1^2 + k (X_2 - X_1)^2}{m_1 X_1^2 + m_2 X_2^2}$$

The value of ω^2 obtained is, of course, a function of the amplitudes X_1 and X_2. Note that it is a fraction, with the numerator a function of potential energy and the denominator a function of the kinetic energy. This is referred to as the *Rayleigh Fraction* or *Rayleigh Quotient*. In the case of a single degree of freedom, ω_n^2 was not a function of amplitude, since ω_n^2 was a *natural frequency*, and the amplitude squared appeared in both the kinetic and potential energy terms.

Assuming a proportionate relation between X_1 and X_2 the value of ω^2 would be a function of that proportion, or more specifically, the constraint imposed on the system. We have, in effect, constrained the two-mass system to move in one and only one mode. In doing so, however, we impose forces on m_1 and m_2 through our constraining members. Let us examine this constraint and minimize these forces.

Fig. 9.9

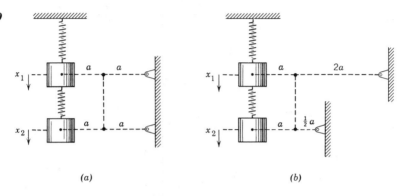

(a)   (b)

As an example, the two-mass system can be constrained by means of rigid linkages, as in Figure 9.9a and b. The problem has also been simplified by letting $m_1 = m_2 = m$ and $k_1 = k_2 = k$. With this simplification, it is easier to see the concept of Rayleigh's principle without an added confusion of different masses and springs. Because of the particular linkage of Figure 9.9a, $X_2 = X_1$ and $\omega^2 = \frac{1}{2}k/m$. The idea of physical constraint helps us accept the idea that the system can move with only one frequency and one mode. If we constrain the system to move so that $X_2 = 2X_1$, as in Figure 9.9b, which constraint is physically more logical, $\omega^2 = \frac{2}{5}k/m$. Each value was determined by substituting in equation 9.26.

Neither of these values is a natural frequency. If we substitute $\omega^2 = \frac{1}{2}k/m$ and its corresponding mode shape $X_2 = X_1$, into the equations of motion, $\Sigma F_1 = m_1\ddot{x}_1$ and $\Sigma F_2 = m_2\ddot{x}_2$, for sin ωt a maximum, or,

$$kX_1 - k(X_2 - X_1) = m_1\omega^2X_1 \qquad (9.27)$$
$$k(X_2 - X_1) = m_2\omega^2X_2$$

The equalities are not satisfied at $X_2 = X_1$. The same is true for $\omega^2 = \frac{2}{5}k/m$ and $X_2 = 2X_1$. The equalities can only be satisfied if ω^2 is a natural frequency, $\omega^2 = \omega_n^2$.

If we constrain the two-mass system such that

$$X_2 = \chi X_1$$

and

$$\dot{X}_2 = \chi \dot{X}_1$$

We can solve for the particular value of χ that will satisfy the equations of motion. That is the value of ω^2 when,

$$\omega^2 = \frac{k + k(\chi - 1)^2}{m + \chi^2 m} = \left[\frac{1 + (\chi - 1)^2}{1 + \chi^2}\right]\frac{k}{m} \qquad (9.28)$$

Plotting the value of ω^2 as a function of χ yields the curve of Figure 9.10.

Fig. 9.10

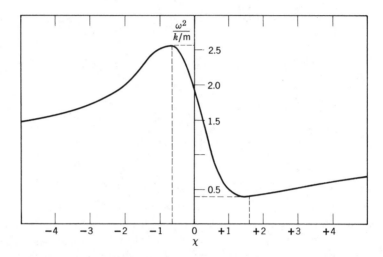

The value of the fundamental natural frequency, $\omega_1^2 = 0.382(k/m)$, corresponds to the minimum of the curve. The value of χ at the minimum, $\chi = 1.62$, yields a mode shape of $X_2 = 1.62X_1$, which does satisfy both equations of motion, (9.27).

If we investigate the minimum of the curve it can be shown to correspond with a natural frequency. Differentiating, with respect to χ,

$$\frac{d\omega^2}{d\chi} = \left[\frac{2(1 + \chi^2)(\chi - 1) - 2\chi(1 + (\chi - 1)^2)}{(1 + \chi^2)^2}\right]\frac{k}{m}$$

Substituting for ω^2 from equation 9.28,

$$\omega^2 = \left[\frac{1 + (\chi - 1)^2}{1 + \chi^2}\right]\frac{k}{m}$$

$$\frac{d\omega^2}{d\chi} = 2\left[\frac{(\chi - 1) - (m/k)\chi\omega^2}{1 + \chi^2}\right]\frac{k}{m} \qquad (9.29)$$

For the constraint, $X_2 = \chi X_1$, the two equations of motion would be

$$k - k(\chi - 1) = m\omega^2$$

and,

$$k(\chi - 1) = m\omega^2\chi$$

The numerator of equation 9.29 is nothing more than the second equation when $\omega^2 = \omega_n^2$, and

$$\frac{d\omega_n^2}{d\chi} = 0 \qquad (9.30)$$

A change in amplitude produces a second-order change in frequency at a natural frequency. The numerator of equation 9.29 is equal to zero if $\omega^2 = \omega_n^2$. At that value, the numerator is nothing more than one of the equations of motion, which must be in a force balance satisfying Newton's second law of motion. In an elementary way, $d\omega_n^2/d\chi = 0$ is a simple statement of Rayleigh's principle.

The value of χ satisfying the requirement $d\omega_n^2/d\chi = 0$ is that ratio of the amplitudes X_2 and X_1, which satisfies the equations of motion without external forces. Note that this ratio places no restraint on the absolute magnitude of X_2 and X_1, but does require that they have a specific ratio, or *eigenvector*.

A more general statement of Rayleigh's principle is that *for a linear system, there will be no change in the natural frequency with change in amplitude or mode shape.* Under these conditions, for the fundamental natural frequency, the mode shape of vibration will be such that the distribution of kinetic and potential energy will make the frequency of the vibration a minimum. These statements are physically logical if one considers that *simple harmonic motion can only exist if frequency and amplitude are independent of each other.*

Rayleigh's principle is valid for any natural frequency, although its greatest utility is in determining the mode and frequency of the fundamental. Attention must again be given to the role that constraint plays in Rayleigh's principle. It is essential that a mode shape exist, which is a function only of the relation of the generalized coordinates with one another. Any approximations of the mode shape for the fundamental will always yield a frequency that is higher than the true natural frequency. Better approximations of the true mode shape will result in lower frequencies. If differentiation is difficult, trial and error will quickly show which approximations are better.

EXAMPLE PROBLEM 9.49

Using Rayleigh's principle, approximate the natural frequency and mode shape for the lowest natural frequency of the double pendulum.

Solution:

θ_1 and θ_2 are the generalized coordinates, which recognize constraint on the double pendulum. The datum position is selected as the position where

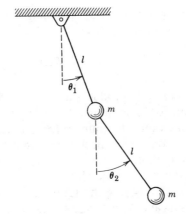

$\theta_1 = \theta_2 = 0$, which is also the equilibrium position. At any position, the change in potential energy is

$$\Delta V = mg(l - l \cos \theta_1) + mg(l - l \cos \theta_1 + l - l \cos \theta_2)$$

Using the approximation,

$$1 - \cos \theta_1 = \frac{\theta_1^2}{2}$$

and,

$$1 - \cos \theta_2 = \frac{\theta_2^2}{2}$$

$$\Delta V = \left(mg\theta_1^2 + mg\frac{\theta_2^2}{2} \right) l$$

The change in kinetic energy is

$$\Delta T = \tfrac{1}{2} mv_1^2 + \tfrac{1}{2} mv_2^2$$

With the aid of elementary kinematics, $v_1 = l\dot{\theta}_1$ and $v_2 = l\dot{\theta}_1 + l\dot{\theta}_2$. For small oscillations

$$\Delta T = \tfrac{1}{2}ml^2\dot{\theta}_1^2 + \tfrac{1}{2}ml^2(\dot{\theta}_1 + \dot{\theta}_2)^2$$

Assuming simple harmonic motion of the generalized coordinates θ_1 and θ_2.

$$\theta_1 = \Theta_1 \sin \omega_n t$$

$$\theta_2 = \Theta_2 \sin \omega_n t$$

Setting the maximum kinetic energy change to be equal to the maximum potential energy change,

$$\Delta T_{max} = \Delta V_{max}$$

$$\tfrac{1}{2}ml^2\Theta_1^2\omega_n^2 + \tfrac{1}{2}ml^2(\Theta_1 + \Theta_2)^2\omega_n^2 = mgl\Theta_1^2 + mgl\frac{\Theta_2^2}{2}$$

At this point, let us define the mode shape,

$$\chi(\theta) = \frac{\Theta_2}{\Theta_1}$$

and substitute this expression in the last equation. This yields the Rayleigh fraction that is an equation for ω_n^2.

$$\omega_n^2 = \frac{g}{l}\left(\frac{2+\chi^2}{1+(1+\chi)^2}\right)$$

Applying Rayleigh's principle, $d\omega^2/d\chi = 0$, will yield a quadratic equation for

$$\chi^2 - 2 = 0$$

The roots of this equation are the characteristic vectors for the natural frequencies, $\chi = \pm\sqrt{2}$. The positive root represents the lowest mode shape, for which

$$\omega_1^2 = 0.586\frac{g}{l}$$

The negative root and negative mode shape represent the second natural frequency, for which

$$\omega_2^2 = 3.414\frac{g}{l}$$

PROBLEM 9.50 Using Rayleigh's principle, determine the natural frequency and mode shape of the two-mass system consisting of the automobile and trailer of Problem 9.4. Note that this is a degenerate two degree of freedom system and one natural frequency is $\omega^2 = 0$. Find the other natural frequency.

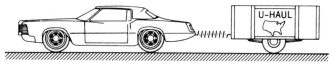

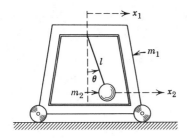

PROBLEM 9.51 Using Rayleigh's principle, determine the two natural frequencies and mode shapes for the frame and pendulum of Problem 9.5. Use coordinates x_1 and x_2.

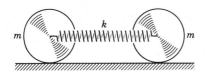

PROBLEM 9.52 Using Rayleigh's principle, determine the two natural frequencies and mode shapes of the two elastically coupled solid cylinders of Problem 9.6.

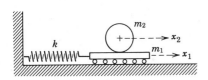

PROBLEM 9.53 Using Rayleigh's principle, determine the two natural frequencies and mode shapes of the cylinder and platform of Problem 9.7.

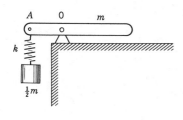

PROBLEM 9.54 A long slender bar of mass m is pivoted at point 0 and supports a mass of $\frac{1}{2} m$ from its left end A. In the equilibrium position that is shown, the bar and mass m are perfectly balanced. The modulus of the spring is k. Using Rayleigh's principle, compute the natural frequency of the system.

Answer: $\omega_n^2 = 3 \dfrac{k}{m}$

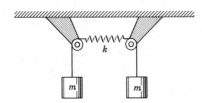

PROBLEM 9.55 Using Rayleigh's principle, approximate the fundamental mode and determine its frequency (ω^2) for the two-mass system shown.

Answer: $\omega_n^2 = \dfrac{2k}{m}$

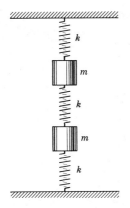

PROBLEM 9.56 Using Rayleigh's principle, determine the lowest natural frequency and mode shape for the two mass system. The springs are all identical and the masses are equal.

Answer: $\omega_n^2 = \dfrac{k}{m}$; $x_2 = x_1$

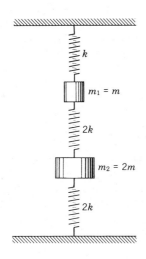

PROBLEM 9.57 Using Rayleigh's principle, determine the two natural frequencies and mode shapes of the two-mass system of Problem 9.9.

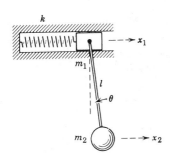

PROBLEM 9.58 Using Rayleigh's principle, determine the two natural frequencies and mode shapes of the two-mass system of Problem 9.13.

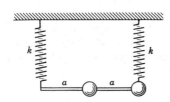

PROBLEM 9.59 Using Rayleigh's principle, determine the lowest natural frequency of the system shown.

$$k_1 = 10 \text{ N/mm}$$
$$k_2 = 20 \text{ N/mm}$$
$$m_1 = 5 \text{ kg}$$
$$m_2 = 3 \text{ kg}$$

Answer: $f_n = 9.02$ Hz

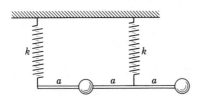

PROBLEM 9.60 Using Rayleigh's principle, determine the two natural frequencies and mode shapes of the two-mass system of Problem 9.16.

PROBLEM 9.61 Using Rayleigh's principle, determine the two natural frequencies and mode shapes of the two-mass system of Problem 9.17.

PROBLEM 9.62 For the slender rod suspended as a pendulum, Problem 9.18, use Rayleigh's principle to determine the two natural frequencies and mode shapes of small vibrations.

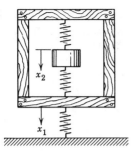

PROBLEM 9.63 Using Rayleigh's principle, determine the two natural frequencies and mode shapes of the tension spring package described in Problem 9.25.

PROBLEM 9.64 Using Rayleigh's principle, determine the two natural frequencies and mode shapes of the long slender bar supported on two springs (see Problem 9.26).

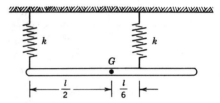

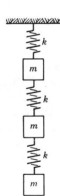

PROBLEM 9.65 Using Rayleigh's fraction, estimate the lowest natural frequency and mode shape for the three-mass system. The springs are all identical and the masses are all equal.

Answer: $\omega_1^2 = 0.198\dfrac{k}{m}$

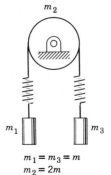

$m_1 = m_3 = m$
$m_2 = 2m$

PROBLEM 9.66 Using Rayleigh's fraction, estimate the lowest natural frequency and the mode shape for the given system. The pulley can be considered as a solid circular cylinder. *Hint:* Consider symmetry in estimating mode shapes.

Answer: $\omega_1^2 = \dfrac{k}{m}$

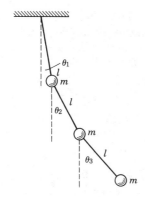

PROBLEM 9.67   Using Rayleigh's fraction, estimate the lowest natural frequency and the mode shape for the triple pendulum.

Answer: $\omega_1^2 = 0.416 \dfrac{g}{l}$

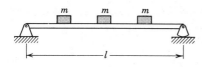

PROBLEM 9.68   A simply supported beam has three equal concentrated masses at its quarter points.

Using Rayleigh's fraction, estimate the lowest natural frequency and the mode shape.

Answer: $\omega_1^2 = 25.11 \dfrac{EI}{ml^3}$

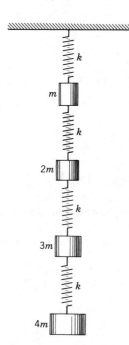

PROBLEM 9.69   Using Rayleigh's fraction, estimate the lowest natural frequency and mode shape for the four-mass system. The springs are all identical. The attachment of the springs to the four masses is symbolic. There is no rotation. The motion of all four masses is only vertical.

Answer: $\omega_1^2 = 0.038 \dfrac{k}{m}$

9.6. FORCED VIBRATION

When a two degree of freedom system is forced by a harmonic forcing function, the system will exhibit a response that is similar in many ways to the response of a single degree of freedom system. Resonance will occur when the forcing frequency equals the natural frequency. Since there are two natural frequencies, resonance can occur twice, once with either natural frequency.

In Figure 9.11 let us consider the system of Figure 9.1 as being forced by a forcing function $F(t)$ applied to the first of the two masses, m_1. The equations of motion are

$$-k_1 x_1 + k_3(x_2 - x_1) + F_1(t) = m_1 \ddot{x}_1$$
$$-k_2 x_2 - k_3(x_2 - x_1) = m_2 \ddot{x}_2$$
(9.31)

If the forcing function is harmonic, $F_1(t) = F_1 \sin \omega t$, the response will be a harmonic displacement at the same frequency as the forcing frequency.

$$x_1 = X_1 \sin \omega t$$

$$x_2 = X_2 \sin \omega t$$

Fig. 9.11

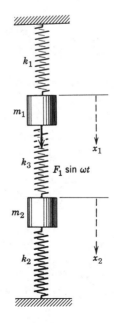

Substituting in equation 9.31, and expressing these two simultaneous equations in matrix form gives

$$\begin{bmatrix} (k_1+k_3-m_1\omega^2) & -k_3 \\ -k_3 & (k_2+k_3-m_2\omega^2) \end{bmatrix} \begin{Bmatrix} X_1 \\ X_2 \end{Bmatrix} = \begin{Bmatrix} F_1 \\ 0 \end{Bmatrix} \qquad (9.32)$$

The value of the determinant is the frequency equation for the system, which we solved earlier (see equation 9.4) for the natural frequencies. Referring to the two roots, ω_1 and ω_2

$$\begin{aligned} \text{Det}(\omega^2) &= (k_1+k_3-m_1\omega^2)(k_2+k_3-m_2\omega^2)-k_3^2 \\ &= m_1m_2(\omega^2-\omega_1^2)(\omega^2-\omega_2^2) \end{aligned} \qquad (9.33)$$

Using Cramer's rule, we can solve for the maximum displacement X_1 or X_2,

$$X_1 = \frac{\begin{vmatrix} F_1 & -k_3 \\ 0 & (k_2+k_3-m_2\omega^2) \end{vmatrix}}{\text{Det}(\omega^2)}$$

$$X_1 = \frac{F_1(k_2+k_3-m_2\omega^2)}{m_1m_2(\omega^2-\omega_1^2)(\omega^2-\omega_2^2)} \qquad (9.34a)$$

$$X_2 = \frac{\begin{vmatrix} (k_1+k_3-m_1\omega^2) & F_1 \\ -k_3 & 0 \end{vmatrix}}{\text{Det}(\omega^2)}$$

$$X_2 = \frac{F_1k_3}{m_1m_2(\omega^2-\omega_1^2)(\omega^2-\omega_2^2)} \qquad (9.34b)$$

It is obvious that the amplitudes X_1 and X_2 are infinite if $\omega^2=\omega_1^2$ or $\omega^2=\omega_2^2$.

Figure 9.12 shows the response for both X_1 and X_2, in terms of the dimensionless parameter ω/ω_1. As in Example Problem 9.1, for symmetry, two equal masses and three equal springs were used, so that

$$\omega_1^2=\frac{k}{m} \qquad \text{or} \qquad \omega_2^2=\frac{3k}{m}$$

In the dimensionless parameter ω/ω_1, ω_1 was selected arbitrarily. It would have been just as easy to select ω_2 but the resulting graphs would have been more difficult to read.

The graphs of the responses show the two resonance conditions at ω_1 and ω_2. At other frequencies, the vibration is finite and could be calculated, if the magnitude of the harmonic force were known. Note that there is a frequency where the vibration of the first mass, to which the forcing function is applied, is reduced to zero. This is the entire basis of the *dynamic vibration absorber*.

Fig. 9.12

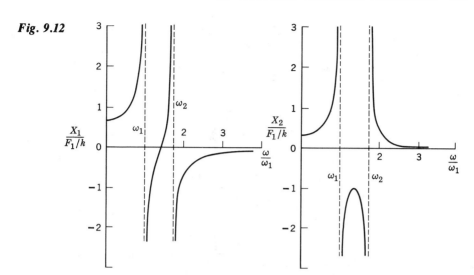

9.7. DYNAMIC VIBRATION ABSORBERS

The dynamic vibration absorber is a mechanical device used to decrease or eliminate unwanted mechanical vibration. Its greatest application is to synchronous machinery, for the dynamic vibration absorber is tuned to one particular frequency and is only effective over a narrow band of frequencies.

Absorbers are used extensively on large reciprocating internal combustion engines, which run at a constant speed for minimum fuel consumption, and in reciprocating tools, such as clippers, shuttles, sanders, and compactors. In all of these applications, the operating frequency is nearly constant and the vibration absorber system balances the reciprocating forces. Without the vibration absorber, the reciprocating forces would make the tool impossible to hold or control. The dumbbell-shaped devices that hang from highest voltage transmission lines are dynamic vibration absorbers used to mitigate the fatiguing effects of wind-induced vibration. A short drive in the open country will verify that there are thousands in use.

In its simplest form, a dynamic vibration absorber consists of one spring and a mass. Such an absorber system is attached to the single degree of freedom system, as shown in Figure 9.13. One effect of adding a dynamic vibration absorber is obvious. Its presence adds an additional degree of freedom to the system.

The equations of motion for the main mass and for the absorber are

$$-k_1x_1+k_2(x_2-x_1)+F_1 \sin \omega t = m_1\ddot{x}_1 \qquad (9.35)$$

$$-k_2(x_2-x_1) = m_2\ddot{x}_2 \qquad (9.36)$$

Fig. 9.13

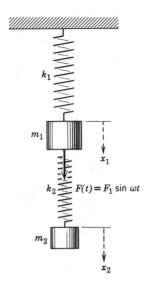

For this forced excitation, each mass will move with the frequency of the forcing function ω, $x_1 = X_1 \sin \omega t$, $x_2 = X_2 \sin \omega t$.

$$\begin{bmatrix} (k_1 + k_2 - m_1\omega^2) & -k_2 \\ -k_2 & (k_2 - m_2\omega^2) \end{bmatrix} \begin{Bmatrix} X_1 \\ X_2 \end{Bmatrix} = \begin{Bmatrix} F_1 \\ 0 \end{Bmatrix} \qquad (9.37)$$

The natural frequencies of the system are the eigenvalues or characteristic roots of the determinant.

$$\text{Det}(\omega^2) = (k_1 + k_2 - m_1\omega^2)(k_2 - m_2\omega^2) - k_2^2 = 0 \qquad (9.38)$$

and the amplitudes X_1 and X_2 are

$$X_1 = \frac{F_1(k_2 - m_2\omega^2)}{(k_1 + k_2 - m_1\omega^2)(k_2 - m_2\omega^2) - k_2^2} \qquad (9.39a)$$

$$X_2 = \frac{F_1 k_2}{(k_1 + k_2 - m_1\omega^2)(k_2 - m_2\omega^2) - k_2^2} \qquad (9.39b)$$

Both X_1 and X_2 are determinate, if the magnitude of the forcing function $F(t) = F_1 \sin \omega t$ is known. For simplification, the following substitutions are made:

$$\omega_{11}^2 = \frac{k_1}{m_1} \qquad \text{the natural frequency of the main system alone}$$

$$\omega_{22}^2 = \frac{k_2}{m_2} \qquad \text{the natural frequency of the absorber system alone}$$

$$\mu = \frac{m_2}{m_1} \qquad \text{the mass ratio}$$

$$\frac{k_2}{k_1} = \frac{\omega_{22}^2}{\omega_{11}^2} \mu$$

Solving for X_1 and X_2

$$X_1 = \frac{F_1}{k_1} \frac{\left(1 - \dfrac{\omega^2}{\omega_{22}^2}\right)}{\left(1 + \mu\dfrac{\omega_{22}^2}{\omega_{11}^2} - \dfrac{\omega^2}{\omega_{11}^2}\right)\left(1 - \dfrac{\omega^2}{\omega_{22}^2}\right) - \mu\dfrac{\omega_{22}^2}{\omega_{11}^2}} \qquad (9.40a)$$

$$X_2 = \frac{F_1}{k_1} \frac{1}{\left(1 + \mu\dfrac{\omega_{22}^2}{\omega_{11}^2} - \dfrac{\omega^2}{\omega_{11}^2}\right)\left(1 - \dfrac{\omega^2}{\omega_{22}^2}\right) - \mu\dfrac{\omega_{22}^2}{\omega_{11}^2}} \qquad (9.40b)$$

At $\omega = \omega_{22}$, the motion of the main mass m_1 does not simply diminish, it ceases altogether. The displacements X_1 and X_2, as related to the arbitrary parameter F_1/k_1 are shown in Figures 9.14a and b. The mass ratio

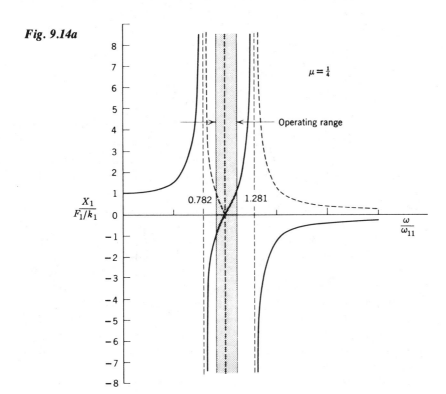

Fig. 9.14a

Fig. 9.14b

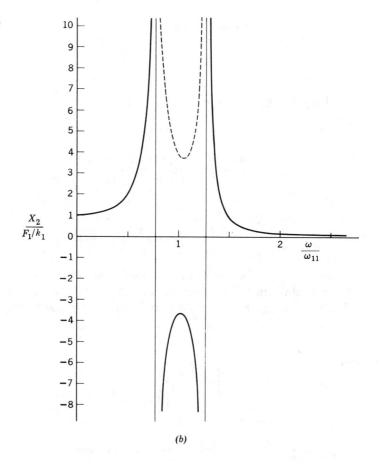

$\dfrac{X_2}{F_1/k_1}$

(b)

$\mu = \frac{1}{4}$, and $\omega_{11} = \omega_{22}$, the absorber being tuned to the natural frequency of the main system. The satisfactory operating range where $X/(F/k_1) < 1$ is shaded.

Two parameters can be varied. One is the mass ratio μ. Obviously a large mass ratio presents a practical problem. An absorber system that matches the original system in size is not a good solution to any vibration problem. At the same time, the smaller the mass ratio, the narrower will be the operating band of the absorber. The second parameter is the frequency ratio $\beta = \omega_{22}/\omega_{11}$. The natural frequency of the absorber system ω_{22} is the frequency at which $X_1 = 0$. It should be selected to best satisfy the operating requirements. It is not necessarily equal to ω_{11}, although the use of a vibration absorber is most warranted when the forcing frequency is close to the natural frequency of the main system, and operating restrictions make it impossible to vary either one.

One disadvantage of the dynamic vibration absorber has already been mentioned. It does add an additional degree of freedom. To find the new natural frequencies of the main system and the absorber, one can go back to the frequency equation 9.37. Setting $\beta = \omega_{22}/\omega_{11}$, for $\text{Det}(\omega^2) = 0$,

$$\left(1 + \mu\beta^2 - \frac{\omega^2}{\omega_{11}^2}\right)\left(1 - \frac{\omega^2}{\omega_{22}^2}\right) - \mu\beta^2 = 0$$

$$\beta^2\left(\frac{\omega^4}{\omega_{22}^4}\right) - \frac{\omega^2}{\omega_{22}^2}[1 + \beta^2(1+\mu)] + 1 = 0$$

(9.41)

This is the frequency equation expressed in terms of the two parameters μ and β.

For $\omega_{11} = \omega_{22}$, $\beta^2 = 1$,

$$\frac{\omega^2}{\omega_{22}^2} = \frac{2+\mu}{2} \pm \frac{1}{2}\sqrt{(2+\mu)^2 - 4}$$

The separation of the two natural frequencies, ω_1 and ω_2 is apparent in Figure 9.15. For a mass ratio of $\mu = \frac{1}{4}$, the two natural frequencies ω_1 and ω_2 are $\omega_1 = 0.782\omega_{11}$ and $\omega_2 = 1.281\omega_{11}$. As the mass ratio increases, the separation of the two natural frequencies increases.

Fig. 9.15

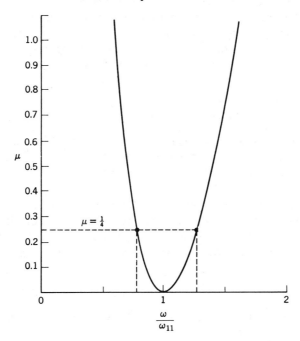

If damping is added to the absorber system, the amplitude of both X_1 and X_2 at resonance are diminished, but not equally. Unfortunately, the lower natural frequency is diminished less than the higher natural frequency, and it is the lower natural frequency that must be passed through in order to reach operating speed. To equalize the maximum amplitudes at resonance, the damped absorber is tuned to a frequency slightly lower than the natural frequency of the main system. Optimum tuning is defined as the ratio ω_{22}/ω_{11}, when the resonant amplitudes are equal. A derivation of the optimum tuning can be found in the vibration books of both S. Timoshenko and J. P. Den Hartog. It suffices here to state the result that at optimum tuning,

$$\beta = \frac{\omega_{22}}{\omega_{11}} = \frac{1}{1+\mu} \tag{9.42}$$

Damping can also be optimized. If it is absent, the amplitude of the main system will be zero at the tuning frequency, $\omega = \omega_{22}$. With damping, the resonant amplitudes of the combined system are diminished but the minimum amplitude of the main system is no longer zero at the tuning frequency. Optimum damping is defined as that amount of damping, which will make the response curve nearly flat between the two natural frequencies ω_1 and ω_2. The resonant amplitudes are decreased and the amplitude at the tuning frequency is increased.

If vibration absorbers are used, they are more often used without damping. Damping defeats the purpose of an absorber, which is to eliminate unwanted vibration, and is only warranted if the frequency band in which an absorber is effective is too narrow for operation.

9.8. TRANSMISSION OF FORCE AND MOTION

For a system with two or more degrees of freedom, the transmission of force and motion is more complex than it is for a single degree of freedom. The modal relationship between coordinates has much to do with force transmitted.

Referring back to Figure 9.11, force can be transmitted through either spring, k_1 or k_2, but only if the springs are stretched or compressed. Using the sign convention in Figure 9.11, a positive force can be transmitted if the spring k_1 is stretched or the spring k_2 is compressed. These forces add. Thus,

$$|F_{TR}| = k_1 x_1 + k_2 x_2 = k(x_1 + x_2) \tag{9.43}$$

If the impressed force $F(t) = F_1 \sin \omega t$ is applied to mass m_1, the transmission ratio is

$$T.R. = \frac{1}{1 - \left(\dfrac{\omega^2}{\omega_1^2}\right)} \tag{9.44}$$

ω_1^2 is the square of the first natural frequency. This could be expected, since $x_1 + x_2$ is a direct measure of the principal coordinate p_1. In the second mode, kx_1 cancels kx_2. This particular result is due to the symmetry of the example, $\chi^{(1)} = 1$, $\chi^{(2)} = -1$. Although it looks like the transmission ratio for a single degree of freedom, do not make this false conclusion. In that case, if we had lumped both masses together, the natural frequency of the single degree of freedom would be $\omega_n^2 = k/2m$, whereas here, $\omega_1^2 = k/m$ and $\omega_2^2 = 3k/m$.

At the same time, the relative displacement between x_1 and x_2 may be more important than the force transmitted to the foundation. As an example, perhaps deflection of the coupling member would be considered to be an engineering failure. Visualize the center coupling member as an egg. In the first mode, the coupling member does not deflect. In the second mode, deflection is at a maximum. In that case,

$$X_2 - X_1 = \frac{F_1/k}{3\left(1 - \dfrac{\omega^2}{\omega_2^2}\right)} \tag{9.45}$$

If the coupling member were an egg, as depicted in Figure 9.16, the second mode would be catastrophic, despite the fact that there would be absolutely no force or motion transmitted to the foundation. It is obvious that the most significant parameter is that in which you are most interested. It could be structural damage or it could be unwanted vibration.

Fig. 9.16

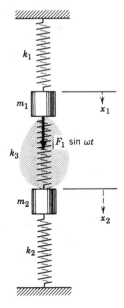

The importance of coupled modes is readily apparent. Suppose a sinusoidal force excited the system of Figure 9.4a. If the excitation were purely vertical, and the geometric center of support coincided with the center of gravity, the second mode would not be forced. If the excitation were a pure couple, in the θ direction, the first mode would not be forced. On the other hand, if the center of gravity were not symmetrically located, the first and second mode would not be independent. It is advantageous to locate the center of gravity symmetrically and make modes independent. In practice, great lengths are taken to ensure that the center of gravity is supported symmetrically and that the plane of support passes through the center of gravity. This is one of those principles of vibration isolation that are so easy to apply that it is considered poor engineering practice to ignore them, even if it is unnecessary. Exciting an unwanted mode, which could have been avoided by locating the center of gravity symmetrically, is something that you learn not to do twice.

EXAMPLE PROBLEM 9.70

The rigid block of Problem 9.32 is now in place. It is a model for a large piece of machinery, so its vital statistics are large, $m = 600$ kg, and $a = 1$ m. For the purposes of this problem, assume that the block is homogeneous, so that the moment of inertia is known, $I = 100$ kg·m², and the center of mass is centrally located and also known. The block sits on four springs, each with an elastic constant $k = 60$ kN/m. Horizontal motion is resisted by a set of four identical springs. What are the linear and angular displacements, if the block is subjected to a horizontal forcing function acting at the mass center, $F(t) = \sin 30t$, expressed in kilo-Newtons.

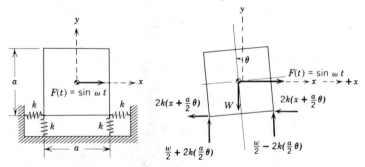

Solution:

This is a problem typical of the vibration isolation of large machinery. In the horizontal direction, the equation of motion would now include the forcing function, $F(t) = F_1 \sin \omega t$.

$$\sum F_x = m\ddot{x}$$

$$F_1 \sin \omega t - 4k\left(x + \frac{a}{2}\theta\right) = m\ddot{x}$$

But, in the θ direction, the equation of motion remains

$$\sum \mathbf{M}_G = I_G \ddot{\theta}$$

$$-4k\left(x+\frac{a}{2}\theta\right)\frac{a}{2} - \left[\frac{W}{2}+2k\left(\frac{a}{2}\theta\right)\right]\frac{a}{2} + \left[\frac{W}{2}-2k\left(\frac{a}{2}\theta\right)\right]\frac{a}{2} = I\ddot{\theta}$$

This will be true if the forcing function acts through the mass center, since there would be no moment of the forcing function about the mass center.

These two equations of motion can be expressed in matrix form. For harmonic motion at the forcing frequency ω, $x = X \sin \omega t$ and $\theta = \Theta \sin \omega t$. Substituting,

$$\begin{bmatrix} (-m\omega^2+4k) & +2ka \\ +2ka & (-I\omega^2+2ka^2) \end{bmatrix}\begin{bmatrix} X \\ \Theta \end{bmatrix} = \begin{Bmatrix} F_1 \\ 0 \end{Bmatrix}$$

Using Cramer's rule, we can solve for the displacements X and Θ.

$$X = \frac{\begin{vmatrix} F_1 & 2ka \\ 0 & (2ka-I\omega^2) \end{vmatrix}}{\begin{vmatrix} (4k-m\omega^2) & 2ka \\ 2ka & (2ka-I\omega^2) \end{vmatrix}}$$

$$\Theta = \frac{\begin{vmatrix} (4k-m\omega^2) & F_1 \\ 2ka & 0 \end{vmatrix}}{\begin{vmatrix} (4k-m\omega^2) & 2ka \\ 2ka & (2ka-I\omega^2) \end{vmatrix}}$$

With the numerical values $m = 600$ kg, $k = 60$ kN/m, $I = 100$ kg·m^2, the forcing frequency $\omega = 30$ s^{-1}, and the magnitude of the exciting force 1 kN or 1000 N, the displacement Θ and X are determinate.

$$X = \frac{\begin{vmatrix} 1000 & 120,000 \\ 0 & [120,000-(100)(900)] \end{vmatrix}}{\begin{vmatrix} [240,000-(600)(900)] & 120,000 \\ 120,000 & [120,000-(100)(900)] \end{vmatrix}} = \frac{3 \times 10^7}{2.34 \times 10^{10}}$$

$$= 1.28 \text{ mm}$$

$$\Theta = \frac{\begin{vmatrix} [240,000-(600)(900)] & 1000 \\ 120,000 & 0 \end{vmatrix}}{\begin{vmatrix} [240,000-(600)(900)] & 120,000 \\ 120,000 & [120,000-(100)(900)] \end{vmatrix}} = \frac{-12 \times 10^7}{2.34 \times 10^{10}}$$

$$= -0.00513 \text{ rad}$$

This is about 0.3°, clockwise. All these values should be acceptable.

At this point, it should be very clear why the forcing function was placed to act through the mass center. If it acted anywhere else, a forcing moment would occur, causing vibration through the second equation of motion. In practice, great effort is spent to direct the exciting force through the mass center.

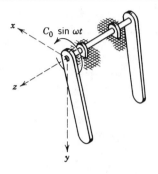

PROBLEM 9.71   An oscillating couple $C = C_0 \sin \omega t$ is applied to one of the two pendula of Problem 9.21. Determine the amplitude of motion of both pendula if the forcing frequency is 200 cpm, and the couple C_0 is 2N·m. Other physical constants are $m = 5$ kg, $K = 9$ N·m/rad, $l = 300$ mm.

Answer: $\theta_1 = -0.0418$ rad;
$\quad\quad\quad \theta_2 = +0.0076$ rad

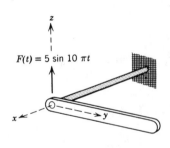

PROBLEM 9.72   Determine the forced amplitudes of motion for the cantilevered shaft of Problem 9.23, if a 5 N force is applied along the z-axis at a frequency of 5 Hz.

Answer: $\Theta = 1.205°$;
$\quad\quad\quad Z = 3.6$ mm

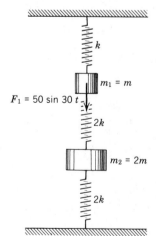

PROBLEM 9.73   The two-mass system of Problem 9.47, has the following system constants:

$$k = 8000 \text{ N/m}; \quad\quad\quad m = 2 \text{ kg}$$

$$\omega_1^2 = \frac{k}{m} = 400 \text{ s}^{-2}; \quad \chi_1 = +1$$

$$\omega_2^2 = 4\frac{k}{m} = 1600 \text{ s}^{-2}; \quad \chi_2 = -\frac{1}{2}$$

What are the amplitudes X_1 and X_2, if the first mass is excited by the harmonic force $F = 50 \sin 30 t$?

Answer: $X_1 = -7.1$ mm, $X_2 = +28.6$ mm

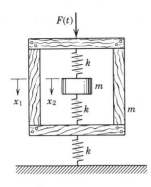

PROBLEM 9.74 A harmonic force $F(t) = 10$ sin 30 t is applied to the outer case of the tension spring packaging system of Problems 9.25, and 9.46. What is the maximum amplitude of the center mass? For this problem, $m_1 = m_2 = 2.5$ kg and $k = 1930$ N/m. For convenience, the natural frequencies and mode shapes are repeated from Problem 9.25.

$$\omega_1^2 = 0.4384 \frac{k}{m}; \ X_1^{(1)} = 1.0000; \ X_2^{(1)} = 1.0000$$

$$\omega_2^2 = 4.5616 \frac{k}{m}; \ X_1^{(2)} = 1.2808; \ X_2^{(2)} = -0.7808$$

Answer: $X_2 = -4.2$ mm

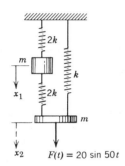

PROBLEM 9.75 For Problem 9.11, determine the forced amplitude of motion of each mass, if the lower mass is forced by $F(t) = 20$ sin 50 t. Use $k = 2000$ N/m and $m = 2$ kg.

Answer: $X_1 = -6.15$ mm; $X_2 = -4.62$ mm

PROBLEM 9.76 What would be the result of Problem 9.75 if the force $F(t) = 20$ sin 50 t were applied to the upper mass rather than the lower mass? Remember, all motion is linear! You do not need to worry about rotation or any mode of motion other than vertical motion.

Answer: $X_1 = -1.54$ mm; $X_2 = -6.15$ mm

PROBLEM 9.77 In Problem 9.14, $m = 2$ kg, $a = 0.25$ m, and $k = 1500$ N/m. Determine the amplitude of motion if a vertical force $F(t) = 10$ sin 10 t is applied to the right mass.

Answer: $X = +6.1$ mm; $\Theta = 2.08°$

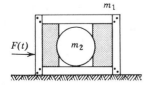

PROBLEM 9.78 A box with a 5-kg mass contains a bowling ball (a sphere) with a 7-kg mass. It has a diameter of 200 mm. Packing supports the ball at the top and bottom, but unfortunately, the box is on its side. Consider the packing to be elastic with a modulus of 1200 N/m. Friction

between the box and the ball cannot be ignored, but for the sake of simplicity, assume that the box rests on a smooth surface and that friction between the box and the horizontal surface can be ignored.

(a) What is the natural frequency or natural frequencies of the box and ball?

(b) Determine the absolute motion of both the ball and the box, if $F(t) = \sin 20\, t$.

Answer: (a) $f_n = 3.62$ Hz

(b) $x_1 = 2.8 \sin 20\, t$, mm
$x_2 = -2.4 \sin 20\, t$, mm

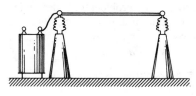

PROBLEM 9.79 Bus bars are hollow tubes used to carry electrical currents in high voltage electrical distribution substations. With long unsupported spans, they often are subjected to wind-induced vibration. In one instance, a 25-m aluminum bus, with a mass of 3.75 kg/m was observed to have a fundamental frequency of 5.0 Hz. Determine the mass and spring constant of a dynamic vibration absorber, placed at midspan, if the bus and absorber system are not to have natural frequencies within the range of 3.5 to 7.0 Hz.

Answer: m = 7.74 kg; k = 7638 N/m

PROBLEM 9.80 A railroad boxcar that has been modified for fragile lading consists of a main car body, m_1, supported near each end by the main springs, k_1. Identical dynamic vibration absorber masses m_2 are mounted at each end of the boxcar. The boxcar has a mass of 25,000-kg empty and without absorbers, and 125,000 kg fully loaded. For AAR standard $1\frac{5}{8}$ in. (41.3 mm) total deflection springs, the minimum vertical movement of the mass m_1 occurs at 100 km/h on standard rails of 39 ft (11.89 m). If each of the absorber masses has a mass of 12,500 kg, determine the two speeds at which resonance would occur.

Answer: 85 km/h; 138 km/h

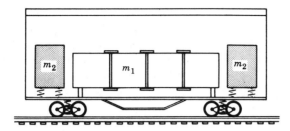

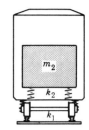

PROBLEM 9.81 For a particular application of a dynamic vibration absorber, $\omega_{11}^2 = \omega_{22}^2$ and $\mu = \frac{1}{4}$. What is the central frequency range for which the transmission ratio is less than unity?

Answer: $0.906\omega_{11} < \omega < 1.12\omega_{11}$

PROBLEM 9.82 A transmission line cable with a mass of 1.5 kg/m is suspended in a long span. Vibration absorbers are attached to the cable, near the suspension tower, to suppress wind-induced vibration, characterized by vertical oscillation of the cable in a series of stationary loops. Each absorber consists of two 2.5-kg weights attached to a short length of flexible cable. The natural frequency of the absorber alone is 15 Hz. Because of the extreme length of the span, the cable is always excited at or near some natural frequency, which can be approximated by $f_n = 75/L$, where L is the length of one loop in meters. Determine the range of frequencies for which one absorber is effective. Note that the mass of a loop of cable decreases as the loop length decreases. What happens if two absorbers are used?

Answer: 10.8 Hz $< f <$ 20.8 Hz

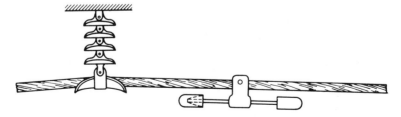

PROBLEM 9.83 An industrial installation of process shakers shows a violent resonance at 456 cpm. As a trial remedy, two 5-kg vibration

absorbers are attached to the process system, which result in two resonant frequencies at 404 and 515 cpm. How many vibration absorbers are required to be assured that no resonance occurs between 350 and 600 cpm?

Answer: 10 absorbers

CHAPTER TEN

TORSIONAL
VIBRATION

10.1. DISCRETE SYSTEMS

Most real systems have many degrees of freedom. Simplifications can be made to approximate a rather complex system with only one or two degrees of freedom, and much can be accomplished using a simplified system, but it is important to know the characteristics of systems with more than two degrees of freedom, and the differences in their vibration characteristics. Some are quite subtle. There will be more than two natural frequencies, one corresponding to each degree of freedom. Each natural frequency will also be characterized by a principal mode, which is descriptive of that natural frequency.

These systems are treated in either of two ways. In one, the elastic and inertial properties of system are separated into discrete springs and masses. The spatial dependence of distributed mass and elasticity is modeled. This method is known as using lumped masses and lumped springs, or more esoterically lumped parameters, or simply as discrete systems. All of the elastic properties are assumed to be exhibited in massless springs and all of the inertial properties are exhibited in point masses. Needless to say, with the advent and widespread use of the modern computer, discretizing a system and using numerical methods to obtain solutions has become the most accepted means of solving mechanical vibration problems. *The process of discretizing a system is also the source of its largest errors.* This fact must be kept in mind when evaluating the results of a computer-aided solution to a discretized problem.

In the second method of attack, the elastic and inertial properties are distributed spatially as a continuous system or distributed system. This method is more exact, but analyses are limited to a narrow selection of problems, such as uniform beams and slender rods. Whichever method of modeling is selected, discrete or distributed, the *practical result is only as good as the model.* This is most important, for we can spend hours on manipulating a mathematical problem, without realizing that inherent errors in the mathematical model do not justify it.

10.2. TORSIONAL VIBRATION

The first discretized problems that were treated as a group arose as a result of the torsional vibration of the crankshafts of large reciprocating steam engines, line shafts, and during and after World War I, the internal combustion engine and motor generators of marine and submarine propulsion systems. A considerable body of literature has been developed about the solution of these problems and it is widely available. It is of interest in the study of mechanical vibration, since the transition from these classical problems, which are easy to visualize, to the more contemporary problems, which are not, is logical and direct.

As examples, let us consider four torsional systems. The two-mass system of Figure 10.1a, has two degrees of freedom, but one of the natural

Fig. 10.1

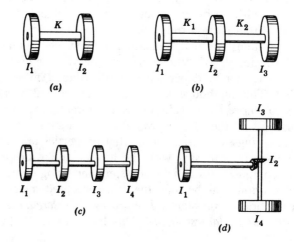

(a)

(b)

(c)

(d)

frequencies is zero, $\omega_1^2 = 0$. More correctly, it is a degenerate two degree of freedom system. The equations of motion are easily written with our knowledge of two degrees of freedom

$$I_1\ddot{\theta}_1 + K(\theta_1 - \theta_2) = 0$$
$$I_2\ddot{\theta}_2 + K(\theta_2 - \theta_1) = 0$$

The frequency equation is found by assuming harmonic vibration in a principal mode with a frequency ω, and eliminating the amplitudes Θ_1 and Θ_2.

$$\theta_1 = \Theta_1 \sin \omega t$$
$$\theta_2 = \Theta_2 \sin \omega t$$

The frequency equation is

$$\omega^4 I_1 I_2 - \omega^2 (I_1 + I_2) K = 0 \tag{10.1}$$

from which the characteristic values are the natural frequencies.

$$\omega_1^2 = 0$$
$$\omega_2^2 = K\left(\frac{I_1 + I_2}{I_1 I_2}\right)$$

The modes corresponding to these natural frequencies are described by substituting ω_1^2 and ω_2^2 in the equations of motion. It is the same physical problem of Example Problem 2.28.

The three-mass systems of Figure 10.1b is a degenerate three degree of freedom system. Considering it in more detail, using three generalized coordinates θ_1, θ_2, and θ_3, for the angular displacements of each mass, the equations of motion are

$$I_1\ddot{\theta}_1 + K_1(\theta_1 - \theta_2) = 0$$
$$I_2\ddot{\theta}_2 + K_2(\theta_2 - \theta_3) - K_1(\theta_1 - \theta_2) = 0 \tag{10.2}$$
$$I_3\ddot{\theta}_3 - K_2(\theta_2 - \theta_3) = 0$$

For harmonic vibration in a principal mode,

$$\theta_1 = \Theta_1 \sin \omega t$$
$$\theta_2 = \Theta_2 \sin \omega t$$
$$\theta_3 = \Theta_3 \sin \omega t$$

and substituting,

$$(K_1 - I_1\omega^2)\Theta_1 - K_1\Theta_2 = 0$$
$$(K_1 + K_2 - I_2\omega^2)\Theta_2 - K_1\Theta_1 - K_2\Theta_3 = 0 \tag{10.3}$$
$$(K_2 - I_3\omega^2)\Theta_3 - K_2\Theta_2 = 0$$

As for two degrees of freedom, the frequency equation for this three degree of freedom system can be found by eliminating Θ_1, Θ_2, and Θ_3

$$\omega^6(I_1I_2I_3) - \omega^4[K_1(I_2I_3 + I_1I_3) + K_2(I_1I_2 + I_1I_3)]$$
$$+ \omega^2 K_1 K_2 (I_1 + I_2 + I_3) = 0 \quad (10.4)$$

This also has three characteristic values, one of which, $\omega_1^2 = 0$.

Substituting the characteristic values of ω^2 in the equations of motion will yield modal relations between Θ_1, Θ_2, Θ_3. These relations will involve ratios. As examples, the first equation of motion will involve Θ_1 and Θ_2, and the third will involve Θ_2 and Θ_3. The unused equation is redundant, since it is impossible to solve for Θ_1, Θ_2, and Θ_3 explicitly. These relations are called the *characteristic vectors*, *modal vectors*, or *eigenvectors*. For each natural frequency, ω^2, each modal vector will have a unique shape, but arbitrary amplitudes. This agrees with our knowledge of Rayleigh's principle, which states that at a natural frequency, amplitudes are independent of the frequency.

To aid calculation of modal vectors, it is conventional to normalize these ratios, using $\Theta_1 = 1$ radian, and expressing all other amplitudes with respect to Θ_1,

$$\frac{\Theta_2}{\Theta_1} = \frac{K_1 - I_1\omega^2}{K_1}$$

$$\frac{\Theta_3}{\Theta_1} = \frac{K_2(K_1 - I_1\omega^2)}{K_1(K_2 - I_3\omega^2)} \quad (10.5)$$

The four-mass system of Figure 10.1c has four natural frequencies, again with $\omega_1^2 = 0$, but it is not the only four-mass torsional system that is possible. The drive train of a conventional automobile, if we include only the rear wheels, the differential, and engine (Figure 10.1d), is a four-mass torsional system. Its arrangement is completely unlike that of the in-line arrangement, due to the geared transmission at I_2, and it has a different frequency equation. This arrangement is called a *branched system,* because of the branching of the elastic system at one of the masses.

For more than four masses, frequency equations become more and more cumbersome. Additional branched spring-mass arrangements are possible, with each additional degree of freedom making the task even more forbidding. For a five-mass torsional system, there are three possible arrangements. For a six-mass torsional system there are four. The task to uniquely determine the characteristic values of the frequency equation becomes impossibly difficult. What is needed is a scheme of determining the characteristic values and mode shapes, without determining the frequency equations. This is particularly true since only one or two modes may be important, and it is senseless to determine all modes explicitly, if you need only one or two to solve a particular problem.

10.3. HOLZER'S METHOD

The method that is attributed to Holzer was actually developed through the work of many of the early investigators of torsional vibration problems. It is basically a trial and error scheme to find natural frequencies, but it is a logical scheme.

At a natural frequency, resonant amplitudes can be maintained without the application of an external force. This is one of the physical meanings of natural frequency. Also, the actual amplitudes are arbitrary. But, if one displacement Θ_1 is assigned a definite value, all the other displacements are uniquely determined. The essence of Holzer's method is to use some convenient value, such as $\Theta_1 = 1$ rad, arbitrarily, and relate all the other amplitudes to that value. It then is only necessary to find the frequencies for which the sum of the inertial forces or couples is zero. These frequencies would have to be the natural frequencies of the system.

In Figure 10.2, a series of torsional masses and torsional springs are in harmonic motion in a principal mode at a frequency ω. For each mass

Fig. 10.2

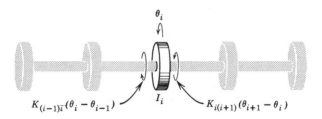

$$K_{(i-1)i}(\theta_i - \theta_{i-1}) \qquad I_i \qquad K_{i(i+1)}(\theta_{i+1} - \theta_i)$$

an equation of motion can be written in terms of the generalized coordinates $\theta_1, \theta_2, \theta_3, \ldots, \theta_i$ where

$$\theta_1 = \Theta_1 \sin \omega t$$

$$\theta_2 = \Theta_2 \sin \omega t$$

$$\cdots \cdots \cdots$$

$$\theta_i = \Theta_i \sin \omega t$$

$$-I_1\omega^2\Theta_1 + K_{12}(\Theta_1 - \Theta_2) = 0, \qquad (10.6)$$

$$-I_2\omega^2\Theta_2 + K_{12}(\Theta_2 - \Theta_1) + K_{23}(\Theta_2 - \Theta_3) = 0$$

$$-I_3\omega^2\Theta_3 + K_{23}(\Theta_3 - \Theta_2) + K_{34}(\Theta_3 - \Theta_4) = 0$$

$$\cdots \cdots \cdots \cdots \cdots \cdots \cdots \cdots \cdots \cdots \cdots$$

$$-I_i\omega^2\Theta_i + K_{(i-1)i}(\Theta_i - \Theta_{i-1}) = 0$$

The amplitude Θ_2 can be expressed from the first equation of motion, in terms of Θ_1,

$$K_{12}(\Theta_2 - \Theta_1) = -I_1 \omega^2 \Theta_1$$

$$\Theta_2 = \Theta_1 - \frac{I_1 \omega^2 \Theta_1}{K_{12}}$$

The amplitude Θ_3 can be expressed from the first and second equations of motion, in terms of Θ_2 and Θ_1,

$$\Theta_3 = \Theta_2 + \frac{K_{12}}{K_{23}}(\Theta_2 - \Theta_1) - \frac{I_2 \omega^2 \Theta_2}{K_{23}}$$

or,

$$\Theta_3 = \Theta_2 - \frac{I_2 \omega^2 \Theta_2}{K_{23}} - \frac{I_1 \omega^2 \Theta_1}{K_{23}}$$

The amplitude Θ_4 can be expressed using the first, second, and third equations, in terms of Θ_3, Θ_2, and Θ_1. It is evident that a series can be written for the amplitude of the nth mass, in terms of $n-1$ equations of motion and the amplitudes of $n-1$ coordinates.

Combining this with the equation of motion for the nth mass, since there is no external force or couple,

$$\sum_{i=1}^{n} I_i \omega^2 \Theta_i = 0 \tag{10.7}$$

This is another form of the frequency equation. The roots of this equation are the characteristic values or eigenvalues of the system.

In practice, the amplitudes and inertia forces or couples are tabulated, by machine or by hand calculation. The amplitudes and inertia forces or couples are used to determine successively the elastic displacement of one mass relative to the next. Beginning with the amplitude of the first mass as $\Theta_1 = 1$ rad, the amplitude of the second mass is found. Moving from one mass to the next, in succession, the amplitudes of all the masses are determined. Calling the sum of the inertial forces and couples $y(\omega^2)$.

$$\sum_{i=1}^{n} I_i \omega^2 \Theta_i = y(\omega^2) \tag{10.8}$$

and plotting $y(\omega^2)$ as a function of ω^2, the roots or eigenvalues can be easily found. Once the location of an approximate root is known, numerical techniques such as Simpson's rule can be used to find the characteristic value more exactly. Example Problem 10.1 shows such a plot and the roots.

10.4. KINETICALLY EQUIVALENT SYSTEMS

Before making a torsional vibration study of a real system, it is necessary to replace reciprocating pistons, machine linkages and parts with equivalent masses and springs that have equivalent kinetic properties. This requires judgment and experience, but there is a substantial history of judgment and experience from which to make these approximations.

Reciprocating engines have already been discussed. In general, one half the mass of the reciprocating parts times the square of the crank radius is added to the actual moment of inertia of rotating parts to obtain the equivalent moment of inertia of a piston and its crank.

$$I_e = I_c + \tfrac{1}{2}m_p r^2$$

where I_c is the moment of inertia of the crankshaft and rotating parts, and m_p is the mass of the reciprocating parts. This approximation is used extensively in seeking the natural frequencies and mode shapes for internal combustion engines.

For machine parts in plane motion, such as a connecting rod, part of the mass is considered to be in rotation and part in translation. To determine the fractions of the mass that are kinetically equivalent to the rod requires two conditions. One is that the mass center remain unchanged and the other is that the center of percussion be unchanged. Usually, you cannot satisfy both conditions simultaneously and a compromise must be made (refer to Section 3.7).

The elastic characteristics of a complex structure such as a crankshaft with webs, journals, varying diameters, keys, and so on, are extensively detailed in such massive works as the *Practical Solution of Torsional Vibration Problems* by W. Ker Wilson. It is impractical to make more than a few simple statements here. Except in the case of hollow shafts, the torsional stiffness of a crankshaft is approximately the same as a solid shaft of the same diameter and same length. This is affected by the number of journals and the stiffness of the crank webs.

In geared systems, the mass moment of inertia and torsional stiffness must include the effect of the increased inertia of the moving parts that have higher velocities. It is conventional to reduce a geared system to an effective system, considering to be moving at the lowest velocity. In such a system, it is easy to see that the effective mass moment of inertia would be the real mass moment of inertia increased by a factor proportional to the velocity squared. The kinetic energy of the effective system must be identical with the real system. In geared systems, if N is the gear ratio,

$$I_e = IN^2$$

An analogy can be made for the effective spring modulus,

$$K_e = KN^2$$

In this case, it would be the potential energy that must be unchanged in both the real and effective systems.

10.5. FORCED VIBRATION OF TORSIONAL SYSTEMS

The steady-state forced vibration of torsional systems can be handled with simplicity if damping is not present. Even if damping is present, damping can be neglected if the frequency of the forced vibration is away from one of the resonant frequencies such as $\pm 20\%$. Many cases of forced vibration consist simply of a single engine or motor driving a shaft with gearing, clutches, propellers or some other driven machinery. In such cases, each mass moves with the same harmonic motion of the forcing frequency of the driver. The constant resistance of a driven propellor or road traction does not constitute anything more than a constant displacement of the elastic member. It has the same effect as a static load. The forcing function comes only from the engine or motor.

If we consider two or more engines or motors in parallel or tandem, or if we consider a multimass system with more than one mass being forced, or if the driven resistance is variable, such as a reciprocating air compressor, we can superpose solutions, provided the system has no nonlinearities. There would probably be a phase difference in the motion of one mass relative to another, and it is more than likely that more than one frequency would be present, but these facts would evolve from the solution.

So many practical problems consist of a single exciting force acting on only one mass, that it is very useful to consider that simple problem. It is only necessary to modify equation 10.7 to include a forcing function, $M(\omega t) = M_i \sin \omega t$. For the particular mass being forced, the torque M_i would add to the inertial torque. For all other $M_i = 0$. It is possible for two or more masses to be forced with the same frequency and in phase, but it is highly unlikely. This is the reason that forcing functions are taken singly and solutions are superposed. It does make the solution for the torsional amplitude of forced vibration in an internal combustion engine rather complex.

$$\sum_{i=1}^{n} (I_i \omega^2 \Theta_i - M_i) = 0 \qquad (10.9)$$

Numerically, this equation can be solved for the individual amplitudes such as Θ_1, Θ_2, and Θ_3.

If the forcing frequency is at or near a resonant frequency, damping cannot be neglected. Damping is the only thing that limits amplitude from being infinite. The energy dissipated through damping must be equal to that injected by the forcing function over a complete cycle.

$$\Delta U_{in} = \Delta U_{out} \qquad (10.10)$$

$$\sum_{i=1}^{n} \pi \Theta_i M_i \sin \phi_i = \sum_{i=1}^{n} \pi c_i \Theta_i^2 \omega$$

Using the modal relations of the resonant mode, since the forcing frequency is at or near resonancy, equation 10.10 can be solved explicitly for the torsional amplitude of each mass. This is useful only near resonance ($\pm 20\%$). If the forcing function is not near resonance or if damping cannot be neglected, the Holzer method can still be used, but it becomes more involved. The forcing function at each mass is

$$M(t) = M_i' \cos \omega t + M_i \sin \omega t$$

and the amplitude of each mass is

$$\theta_i = \kappa_i \cos \omega t + \lambda_i \sin \psi t$$

Using the same procedure as before yields the simultaneous equations

$$\sum_{i=1}^{n} (I_i \omega^2 \lambda_i + c_i \omega \kappa_i - M_i) = 0$$

$$\sum_{i=1}^{n} (I_i \omega^2 \kappa_i - c_i \omega \lambda_i - M_i') = 0$$

(10.11)

These are a conjugate set of tabular calculations and are tedious even with the aid of a computer. Better methods of solution are available.

EXAMPLE PROBLEM 10.1

A four-cylinder gasoline engine is direct coupled to an electric generator. The actual system has been reduced to five rigid masses connected by sections of massless shafts. The effective moment of inertia of the rotating parts of each cylinder is 1 kg · m². The effective moment of inertia of the generator is 2 kg · m². The elastic modulus of the 40-mm crankshaft between each cylinder is 1.5×10^5 N · m/rad. The 50-mm diameter shaft, coupling the engine to the generator, has an effective elastic modulus of 2×10^6 N · m/rad.

Using Holzer's method, determine the frequency and mode shape of the first natural frequency in torsion.

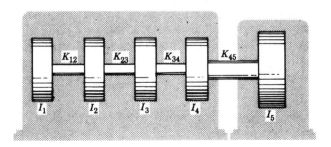

Solution:

(a) The natural frequency of torsional systems can be found in either of two ways using Holzer's method. The first is to seek each natural frequency by a series of trials, each using a frequency that is suspected of being closer to the actual natural frequency. The series is concluded when

$$\sum_{i=1}^{n} I_i \omega^2 \Theta_i = 0$$

In order to begin, we must have some idea of the order of magnitude of the first natural frequency. The five-mass system can be simplified in several ways, but one that is most logical would be to lump all of the mass of the engine in one place, creating a two-mass system. If we locate the effective mass of the four cylinders at the midpoint of the engine, we would have to make an adjustment of the elastic modulus.

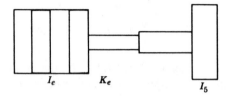

For the engine, the effective moment of inertia is the sum,

$$I_e = I_1 + I_2 + I_3 + I_4 = 4 \text{ kg} \cdot \text{m}^2$$

For the elastic modulus of the shaft,

$$\frac{1}{K_e} = \frac{2}{1.5 \times 10^5} + \frac{1}{1.5 \times 10^5} + \frac{1}{2 \times 10^5}$$

$$K_e = 0.4 \times 10^5 \text{ N} \cdot \text{m/rad}$$

Using, these values, the natural frequency of the simplified system is

$$\omega^2 = K_e \left[\frac{1}{I_e} + \frac{1}{I_5} \right] = 0.4 \times 10^5 \left[\frac{1}{4} + \frac{1}{2} \right] = .3 \times 10^4 \text{ s}^{-2}$$

Fortunately, this is a round number, but if it had not been, we would have rounded it to some convenient trial number.

A tabular calculation is a good way to organize calculations using Holzer's method. In the second column, the moment of inertia of each mass is entered, in succession. In the third column are the amplitudes. As we know, they are arbitrary, so Θ_1 is set at one radian. This is more than just a convenience, it also normalizes the modal vector. The product $I_i \omega^2 \Theta_i$ is the inertial torque contributed at each mass. The sum $\sum_{i=1}^{n} I_i \omega^2 \Theta_i$ is the total inertial torque at mass i, which will deflect the elastic shaft between mass i and mass $i+1$. In the first row, this entry would be the inertial torque causing the deflection of the crankshaft between cylinder 1 and 2. If we divide by the modulus of the crankshaft, we will have the relative angular displacement between Θ_1 and Θ_2. Subtracting $\Theta_1 - \Theta_2$ from Θ_1 gives the value of Θ_2. This is repeated for Θ_3, Θ_4, and Θ_5.

Try $\omega^2 = 3 \times 10^4$ s^{-2}:

i	I_i	Θ_i	$I_i\omega^2\Theta_i$	$\Sigma I_i\omega^2\Theta_i$	$K_{i(i+1)}$	$\Theta_i - \Theta_{i+1}$
1	1	1.0000	3.0000×10^4	3.0000×10^4	1.5×10^5	0.2000
2	1	0.8000	2.4000×10^4	5.4000×10^4	1.5×10^5	0.3600
3	1	0.4400	1.3200×10^4	6.7200×10^4	1.5×10^5	0.4480
4	1	-0.0080	-0.0240×10^4	6.6960×10^4	2×10^5	0.3348
5	2	-0.3428	-2.0568×10^4	4.6392×10^4	—	—

If $\omega^2 = 3 \times 10^4$ s^{-2} were a natural frequency, $\Sigma_{i=1}^5 I_i\omega^2\Theta_i = 0$, which is *not* the case. It would require an external torque of 4.6392×10^4 N $\cdot$ m to drive this five-mass system with the modal vector of column three, Θ_1 having an amplitude of 1 rad, in harmonic motion at a frequency of $\omega = 173.2$ s^{-1}. This frequency is less than the first natural frequency. If it were slightly above it, the sum $\Sigma_{i=1}^5 I_i\omega^2\Theta_i$ would have been negative.

Try $\omega^2 = 4 \times 10^4$ s^{-2}:

i	I_i	Θ_i	$I_i\omega^2\Theta_i$	$\Sigma I_i\omega^2\Theta_i$	$K_{i(i+1)}$	$\Theta_i - \Theta_{i+1}$
1	1	1.0000	4.0000×10^4	4.0000×10^4	1.5×10^5	0.2666
2	1	0.7333	2.9333×10^4	6.9333×10^4	1.5×10^5	0.4622
3	1	0.2711	1.0844×10^4	8.0177×10^4	1.5×10^5	0.5345
4	1	-0.2634	-1.0536×10^4	6.9641×10^4	2×10^5	0.3482
5	2	-0.6116	-4.8928×10^4	2.0713×10^4	—	

This is again too low, but it is closer to the first natural frequency, since the remainder of column 5 is less than it was when we assumed $\omega^2 = 3 \times 10^4$ s^{-2}.

Try $\omega^2 = 5 \times 10^4$ s^{-2}:

i	I_i	Θ_i	$I_i\omega^2\Theta_i$	$\Sigma I_i\omega^2\Theta_i$	$K_{i(i+1)}$	$\Theta_i - \Theta_{i+1}$
1	1	1.0000	5.0000×10^4	5.0000×10^4	1.5×10^5	0.3333
2	1	0.6667	3.3333×10^4	8.3333×10^4	1.5×10^5	0.5555
3	1	0.1111	0.5556×10^4	8.8889×10^4	1.5×10^5	0.5926
4	1	-0.4815	-2.4075×10^4	6.4814×10^4	2×10^5	0.3241
5	2	-0.8056	-8.0560×10^4	-1.5746×10^4	—	—

This trial is too high, $\Sigma_{i=1}^{5}\ I_i\omega^2\Theta_i = -1.5746 \times 10^4$ N · m/rad. A simple plot of the external torque, $y(\omega^2) = \Sigma_{i=1}^{5}\ I_i\omega^2\Theta_i$ versus ω^2 will permit interpolation of closer and closer approximations.

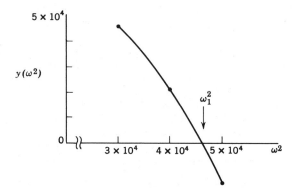

Successive trials will converge on the actual first natural frequency, $\omega_1^2 = 4.595 \times 10^4$ s^{-2}.

i	I_i	Θ_i	$I_i\omega^2\Theta_i$	$\Sigma I_i\omega^2\Theta_i$	$K_{i(i+1)}$	$\Theta_i - \Theta_{i+1}$
1	1	1.0000	4.5950×10^4	4.5950×10^4	1.5×10^5	0.3063
2	1	0.6937	3.1873×10^4	7.7824×10^4	1.5×10^5	0.5188
3	1	0.1748	0.8034×10^4	8.5858×10^4	1.5×10^5	0.5724
4	1	−0.3976	-1.8267×10^4	6.7591×10^4	2×10^5	0.3380
5	2	−0.7355	-6.7592×10^4	0	—	—

This is the first natural frequency, since $\Sigma_{i=1}^{5}\ I_i\omega^2\Theta_i = 0$.

Although this is a cumbersome way of finding natural frequencies, there are two dividends. One is the modal vector, column three, and the other is an evaluation of the shear moment in the elastic shaft, column five. Both are determined along with the natural frequency.

From the shear moment it is only a single step to determine the nominal shear stress in the shaft.

$$\tau = \frac{16M}{\pi d^3} = \frac{16(8.5857 \times 10^4)}{\pi (0.04)^3} = 6832 \text{ MN} \cdot \text{m}^{-2} \cdot \text{rad}^{-1}$$

For the first mode, the maximum shear stress occurs within the engine between cylinders three and four and would be 6832 MN/m^2 for 1-rad displacement at cylinder number 1. Obviously, the displacement cannot

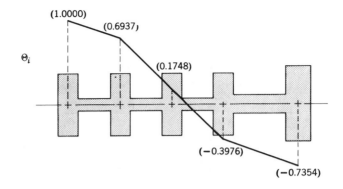

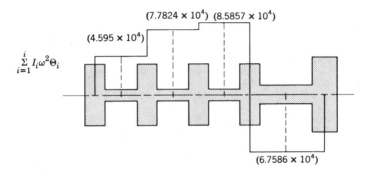

be 1 rad, but a torsiograph could be made of the actual displacement at cylinder one and the displacement measured. If it were only 0.001 rad at the natural frequency, this would indicate a nominal maximum shear stress of 6.832 MN/m^2 (991 lb/in.^{2}), which is entirely satisfactory for a steel crankshaft, even considering stress concentrations, such as oil holes, and so on.

(b) The second way to use Holzer's method is to set $\Sigma_{i=1}^5 I_i\omega^2\Theta_i = y(\omega^2)$ and plot the remainder as a function of ω^2. Since the equation is another form of the frequency equation, the roots will be the natural frequencies. Plotted in the following figure is $y(\omega^2)$ versus ω^2, showing the four natural frequencies. Note that the function passes through the origin, confirming that there is a degenerate natural frequency at $\omega^2 = 0$.

Comparing the modal vector and the value of $y(\omega^2)$, it is possible to locate whether a particular value of ω^2 is above or below a natural frequency. For example, in (a), for $\omega^2 = 3 \times 10^4$ s^{-2} and $\omega^2 = 4 \times 10^4$ s^{-2}, the value of $y(\omega^2)$ is positive. We knew that $\omega^2 = 5 \times 10^4$ s^{-2} was above ω_1^2 because $y(\omega^2)$ was negative and the modal vector had only one node, or one change in sign. If we had tried $\omega^2 = 50 \times 10^4$ s^{-2}, $y(\omega^2)$ would also be negative, but the modal vector would have had three nodes.

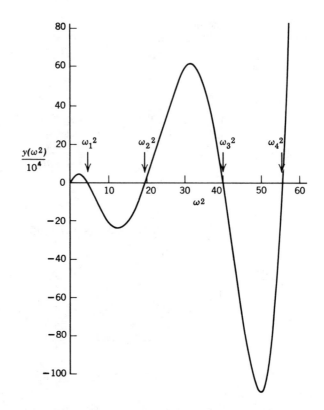

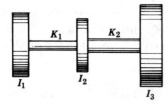

PROBLEM 10.2 Determine the two natural frequencies and the mode shapes for those frequencies, if

$$I_1 = 0.2 \text{ kg} \cdot \text{m}^2 \qquad K_1 = 100 \text{ N} \cdot \text{m/rad}$$
$$I_2 = 0.1 \text{ kg} \cdot \text{m}^2 \qquad K_2 = 200 \text{ N} \cdot \text{m/rad}$$
$$I_3 = 0.4 \text{ kg} \cdot \text{m}^2$$

Answers: $\omega_1^2 = 500 \text{ s}^{-2} \qquad \omega_2^2 = 3500 \text{ s}^{-2}$

$$\Theta_1^{(1)} = 1.00 \qquad \Theta_1^{(2)} = 1.00$$
$$\Theta_2^{(1)} = 0 \qquad \Theta_2^{(2)} = -6.00$$
$$\Theta_3^{(1)} = -0.50 \qquad \Theta_3^{(2)} = 1.00$$

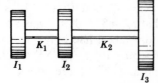

PROBLEM 10.3 Determine the two natural frequencies and the mode shapes for those frequencies, if

$$I_1 = 1 \text{ kg} \cdot \text{m}^2 \qquad K_1 = 10\,000 \text{ N} \cdot \text{m/rad}$$
$$I_2 = 1 \text{ kg} \cdot \text{m}^2 \qquad K_2 = 5000 \text{ N} \cdot \text{m/rad}$$
$$I_3 = 2 \text{ kg} \cdot \text{m}^2$$

Answers: $\omega_1^2 = 4313 \text{ s}^{-2}$ $\omega_2^2 = 23\ 187 \text{ s}^{-2}$

$\Theta_1^{(1)} = 1.0000$ $\Theta_1^{(2)} = 1.0000$

$\Theta_2^{(1)} = -0.5687$ $\Theta_2^{(2)} = -1.3187$

$\Theta_3^{(1)} = -0.7844$ $\Theta_3^{(2)} = 0.1594$

PROBLEM 10.4 Determine the two natural frequencies and the mode shapes for those frequencies, if

$I_1 = 1 \text{ kg} \cdot \text{m}^2$ $K_1 = 10\ 000 \text{ N} \cdot \text{m/rad}$
$I_2 = 2 \text{ kg} \cdot \text{m}^2$ $K_2 = 20\ 000 \text{ N} \cdot \text{m/rad}$
$I_3 = 10 \text{ kg} \cdot \text{m}^2$

PROBLEM 10.5 Repeat Problem 10.3, using Holzer's method.

PROBLEM 10.6 Repeat Problem 10.4, using Holzer's method.

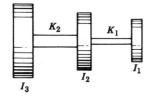

PROBLEM 10.7 A motor generator set consists of a six-cylinder, in-line diesel engine and generator. Each cylinder of the engine has an effective polar mass moment of inertia of 15 kg · m². The crankshaft between each cylinder has a torsional rigidity of 200×10^5 N · m/rad. The generator rotor has a polar moment of inertia of 50 kg · m², and the shaft between the generator and the engine has a torsional rigidity of 300×10^5 N · m/rad. Determine the natural frequency and the mode shape of the first mode.

PROBLEM 10.8 Determine the second natural frequency and mode shape for Problem 10.7.

PROBLEM 10.9 Setting $y(\omega^2) = \Sigma_{i=1}^{7} I_i \omega^2 \Theta_i$, plot $y(\omega^2)$ up to $\omega^2 = 10^6 \text{ s}^{-2}$ for Problem 10.7.

PROBLEM 10.10 If the normal crankshaft diameter is 100 mm, determine the shear stress per degree of amplitude at cylinder number 1.

Answer: 659.6 MN · m^{-2} · deg^{-1}

PROBLEM 10.11 A gas turbine rotor is arranged schematically here. Neglect the inertia of the shafts and determine the fundamental frequency and mode shape of the system in torsional vibration.

I (compressor rotor) $= 10 \text{ kg} \cdot \text{m}^2$

I (turbine rotor) $= 5 \text{ kg} \cdot \text{m}^2$

I (generator armature) $= 5 \text{ kg} \cdot \text{m}^2$

I (coupling) $= 1 \text{ kg} \cdot \text{m}^2$

$$K_1 = 2 \times 10^6 \text{ N} \cdot \text{m}/\text{rad}$$

$$K_2 = 10^6 \text{ N} \cdot \text{m}/\text{rad}$$

Answer: $\omega_1^2 = 0.9586 \times 10^5 \text{ s}^{-2}$

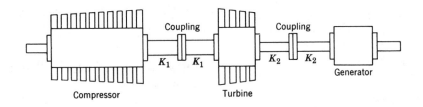

PROBLEM 10.12 Determine the second natural frequency and mode shape for the gas turbine system of Problem 10.11.

Answer: $\omega_2^2 = 3.866 \times 10^5 \text{ s}^{-2}$

PROBLEM 10.13 For the characteristic, plot the torsional moment as a function of axial position along the shaft. (For a uniform shaft, where is the maximum shear stress?)

PROBLEM 10.14 A shipboard diesel engine and pump unit consists of a six-cylinder, two-cycle marine engine rated at 240 BHP at 400 rpm, a fly-wheel, coupling, speed increaser and centrifugal pump. Using Holzer's method, deter-

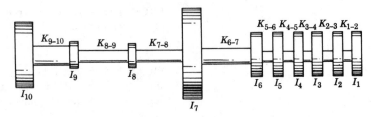

mine the first natural frequency and mode shape. The system constants have been corrected to the speed of the diesel engine.

$$\text{Cylinders, } I_{1-6} = 36 \text{ kg} \cdot \text{m}^2$$

$$\text{Fly-wheel, } I_7 = 80 \text{ kg} \cdot \text{m}^2$$

$$\text{Coupling, } I_8 = 9 \text{ kg} \cdot \text{m}^2$$

$$\text{Gears, } I_9 = 12 \text{ kg} \cdot \text{m}^2$$

$$\text{Pump, } I_{10} = 60 \text{ kg} \cdot \text{m}^2$$

$$K_{1-2 \text{ to } 5-6} = 1\ 850\ 000 \text{ N} \cdot \text{m/rad}$$

$$K_{6-7} = 3\ 000\ 000 \text{ N} \cdot \text{m/rad}$$

$$K_{7-8} = 2\ 500\ 000 \text{ N} \cdot \text{m/rad}$$

$$K_{8-9} = 2\ 650\ 000 \text{ N} \cdot \text{m/rad}$$

$$K_{9-10} = 4\ 000\ 000 \text{ N} \cdot \text{m/rad}$$

Answer: $f_1 = 12.04$ Hz

PROBLEM 10.15   Determine the second natural frequency and mode shape.

PROBLEM 10.16   For the system in Problem 10.14 determine the nominal maximum shear stress per degree amplitude for the first and second mode, if the crankshaft has a nominal diameter of 100 mm. Note that this will be in terms of a degree of amplitude at mass number 1.

Answer: $55.4 \text{ MN} \cdot \text{m}^{-2} \cdot \text{deg}^{-1}$
$-183.3 \text{ MN} \cdot \text{m}^{-2} \cdot \text{deg}^{-1}$

PROBLEM 10.17   A six-cylinder marine oil engine drive is shown. Determine the first natural frequency of the installation.

Effective Moments of Inertia
Propeller, $I_1 = 9.5 \text{ kg} \cdot \text{m}^2$
Slip coupling, $I_2 = 2.5 \text{ kg} \cdot \text{m}^2$
Each cylinder, $I_3 = 8.0 \text{ kg} \cdot \text{m}^2$
Air compressor, $I_4 = 6.0 \text{ kg} \cdot \text{m}^2$

Effective Shaft Moduli
Air compressor shaft = 600 000 N · m/rad
Crankshaft between cylinders

$$= 500\ 000 \text{ N} \cdot \text{m/rad}$$

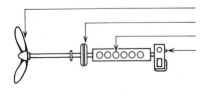

Slip coupling shaft $= 1\,000\,000$ N · m/rad
Propeller shaft $= 10\,000$ N · m/rad

Answer: $f_1 = 5.48$ Hz

PROBLEM 10.18 Determine the second natural frequency for Problem 10.17

Answer: $f_2 = 17.9$ Hz

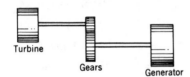

PROBLEM 10.19 A turbo-electric drive has the following characteristics:

I (turbine rotor)	$= 3500$ kg · m²
I (turbine pinion)	$= 50$ kg · m²
I (generator gear)	$= 3000$ kg · m²
I (generator armature)	$= 5000$ kg · m²

The elastic modulus of the turbine shaft is 1.2×10^6 N · m/rad. The elastic modulus of the generator shaft is 2×10^6 N · m/rad. The speed of the turbine is 5400 rpm. The speed of the generator is 1800 rpm. Neglect inertia of shafts and determine the natural frequencies corresponding to the one- and two-node modes of vibration.

Answer: $f_1 = 3.075$ Hz; $f_2 = 10.17$ Hz

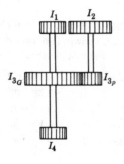

PROBLEM 10.20 Determine the lower natural frequency of the torsional system shown here. Shaft 2–3 is geared to run at three times the speed of shafts 3–4 and 1–3.

$$I_1 = 1.0 \text{ kg · m}^2$$
$$I_4 = 1.0 \text{ kg · m}^2$$
$$I_2 = 1.5 \text{ kg · m}^2$$
$$I \text{ pinion} = 0.5 \text{ kg · m}^2$$
$$I \text{ gear} = 3.0 \text{ kg · m}^2$$
$$K_{13} = 4000 \text{ N · m/rad}$$
$$K_{34} = 4000 \text{ N · m/rad}$$
$$K_{23} = 10\,000 \text{ N · m/rad}$$

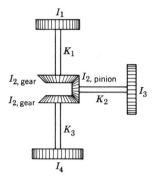

PROBLEM 10.21 Determine the lowest natural frequency of the torsional system shown here. I_3 runs at twice the speed of I_1 and I_4. I_1 and I_4 run at the same speed.

$$K_1 = K_2 = K_3 = 10^4 \text{ N} \cdot \text{m/rad}$$

$$I_3 = I_1 = 0.4 \text{ kg} \cdot \text{m}^2$$

$$I_2 \text{ (gear)} = 0.2 \text{ kg} \cdot \text{m}^2$$

$$I_2 \text{ (pinion)} = 0.025 \text{ kg} \cdot \text{m}^2$$

$$I_4 = 0.5 \text{ kg} \cdot \text{m}^2$$

Hint: As a first approximation, consider the gears locked, that is, a node at the gears.

Answer: $\omega_1^2 = 2.0804 \times 10^4 \text{ s}^{-2}$;

$$\Theta_1^{(1)} = 1.00 \qquad \Theta_2^{(1)} = 0.16$$

$$\Theta_3^{(1)} = 1.00 \qquad \Theta_4^{(1)} = -4.18$$

PROBLEM 10.22 Determine the second natural frequency of the torsional system of Problem 10.21.

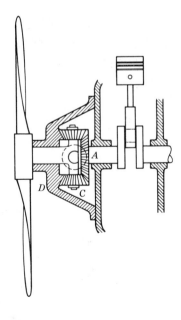

PROBLEM 10.23 An epicyclic transmission system for a radial aircraft engine consists of a bevel gear A secured to the engine crankshaft and four bevel planetary gears C, which are free to rotate on stub arms that are integral with the propeller shaft. The speed reduction between the crankshaft and propeller shaft is 2:1. The crankshaft bevel gear A and the fixed bevel gear D have the same number of teeth, which is three times the number for each planetary gear. Each planetary gear has a mass of 1 kg, a polar moment of inertia of 0.003 kg · m², and is located 100 mm from the centerline of the propeller shaft. The effective polar moment of inertia of the planetary stub arms is 0.03 kg · m². The effective moment of inertia of the crankshaft bevel gear is 0.02 kg · m². Determine the effective mass moment of inertia of the epicyclic gear system, referred to the speed of the propeller shaft.

Answer: 0.2395 kg · m²

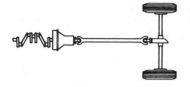

PROBLEM 10.24 Determine the first two natural frequencies and mode shapes for the automobile transmission system. The system constants are given. Note that the moment of inertia of the rear wheels is unusually large since they cannot vibrate independently of the body, unless the vehicle is supported off the ground.

$$\text{Engine} = 0.20 \text{ kg} \cdot \text{m}^2$$
$$\text{Gears} = 0.25 \text{ kg} \cdot \text{m}^2$$
$$\text{Wheels and body} = 10.0 \text{ kg} \cdot \text{m}^2$$
$$\text{Crank shaft} = 2 \times 10^5 \text{ N} \cdot \text{m/rad}$$
$$\text{Drive shaft} = 8 \times 10^2 \text{ N} \cdot \text{m/rad}$$

PROBLEM 10.25 Using Holzer's method, determine the fundamental frequency of the General Motors G-71 Quad arrangement.

Flywheel,	$I_1 = 3.3472 \text{ kg} \cdot \text{m}^2$
Cylinder 6,	$I_2 = 0.0776 \text{ kg} \cdot \text{m}^2$
Cylinders 2–5,	$I_3–I_6 = 0.0431 \text{ kg} \cdot \text{m}^2$
Cylinder 1,	$I_7 = 0.0756 \text{ kg} \cdot \text{m}^2$
Damper hub, disk and rubber,	$I_8 = 0.0442 \text{ kg} \cdot \text{m}^2$
Light damper,	$I_9 = 0.0442 \text{ kg} \cdot \text{m}^2$
Heavy damper,	$I_{10} = 0.771 \text{ kg} \cdot \text{m}^2$

$$K_1 = 2225 \text{ kN} \cdot \text{m/rad}$$
$$K_{2–6} = 1446 \text{ kN} \cdot \text{m/rad}$$
$$K_7 = 1175 \text{ kN} \cdot \text{m/rad}$$
$$K_8 = 746 \text{ kN} \cdot \text{m/rad}$$
$$K_9 = 565 \text{ kN} \cdot \text{m/rad}$$

Answer: $\omega_1^2 = 8.375 \times 10^5 \text{ s}^{-2}$

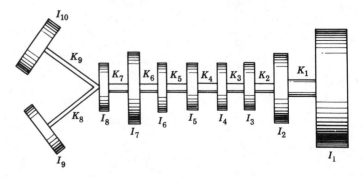

PROBLEM 10.26 For a harmonic torque of $100 \sin \omega t$, in N $\cdot$ m, applied to mass I_2 of Problem 10.4, and $\omega = 1080$ rpm, determine the forced amplitudes of vibration.

$$\textit{Answer:} \ \Theta_1 = -0.00815 \text{ rad;}$$
$$\Theta_2 = \ \ \ 0.00227 \text{ rad;}$$
$$\Theta_3 = -0.000422 \text{ rad}$$

PROBLEM 10.27 A harmonic couple $M = 10 \sin \omega t$, where $\omega^2 = 15{,}790 \text{ s}^{-2}$ or 20 rps and M is in N $\cdot$ m, is applied to I_2 of Problem 10.3. Determine the forced amplitude of each mass.

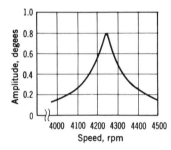

PROBLEM 10.28 At 1416 rpm the third harmonic of the torque applied at each cylinder is near resonance with the first mode of the engine in Problem 10.7. Determine the resonant amplitude for a third harmonic that has separately been determined to have a maximum value of 900 N $\cdot$ m. The resonance curve of the engine, as determined by a torsiograph, is shown. All cylinders are assumed to have the same turning effort and similar damping characteristics. Note that these displacements are not the total torsional displacement, since other harmonics will be present, but the other harmonics will be away from resonance.

Answer: $\Theta_1 = 0.51°$

CHAPTER ELEVEN

DISCRETE SYSTEMS: MATRIX METHODS

11.1. MATRIX EQUATIONS OF MOTION

There are other numerical methods of finding the characteristic values and mode shapes for a vibrating system with many degrees of freedom. One of the most useful is the numerical iteration of a square matrix derived from the equations of motion. There is one equation of motion for each degree of freedom, and if generalized coordinates are used, for each degree of freedom there is one generalized coordinate. This makes the matrix of the equations of motion have an equal number of rows and columns or *square*. The iteration technique relies on the behavior properties of a square matrix to converge on the characteristic vector through successive multiplications of the matrix by itself. Knowing the characteristic vector (or mode shape) also determines the characteristic values.

To show an example of the process of matrix iteration, let us consider a three-mass system. Such a system is shown in both torsional and rectilinear form in Figure 11.1. Mathematically, the two problems are identical.

Writing the three equations of motion,

$$m_1 \ddot{x}_1 = -k_{12}(x_1 - x_2) - k_1 x_1$$
$$m_2 \ddot{x}_2 = -k_{21}(x_2 - x_1) - k_{23}(x_2 - x_3) \tag{11.1}$$
$$m_3 \ddot{x}_3 = -k_{32}(x_3 - x_2)$$

For the torsional equations, substitute $\ddot{\theta}$ and θ for $\ddot{x}$ and x, and I for m.

Fig. 11.1

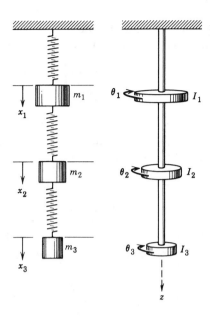

For harmonic motion in a principal mode, all parts of the system move in one mode with the same frequency.

$$x_1 = X_1 \sin \omega t$$

$$x_2 = X_2 \sin \omega t$$

$$x_3 = X_3 \sin \omega t$$

Substituting in the equations of motion and stating them in matrix form,

$$\begin{bmatrix} (k_{12}+k_1-m_1\omega^2) & -k_{12} & 0 \\ -k_{21} & (k_{21}+k_{23}-m_2\omega^2) & -k_{23} \\ 0 & -k_{32} & (k_{32}-m_3\omega^2) \end{bmatrix} \begin{Bmatrix} X_1 \\ X_2 \\ X_3 \end{Bmatrix} = \begin{Bmatrix} 0 \\ 0 \\ 0 \end{Bmatrix}$$

(11.2)

To avoid a trivial solution, this equation can be satisfied only if the determinant

$$\mathrm{Det}(\omega^2) = \begin{bmatrix} (k_{12}+k_1-m_1\omega^2) & -k_{12} & 0 \\ -k_{21} & (k_{21}+k_{23}-m_2\omega^2) & -k_{23} \\ 0 & -k_{32} & (k_{32}-m_3\omega^2) \end{bmatrix} = 0$$

(11.3)

Some will instantly recognize this determinant as an old friend, the frequency equation. The roots ω_1^2, ω_2^2, and ω_3^2 are the characteristic values of the determinant, that is, the values of ω^2 that will make the $\text{Det}(\omega^2) = 0$. Expanding the equations of motion, another matrix statement would be

$$\begin{bmatrix} +k_{12}+k_1 & -k_{12} & 0 \\ -k_{21} & k_{21}+k_{23} & -k_{23} \\ 0 & -k_{32} & k_{32} \end{bmatrix} \begin{Bmatrix} X_1 \\ X_2 \\ X_3 \end{Bmatrix} = \omega^2 \begin{bmatrix} m_1 & 0 & 0 \\ 0 & m_2 & 0 \\ 0 & 0 & m_3 \end{bmatrix} \begin{Bmatrix} X_1 \\ X_2 \\ X_3 \end{Bmatrix} \quad (11.4)$$

The first matrix is the stiffness matrix $[K]$. The second is the mass matrix $[M]$. The equation can be written as

$$[K]\{X\} = \omega^2[M]\{X\} \quad (11.5)$$

where each matrix is symbolized. The column matrices $\{X\}$ are symbolized by a different set of brackets from the square matrices $[M]$ and $[K]$.

Multiplying each side of the matrix equation 11.5 by the inverse matrix $[M^{-1}]$, which defines the matrix for which the product $[M^{-1}][M]$, is the identity matrix $[I]$, defined as

$$[I] = \begin{bmatrix} 1 & 0 & 0 \\ 0 & 1 & 0 \\ 0 & 0 & 1 \end{bmatrix} \quad (11.6)$$

the product becomes

$$[M^{-1}][K]\{X\} = \omega^2[M^{-1}][M]\{X\} = \omega^2\{X\}$$

Now, if we define a matrix $[A]$ as the product of multiplying the inverse matrix $[M^{-1}]$ and the stiffness matrix $[K]$,

$$[M^{-1}][K] = [A]$$

$$[A]\{X\} = \omega^2\{X\} \quad (11.7)$$

This is the eigenvalue or characteristic problem, which is satisfied for every characteristic vector $\{X^{(r)}\}$ with a corresponding natural frequency ω_r^2.

11.2. INFLUENCE COEFFICIENTS

The problem can also be solved using influence coefficients, adopted extensively in structural engineering.

Influence coefficients describe the displacement along a coordinate when loading is in the direction of that or any other coordinate. For the example of Figure 11.2, a_{11} is the displacement in the direction of x_1 for a unit load applied in the direction of x_1; a_{12} is the displacement in the direction of x_1 for a unit load applied in the direction of x_2. . . a_{ij} is the displacement

Fig. 11.2

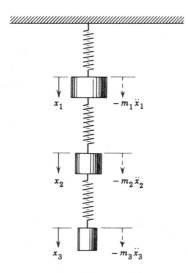

of coordinate x_i for a unit load applied in the direction of x_j. The total displacement at x_i can then be expressed as the sum of the displacements caused by all the applied loads. It is presumed that there are sufficient degrees of freedom to describe completely the total displacement at each discrete point by the applied forces or their component forces. Unit loads are used and influence coefficients are expressed as displacement per unit force (N/m, or rad/N · m).

$$x_i = a_{i1} F_1 + a_{i2} F_2 + a_{i3} F_3 + \cdots$$

For dynamic loads, the static forces are replaced by inertial forces. The result is a set of equations of motion that can also be expressed in matrix form.

$$x_1 = -a_{11} m_1 \ddot{x}_1 - a_{12} m_2 \ddot{x}_2 - a_{13} m_3 \ddot{x}_3$$
$$x_2 = -a_{21} m_1 \ddot{x}_1 - a_{22} m_2 \ddot{x}_2 - a_{23} m_3 \ddot{x}_3$$
$$x_3 = -a_{31} m_1 \ddot{x}_1 - a_{32} m_2 \ddot{x}_2 - a_{33} m_3 \ddot{x}_3$$

With a positive convention downward as shown, inertial forces will be upward. These are reflected in the negative signs. For harmonic motion in a principal mode, acceleration has the opposite sign to displacement and

$$X_1 = a_{11} m_1 \omega^2 X_1 + a_{12} m_2 \omega^2 X_2 + a_{13} m_3 \omega^2 X_3$$
$$X_2 = a_{21} m_1 \omega^2 X_1 + a_{22} m_2 \omega^2 X_2 + a_{23} m_3 \omega^2 X_3 \qquad (11.8)$$
$$X_3 = a_{31} m_1 \omega^2 X_1 + a_{32} m_2 \omega^2 X_2 + a_{33} m_3 \omega^2 X_3$$

Rearranging terms and, in particular, dividing by ω^2, the set of equations can be stated

$$\begin{bmatrix} a_{11}m_1 & a_{12}m_2 & a_{13}m_3 \\ a_{21}m_1 & a_{22}m_2 & a_{23}m_3 \\ a_{31}m_1 & a_{32}m_2 & a_{33}m_3 \end{bmatrix} \begin{Bmatrix} X_1 \\ X_2 \\ X_3 \end{Bmatrix} = \frac{1}{\omega^2} \begin{Bmatrix} X_1 \\ X_2 \\ X_3 \end{Bmatrix} \qquad (11.9)$$

These equations are of the form

$$[B]\{X\} = \frac{1}{\omega^2}\{X\} \qquad (11.10)$$

which is similar to equation 11.7, except for a matrix $[B]$ in place of $[A]$ and a modulus $1/\omega^2$ in place of ω^2. This is also a characteristic value problem, but this time it can be satisfied by a characteristic vector $\{X^{(r)}\}$ for a corresponding modulus $1/\omega_r^2$, which is the inverse of the natural frequency.

This is not surprising, since the equations of motion (11.9) can be defined still further

$$\begin{bmatrix} a_{11} & a_{12} & a_{13} \\ a_{21} & a_{22} & a_{23} \\ a_{31} & a_{32} & a_{33} \end{bmatrix} \begin{bmatrix} m_1 & 0 & 0 \\ 0 & m_2 & 0 \\ 0 & 0 & m_3 \end{bmatrix} \begin{Bmatrix} X_1 \\ X_2 \\ X_3 \end{Bmatrix} = \frac{1}{\omega^2} \begin{Bmatrix} X_1 \\ X_2 \\ X_3 \end{Bmatrix} \qquad (11.11)$$

or

$$[K^{-1}][M]\{X\} = \frac{1}{\omega^2}\{X\} \qquad (11.12)$$

The first matrix is the flexibility matrix, which is the inverse of the stiffness matrix, and the second is the mass matrix.

The academic preparation of most mechanical engineers enables them to write equations of motion without difficulty, but their knowledge of influence coefficients is short lived. The preparation of most civil or structural engineers includes a facility with influence coefficients and less preparation in dynamics.

Few recognize that equation 11.11 can be written directly from equation 11.4, which shows how interchangeable the two methods of iteration are. The inversion of matrices is a computational problem, and may be a minor problem or a major problem, depending on the matrix. The inverse matrix of a matrix $[A]$ is defined as the adjoint matrix $[A^+]$ divided by the value of the determinant of $[A]$. The adjoint matrix is a matrix of which the elements are the cofactors of the matrix $[A]$ with indices interchanged (in other words, the transpose of the matrix of cofactors).

$$\frac{[A^+]}{|A|} = [A^{-1}] \qquad (11.13)$$

11.3. MATRIX ITERATION

A numerical iteration process is used to determine the characteristic values and characteristic vectors associated with the eigenvalue problem.

For example, let us consider characteristic equation 11.10

$$[B]\{X\} = \frac{1}{\omega^2}\{X\}$$

By assuming an arbitrary vector $\{V_0\}$ in place of $\{X\}$ and multiplying $\{V_0\}$ by $[B]$, we will have a first approximation of the characteristic value $1/\omega^2$ and a first approximation of the characteristic vector $\{V_1\}$. If the arbitrary vector is not the characteristic vector, the new vector $\{V_1\}$ can be used as an improved trial vector. A sequence of linear transformations of successive trial vectors will lead to a vector which, when multiplied by $[B]$, will reproduce itself. Since the vector satisfies the characteristic equations, it must be a characteristic vector. The constant of proportionality will be the dominant characteristic value. In this case, it will be the inverse of the lowest natural frequency, ω_1^2.

If we were to begin with equation 11.7,

$$[A]\{X\} = \omega^2\{X\}$$

the iterative process will converge on the largest characteristic value of ω^2. The corresponding vector will be the characteristic vector for the highest natural frequency.

The explanation of this phenomenon requires a knowledge of matrix behavior, but briefly, the arbitrary vector $\{V_0\}$ can be written as a linear expression,

$$\{V_0\} = c_1\{X^{(1)}\} + c_2\{X^{(2)}\} + \cdots + c_r\{X^{(r)}\}$$

where $X^{(1)}$, $X^{(2)}$, . . . , $X^{(r)}$ are the characteristic modal vectors for the corresponding modes and c_1, c_2, . . . , c_r are constants, but unknown. Multiplying by the matrix $[B]$, yields the expression

$$[B]\{V_0\} = c_1[B]\{X^{(1)}\} + c_2[B]\{X^{(2)}\} + \cdots + c_r[B]\{X^{(r)}\}$$

Since each term satisfies the characteristic equation, the expression can be rewritten as

$$[B]\{V_0\} = c_1\left(\frac{1}{\omega_1^2}\right)\{X^{(1)}\} + c_2\left(\frac{1}{\omega_2^2}\right)\{X^{(2)}\} + \cdots + c_r\left(\frac{1}{\omega_r^2}\right)\{X^{(r)}\}$$

It is also a new vector $\{V_1\}$

$$\{V_1\} = c_1'\{X^{(1)}\} + c_2'\{X^{(2)}\} + \cdots + c_r'\{X^{(r)}\}$$

The constants are still unknown, but they are proportional to the constants for $\{V_0\}$.

If we multiply by $[B]$ again

$$[B]\{V_1\} = [B]^2\{V_0\} = c_1\left(\frac{1}{\omega_1^2}\right)^2\{X^{(1)}\} + c_2\left(\frac{1}{\omega_2^2}\right)^2\{X^{(2)}\} + \cdots + c_r\left(\frac{1}{\omega_r^2}\right)^2\{X^{(r)}\}$$

It is obvious that in successive multiplications, the largest modulus will dominate the summation. Since we are using the reciprocal $1/\omega^2$, the dominant modulus will be the reciprocal of the first natural frequency and the trial vector will resemble the first characteristic vector. After n successive iterations,

$$\lim\left\{[B]\{V_{n-1}^{(1)}\}\right\}_{n\to\infty} = \frac{1}{\omega_1^2}\{V_n^{(1)}\} \tag{11.14}$$

Each element of the resulting characteristic vector will have the same constant of proportionality with the corresponding element in the trial vector.

In practice, a finite number of iterations will yield a good estimate of the first mode, with the number of iterations depending on the desired accuracy. The real utility of the method can only be shown through examples.

11.4. ORTHOGONALITY OF PRINCIPAL MODES

A dominant characteristic modulus such as ω^2 or the reciprocal $1/\omega^2$, can be found through iteration of a square matrix. There still remains the problem of determining the characteristic values and vectors that do not correspond to the maximum or minimum natural frequencies. These are found by withholding convergence of the dominant modulus by using the *orthogonality of principal modes*.

The principle is simple to explain by referring to our original example. Designating the first principal mode corresponding to ω_1^2 by a superscript (1), and the second principal mode corresponding to ω_2^2 by a superscript (2), each principal mode must satisfy the equations of motion.

$$m_1 X_1^{(1)}\omega_1^2 = k_1 X_1^{(1)} + k_{12}(X_1^{(1)} - X_2^{(1)})$$
$$m_2 X_2^{(1)}\omega_1^2 = k_{21}(X_2^{(1)} - X_1^{(1)}) + k_{23}(X_2^{(1)} - X_3^{(1)})$$
$$m_3 X_3^{(1)}\omega_1^2 = k_{32}(X_3^{(1)} - X_2^{(1)})$$

and

$$m_1 X_1^{(2)}\omega_2^2 = k_1 X_1^{(2)} + k_{12}(X_1^{(2)} - X_2^{(2)})$$
$$m_2 X_2^{(2)}\omega_2^2 = k_{21}(X_2^{(2)} - X_1^{(2)}) + k_{23}(X_2^{(2)} - X_3^{(2)})$$
$$m_3 X_3^{(2)}\omega_2^2 = k_{32}(X_3^{(2)} - X_2^{(2)})$$

Multiplying the first set of equations by $X_1^{(2)}$, $X_2^{(2)}$, and $X_3^{(2)}$ and the second set of equations by $X_1^{(1)}$, $X_2^{(1)}$, and $X_3^{(1)}$, respectively, and subtracting one set from the other

$$(\omega_1^2 - \omega_2^2)\{m_1 X_1^{(1)} X_1^{(2)} + m_2 X_2^{(1)} X_2^{(2)} + m_3 X_3^{(1)} X_3^{(2)}\} = 0$$

If $\omega_1^2 \neq \omega_2^2$, this can only be true if

$$\sum_{i=1}^{n} m_i X_i^{(j)} X_i^{(k)} = 0 \qquad (11.15)$$

This equation is an expression of a principle known as the orthogonality of principal modes. It is derived directly from the equations of motion and represents a physical relation between principal modes. Care must be taken to see that the correct expression of the principle of orthogonality is used, for example, a cylinder rolling with a speed $\dot{x}$, the inertial term includes rotation and is not simply $m\ddot{x}$. The term orthogonality is a geometric interpretation referring to orthogonal vectors A and B, where

$$\mathbf{A \cdot B} = 0$$

In matrix form, the statement of the principle of orthogonality is

$$\{X^{(j)}\}^T [M]\{X^{(k)}\} = 0 \qquad (11.16)$$

where $j \neq k$.

$\{X^{(j)}\}^T$ is a row matrix, which is the transpose of the normalized column matrix $\{X^{(j)}\}$. If the mass matrix is symmetric, it follows that

$$\{X^{(k)}\}^T [M]\{X^{(j)}\} = 0 \qquad (11.17)$$

It can also be shown that the modal column matrices are orthogonal with respect to the stiffness matrix, provided that it is symmetric. Proofs of these statements are not difficult.

If we combine the principle of orthogonality with equation 11.7, the matrix equation will converge on the largest value of ω^2, which is orthogonal to the highest value of ω^2. Conversely, if we had used influence coefficients, and combined equation 11.10 with the principle of orthogonality, the matrix equation will converge on the largest value of $1/\omega^2$, which has characteristic vectors orthogonal to those of the lowest natural frequency. This would be the second lowest natural frequency and its corresponding characteristic vector.

Repeating this process, we can successively determine each natural frequency and characteristic vector starting with either the highest frequency or the lowest frequency.

11.5. PRINCIPAL COORDINATES

As we learned from two degrees of freedom, it is possible to describe the motion of a system with two degrees of freedom by using principal coordinates. The same is true for systems with more than two degrees of freedom, and the same definitions hold. If the motion of all parts of the system can be described by a single coordinate without reference to any other, that coordinate is a *principal coordinate*. A principal coordinate is also a *generalized coordinate*, which is the term given to a set of coordinates that recognize constraint. Principal coordinates are symbolized by the letter p, that is, $p_1, p_2, p_3, \ldots, p_n$. There are as many principal coordinates as there are degrees of freedom.

Each generalized coordinate is a linear combination of all principal coordinates, unless the generalized coordinate is itself a principal coordinate. For the sake of clarity, let us use q_1, q_2, q_3 as the generalized coordinates, instead of x_1, x_2, x_3. This will avoid any confusion between coordinates and the modal vectors.

$$
\begin{aligned}
x_1 &= q_1 = X_1^{(1)}p_1 + X_1^{(2)}p_2 + X_1^{(3)}p_3 \\
x_2 &= q_2 = X_2^{(1)}p_1 + X_2^{(2)}p_2 + X_2^{(3)}p_3 \\
x_3 &= q_3 = X_3^{(1)}p_1 + X_3^{(2)}p_2 + X_3^{(3)}p_3
\end{aligned}
\tag{11.18}
$$

This means that the first mode, represented by the principal coordinate p_1, is present in the generalized coordinates q_1, q_2, q_3 in proportion to the characteristic vector of the first mode. The second mode is also present in proportion to its characteristic vector, as is the third, fourth, and however many degrees of freedom there may be.

In matrix form, equations 11.18 are

$$
\begin{Bmatrix} q_1 \\ q_2 \\ q_3 \end{Bmatrix} =
\begin{bmatrix} X_1^{(1)} & X_1^{(2)} & X_1^{(3)} \\ X_2^{(1)} & X_2^{(2)} & X_2^{(3)} \\ X_3^{(1)} & X_3^{(2)} & X_3^{(3)} \end{bmatrix}
\begin{Bmatrix} p_1 \\ p_2 \\ p_3 \end{Bmatrix}
\tag{11.19}
$$

It is quickly apparent that the matrix is the modal matrix, with each column being the modal column for one of the natural modes of vibration. This fact is not too difficult to accept. One definition of a natural frequency is that all parts of the vibrating system pass through a minimum displacement at one time, and a maximum displacement at another time. These are exactly the same words that have been used to define principal coordinates. Principal modes and principal coordinates correspond, each coordinate being associated with a separate mode. The principal coordinate p_1 is associated with the natural frequency ω_1, p_2 is associated with the natural frequency ω_2, and so on.

To find p_1, p_2, p_3 in terms of the generalized coordinates only requires replacing q_1, q_2, q_3 in the proper modal column.

$$p_1 = \frac{\begin{bmatrix} q_1 & X_1^{(2)} & X_1^{(3)} \\ q_2 & X_2^{(2)} & X_2^{(3)} \\ q_3 & X_3^{(2)} & X_3^{(3)} \end{bmatrix}}{\begin{bmatrix} X_1^{(1)} & X_1^{(2)} & X_1^{(3)} \\ X_2^{(1)} & X_2^{(2)} & X_2^{(3)} \\ X_3^{(1)} & X_3^{(2)} & X_3^{(3)} \end{bmatrix}} = A_1 \cos \omega_1 t + B_1 \sin \omega_1 t \quad (11.20a)$$

$$p_2 = \frac{\begin{bmatrix} X_1^{(1)} & q_1 & X_1^{(3)} \\ X_2^{(1)} & q_2 & X_2^{(3)} \\ X_3^{(1)} & q_3 & X_3^{(3)} \end{bmatrix}}{\begin{bmatrix} X_1^{(1)} & X_1^{(2)} & X_1^{(3)} \\ X_2^{(1)} & X_2^{(2)} & X_2^{(3)} \\ X_3^{(1)} & X_3^{(2)} & X_3^{(3)} \end{bmatrix}} = A_2 \cos \omega_2 t + B_2 \sin \omega_2 t \quad (11.20b)$$

$$p_3 = \frac{\begin{bmatrix} X_1^{(1)} & X_1^{(2)} & q_1 \\ X_2^{(1)} & X_2^{(2)} & q_2 \\ X_3^{(1)} & X_3^{(2)} & q_3 \end{bmatrix}}{\begin{bmatrix} X_1^{(1)} & X_1^{(2)} & X_1^{(3)} \\ X_2^{(1)} & X_2^{(2)} & X_2^{(3)} \\ X_3^{(1)} & X_3^{(2)} & X_3^{(3)} \end{bmatrix}} = A_3 \cos \omega_3 t + B_3 \sin \omega_3 t \quad (11.20c)$$

With a statement of initial conditions, it is simply a matter of algebra to arrive at the equations of motion, in terms of the principal coordinates. To find the equations of motion in terms of generalized coordinates, we must go back to equations 11.18. To solve completely, we will need six initial conditions to solve for the six arbitrary constants, A_1, A_2, A_3, B_1, B_2, and B_3, if there are three degrees of freedom.

Two things should be quite clear at this point. One is that you cannot set initial conditions for generalized coordinates without solving for principal coordinates. The initial condition that $q_1(0) = 0$ at $t = 0$ means nothing. All possible modes may be present in the coordinate q_1. We cannot evaluate arbitrary constants for q_1 without simultaneously evaluating all the arbitrary constants. The second point is the observation of why we have placed so much emphasis on characteristic values and characteristic vectors. With many degrees of freedom, characteristic values and characteristic vectors are the keys to a general description of motion.

EXAMPLE PROBLEM 11.1

Determine the natural frequencies and mode shapes for the three-mass system using matrix iteration.

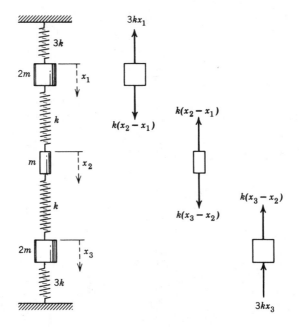

Solution:

In order to make more sense of our iteration process, let us first determine the exact values of the natural frequencies and mode shapes by finding the characteristic values of the frequency equation.

Using the free body diagrams, the equations of motion are

$$-3kx_1 - k(x_1 - x_2) = 2m\ddot{x}_1$$

$$-k(x_2 - x_1) - k(x_2 - x_3) = m\ddot{x}_2$$

$$-3kx_3 - k(x_3 - x_2) = 2m\ddot{x}_3$$

For harmonic motion in a principal mode, $x_1 = X_1 \sin \omega t$, $x_2 = X_2 \sin \omega t$, and $x_3 = X_3 \sin \omega t$. Placing the equations of motion in matrix form

$$\begin{bmatrix} (4k-2m\omega^2) & -k & 0 \\ -k & (2k-m\omega^2) & -k \\ 0 & -k & (4k-2m\omega^2) \end{bmatrix} \begin{Bmatrix} X_1 \\ X_2 \\ X_3 \end{Bmatrix} = 0$$

Setting the determinant equal to zero,

$$(4k-2m\omega^2)^2(2k-m\omega^2) - 2k^2(4k-2m\omega^2) = 0$$

from which the three natural frequencies are

$$\omega_1^2 = \frac{k}{m}$$

$$\omega_2^2 = \frac{2k}{m}$$

$$\omega_3^2 = \frac{3k}{m}$$

The corresponding mode shapes are, normalizing on X_1

for $\omega_1^2 = \dfrac{k}{m}$ $X_1^{(1)} = 1$ $X_2^{(1)} = 2$ $X_3^{(1)} = 1$

for $\omega_2^2 = \dfrac{2k}{m}$ $X_1^{(2)} = 1$ $X_2^{(2)} = 0$ $X_3^{(2)} = -1$

for $\omega_3^2 = \dfrac{3k}{m}$ $X_1^{(3)} = 1$ $X_2^{(3)} = -2$ $X_3^{(3)} = 1$

Matrix Iteration Using Influence Coefficients. After a little thought and deduction, the influence coefficients for this system can be found to be

$$a_{11} = \frac{7}{24k} \qquad a_{21} = \frac{1}{6k} \qquad a_{31} = \frac{1}{24k}$$

$$a_{12} = \frac{1}{6k} \qquad a_{22} = \frac{2}{3k} \qquad a_{32} = \frac{1}{6k}$$

$$a_{13} = \frac{1}{24k} \qquad a_{23} = \frac{1}{6k} \qquad a_{33} = \frac{7}{24k}$$

See p 330
definition
for a_{ij}

As a line of deduction, consider the static deflection at each mass, ignoring the other two masses, indeed, removing the other two from the system. Find a_{22} and then a_{11}, which is also a_{33}. Next find a_{12}, which is also a_{32}. Last, find a_{31}.

Using equation 11.9, and factoring the constants m/k,

$$\begin{bmatrix} \dfrac{7}{12}\dfrac{m}{k} & \dfrac{1}{6}\dfrac{m}{k} & \dfrac{1}{12}\dfrac{m}{k} \\[2mm] \dfrac{1}{3}\dfrac{m}{k} & \dfrac{2}{3}\dfrac{m}{k} & \dfrac{1}{3}\dfrac{m}{k} \\[2mm] \dfrac{1}{12}\dfrac{m}{k} & \dfrac{1}{6}\dfrac{m}{k} & \dfrac{7}{12}\dfrac{m}{k} \end{bmatrix} \begin{Bmatrix} X_1 \\ X_2 \\ X_3 \end{Bmatrix} = \dfrac{1}{\omega^2}\begin{Bmatrix} X_1 \\ X_2 \\ X_3 \end{Bmatrix}$$

the matrix equation is now ready for iteration.

$$\begin{bmatrix} \frac{7}{12} & \frac{1}{6} & \frac{1}{12} \\[1mm] \frac{1}{3} & \frac{2}{3} & \frac{1}{3} \\[1mm] \frac{1}{12} & \frac{1}{6} & \frac{7}{12} \end{bmatrix} \begin{Bmatrix} X_1 \\ X_2 \\ X_3 \end{Bmatrix} = \dfrac{k}{m\omega^2}\begin{Bmatrix} X_1 \\ X_2 \\ X_3 \end{Bmatrix}$$

As a trial vector, let us assume the arbitrary vector $X_1=1$, $X_2=1$, $X_3=1$

$$\begin{bmatrix} \frac{7}{12} & \frac{1}{6} & \frac{1}{12} \\[1mm] \frac{1}{3} & \frac{2}{3} & \frac{1}{3} \\[1mm] \frac{1}{12} & \frac{1}{6} & \frac{7}{12} \end{bmatrix} \begin{Bmatrix} 1 \\ 1 \\ 1 \end{Bmatrix} = \begin{Bmatrix} \frac{5}{6} \\ \frac{4}{3} \\ \frac{5}{6} \end{Bmatrix} = \frac{5}{6}\begin{Bmatrix} 1 \\ \frac{8}{5} \\ 1 \end{Bmatrix}$$

This states that an improved trial vector would be $X=1$, $X_2=\frac{8}{5}$, and $X_3=1$, and the first approximation of the modulus $k/m\omega^2=\frac{5}{6}$.

Using the improved trial vector and successive iterations,

$$\begin{bmatrix} \frac{7}{12} & \frac{1}{6} & \frac{1}{12} \\[1mm] \frac{1}{3} & \frac{2}{3} & \frac{1}{3} \\[1mm] \frac{1}{12} & \frac{1}{6} & \frac{7}{12} \end{bmatrix} \begin{Bmatrix} 1 \\ \frac{8}{5} \\ 1 \end{Bmatrix} = \begin{Bmatrix} \frac{14}{15} \\ \frac{26}{15} \\ \frac{14}{15} \end{Bmatrix} = \frac{14}{15}\begin{Bmatrix} 1 \\ \frac{13}{7} \\ 1 \end{Bmatrix}$$

$$\begin{bmatrix} \frac{7}{12} & \frac{1}{6} & \frac{1}{12} \\[1mm] \frac{1}{3} & \frac{2}{3} & \frac{1}{3} \\[1mm] \frac{1}{12} & \frac{1}{6} & \frac{7}{12} \end{bmatrix} \begin{Bmatrix} 1 \\ \frac{13}{7} \\ 1 \end{Bmatrix} = \begin{Bmatrix} \frac{41}{42} \\ \frac{80}{42} \\ \frac{41}{42} \end{Bmatrix} = \frac{41}{42}\begin{Bmatrix} 1 \\ \frac{80}{41} \\ 1 \end{Bmatrix}$$

$$\begin{bmatrix} \frac{7}{12} & \frac{1}{6} & \frac{1}{12} \\[1mm] \frac{1}{3} & \frac{2}{3} & \frac{1}{3} \\[1mm] \frac{1}{12} & \frac{1}{6} & \frac{1}{12} \end{bmatrix} \begin{Bmatrix} 1 \\ \frac{80}{41} \\ 1 \end{Bmatrix} = \begin{Bmatrix} \frac{122}{123} \\ \frac{242}{123} \\ \frac{122}{123} \end{Bmatrix} = \frac{122}{123}\begin{Bmatrix} 1 \\ \frac{121}{61} \\ 1 \end{Bmatrix}$$

$$\begin{bmatrix} \frac{7}{12} & \frac{1}{6} & \frac{1}{12} \\[1mm] \frac{1}{3} & \frac{2}{3} & \frac{1}{3} \\[1mm] \frac{1}{12} & \frac{1}{6} & \frac{7}{12} \end{bmatrix} \begin{Bmatrix} 1 \\ \frac{121}{61} \\ 1 \end{Bmatrix} = \begin{Bmatrix} \frac{365}{366} \\ \frac{728}{366} \\ \frac{365}{366} \end{Bmatrix} = \frac{365}{366}\begin{Bmatrix} 1 \\ \frac{728}{365} \\ 1 \end{Bmatrix}$$

At this point it should be quite clear that the modulus $k/m\omega^2$ is converging on 1.000, from which $\omega_1^2 = k/m$.

The trial vector is also converging on $X_1^{(1)} = 1$, $X_2^{(1)} = 2$, and $X_3^{(1)} = 1$. After five trials, the error is less than 0.3%.

Matrix Iteration Using Equations of Motion. Writing the equations of motion in the matrix form of equation 11.4

$$
\begin{bmatrix} 4k & -k & 0 \\ -k & 2k & -1 \\ 0 & -k & 4k \end{bmatrix} \begin{Bmatrix} X_1 \\ X_2 \\ X_3 \end{Bmatrix} = \omega^2 \begin{bmatrix} 2m & 0 & 0 \\ 0 & m & 0 \\ 0 & 0 & 2m \end{bmatrix} \begin{Bmatrix} X_1 \\ X_2 \\ X_3 \end{Bmatrix}
$$

The inverse of the mass matrix is

$$
[M^{-1}] = \frac{1}{4m} \begin{bmatrix} 2 & 0 & 0 \\ 0 & 4 & 0 \\ 0 & 0 & 2 \end{bmatrix}
$$

Multiplying both sides of the matrix equation of motion by $[M^{-1}]$,

$$
\frac{1}{4m} \begin{bmatrix} 2 & 0 & 0 \\ 0 & 4 & 0 \\ 0 & 0 & 2 \end{bmatrix} \begin{bmatrix} 4k & -k & 0 \\ -k & 2k & -k \\ 0 & -k & 4k \end{bmatrix} \begin{Bmatrix} X_1 \\ X_2 \\ X_3 \end{Bmatrix} = \omega^2 \begin{Bmatrix} X_1 \\ X_2 \\ X_3 \end{Bmatrix}
$$

Factoring the constants k/m and multiplying the matrices,

$$
\begin{bmatrix} 2 & -\tfrac{1}{2} & 0 \\ -1 & 2 & -1 \\ 0 & -\tfrac{1}{2} & 2 \end{bmatrix} \begin{Bmatrix} X_1 \\ X_2 \\ X_3 \end{Bmatrix} = \frac{m\omega^2}{k} \begin{Bmatrix} X_1 \\ X_2 \\ X_3 \end{Bmatrix}
$$

This matrix statement is now ready for iteration. As a trial vector, we could use the arbitrary vector $X_1 = 1$, $X_2 = 1$, and $X_3 = 1$, as we did earlier, but since this iterative process will converge on the highest natural frequency, let us use an arbitrary vector that has two changes in sign, $X_1 = 1$, $X_2 = -1$, $X_3 = 1$.

$$
\begin{bmatrix} 2 & -\tfrac{1}{2} & 0 \\ -1 & 2 & -1 \\ 0 & -\tfrac{1}{2} & 2 \end{bmatrix} \begin{Bmatrix} 1 \\ -1 \\ 1 \end{Bmatrix} = \begin{Bmatrix} \tfrac{5}{2} \\ -4 \\ \tfrac{5}{2} \end{Bmatrix} = \tfrac{5}{2} \begin{Bmatrix} 1 \\ -\tfrac{8}{5} \\ 1 \end{Bmatrix}
$$

Using the improved trial vector and successive iteration,

$$
\begin{bmatrix} 2 & -\frac{1}{2} & 0 \\ -1 & 2 & -1 \\ 0 & -\frac{1}{2} & 0 \end{bmatrix} \left\{ \begin{array}{c} 1 \\ -\frac{8}{5} \\ 1 \end{array} \right\} = \left\{ \begin{array}{c} \frac{14}{5} \\ -\frac{26}{5} \\ \frac{14}{5} \end{array} \right\} = \frac{14}{5} \left\{ \begin{array}{c} 1 \\ -\frac{13}{7} \\ 1 \end{array} \right\}
$$

$$
\begin{bmatrix} 2 & -\frac{1}{2} & 0 \\ -1 & 2 & -1 \\ 0 & -\frac{1}{2} & 0 \end{bmatrix} \left\{ \begin{array}{c} 1 \\ -\frac{13}{7} \\ 1 \end{array} \right\} = \left\{ \begin{array}{c} \frac{41}{14} \\ -\frac{40}{7} \\ \frac{41}{14} \end{array} \right\} = \frac{41}{14} \left\{ \begin{array}{c} 1 \\ -\frac{80}{41} \\ 1 \end{array} \right\}
$$

$$
\begin{bmatrix} 2 & -\frac{1}{2} & 0 \\ -1 & 2 & -1 \\ 0 & -\frac{1}{2} & 0 \end{bmatrix} \left\{ \begin{array}{c} 1 \\ -\frac{80}{41} \\ 1 \end{array} \right\} = \left\{ \begin{array}{c} \frac{122}{41} \\ -\frac{242}{41} \\ \frac{122}{41} \end{array} \right\} = \frac{122}{41} \left\{ \begin{array}{c} 1 \\ -\frac{121}{61} \\ 1 \end{array} \right\}
$$

$$
\begin{bmatrix} 2 & -\frac{1}{2} & 0 \\ -1 & 2 & -1 \\ 0 & -\frac{1}{2} & 0 \end{bmatrix} \left\{ \begin{array}{c} 1 \\ -\frac{121}{61} \\ 1 \end{array} \right\} = \left\{ \begin{array}{c} \frac{365}{122} \\ -\frac{728}{122} \\ \frac{365}{122} \end{array} \right\} = \frac{365}{122} \left\{ \begin{array}{c} 1 \\ -\frac{728}{365} \\ 1 \end{array} \right\}
$$

This time, $m\omega^2/k = \frac{365}{122}$ or $\omega_3^2 = 2.992(k/m)$, which is again very close to the true value $\omega_3^2 = 3(k/m)$.

EXAMPLE PROBLEM 11.2

Using the results of the previous problem, determine the displacements x_1, x_2, and x_3 as a function of time, if at time $t = 0$, the first mass is displaced a distance of 10 mm from rest and then released. The second and third masses are not displaced at time $t = 0$, but remain at rest until $t > 0$.

Solution:
The generalized coordinates are

$$
x_1 = q_1 = X_1^{(1)} p_1 + X_1^{(2)} p_2 + X_1^{(3)} p_3
$$
$$
x_2 = q_2 = X_2^{(1)} p_1 + X_2^{(2)} p_2 + X_2^{(3)} p_3
$$
$$
x_3 = q_3 = X_3^{(1)} p_1 + X_3^{(2)} p_2 + X_3^{(3)} p_3
$$

or, in matrix form,

$$
\begin{Bmatrix} x_1 \\ x_2 \\ x_3 \end{Bmatrix} = \begin{Bmatrix} q_1 \\ q_2 \\ q_3 \end{Bmatrix} = \begin{bmatrix} X_1^{(1)} & X_1^{(2)} & X_1^{(3)} \\ X_2^{(1)} & X_2^{(2)} & X_2^{(3)} \\ X_3^{(1)} & X_3^{(2)} & X_3^{(3)} \end{bmatrix} \begin{Bmatrix} p_1 \\ p_2 \\ p_3 \end{Bmatrix}
$$

where the modal matrix is

$$
\begin{bmatrix} X_1^{(1)} & X_1^{(2)} & X_1^{(3)} \\ X_2^{(1)} & X_2^{(2)} & X_2^{(3)} \\ X_3^{(1)} & X_3^{(2)} & X_3^{(3)} \end{bmatrix} = \begin{bmatrix} 1 & 1 & 1 \\ 2 & 0 & -2 \\ 1 & -1 & 1 \end{bmatrix}
$$

Each principal coordinate is a harmonic function of time

$$
p_1 = A_1 \cos \omega_1 t + B_1 \sin \omega_1 t
$$
$$
p_2 = A_2 \cos \omega_2 t + B_2 \sin \omega_2 t
$$
$$
p_3 = A_3 \cos \omega_3 t + B_3 \sin \omega_3 t
$$

For the initial conditions $x_1(0) = 10$, $x_2(0) = 0$, and $x_3(0) = 0$. Using equation 11.20

$$
A_1 = \dfrac{\begin{bmatrix} 10 & 1 & 1 \\ 0 & 0 & -2 \\ 0 & -1 & 1 \end{bmatrix}}{\begin{bmatrix} 1 & 1 & 1 \\ 2 & 0 & -2 \\ 1 & -1 & 1 \end{bmatrix}} = \dfrac{-20}{-8} = 2.5
$$

$$
A_2 = \dfrac{\begin{bmatrix} 1 & 10 & 1 \\ 2 & 0 & -2 \\ 1 & 0 & 1 \end{bmatrix}}{\begin{bmatrix} 1 & 1 & 1 \\ 2 & 0 & -2 \\ 1 & -1 & 1 \end{bmatrix}} = \dfrac{-40}{-8} = 5
$$

$$A_3 = \frac{\begin{bmatrix} 1 & 1 & 10 \\ 2 & 0 & 0 \\ 1 & -1 & 0 \end{bmatrix}}{\begin{bmatrix} 1 & 1 & 1 \\ 2 & 0 & 2 \\ 1 & -1 & 1 \end{bmatrix}} = \frac{-20}{-8} = 2.5$$

For the initial conditions $\dot{x}_1(0) = 0$, $\dot{x}_2(0) = 0$, $\dot{x}_3 = (0)$, $B_1 = B_2 = B_3 = 0$. For the natural frequencies,

$$\omega_1^2 = \frac{k}{m}$$

$$\omega_2^2 = 2\frac{k}{m}$$

$$\omega_3^2 = 3\frac{k}{m}$$

the principal coordinates are

$$p_1 = 2.5 \cos\sqrt{\frac{k}{m}}t$$

$$p_2 = 5 \cos\sqrt{2\frac{k}{m}}t$$

$$p_3 = 2.5 \cos\sqrt{3\frac{k}{m}}t$$

and the generalized coordinates x_1, x_2, and x_3 are

$$x_1 = 2.5 \cos\sqrt{\frac{k}{m}}t + 5 \cos\sqrt{2\frac{k}{m}}t + 2.5 \cos\sqrt{3\frac{k}{m}}t$$

$$x_2 = 5 \cos\sqrt{\frac{k}{m}}t - 5 \cos\sqrt{3\frac{k}{m}}t$$

$$x_3 = 2.5 \cos\sqrt{\frac{k}{m}}t - 5 \cos\sqrt{2\frac{k}{m}}t + 2.5 \cos\sqrt{3\frac{k}{m}}t$$

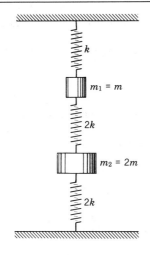

PROBLEM 11.3 Set up the matrix equations for the two-mass system of Problem 9.9 using influence coefficents. Determine the frequency and mode shape of the lowest natural frequency.

Answer: $\omega_1^2 = \dfrac{k}{m}$; $\chi_1 = 1$

PROBLEM 11.4 Set up the matrix equations for the two-mass system of Problem 9.9 using equations of motion. Determine the frequency and mode shape of the highest natural frequency.

Answer: $\omega_2^2 = \dfrac{4k}{m}$; $\chi_2 = -\dfrac{1}{2}$

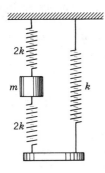

PROBLEM 11.5 Set up an iteration equation for the system using influence coefficients. Determine the frequency and mode shape of the lowest natural mode. Assume linear motion only (see also Problem 11.6).

Answer: $\omega_1^2 = 1.4385\dfrac{k}{m}$; $\chi_1 = 1.2808$

PROBLEM 11.6 Set up an iteration equation for the system of Problem 11.5 using the equations of motion. Determine the frequency and mode shape of the highest natural mode. Assume linear motion (see also Problem 11.5).

Answer: $\omega_2^2 = 5.562 \dfrac{k}{m}$; $\chi_2 = -0.7808$

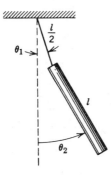

PROBLEM 11.7 The following matrix equation is the matrix equation of motion for Problem 9.18. Using matrix iteration, determine the highest natural frequency and mode shape.

$$\begin{bmatrix} 2 & 0 \\ -6 & 6 \end{bmatrix} \begin{Bmatrix} \Theta_1 \\ \Theta_2 \end{Bmatrix} = \frac{l\omega^2}{g} \begin{bmatrix} 1 & 1 \\ 0 & 1 \end{bmatrix} \begin{Bmatrix} \Theta_1 \\ \Theta_2 \end{Bmatrix}$$

Answer: $\omega_2^2 = 13.083 \dfrac{g}{l}$; $\Theta_1^{(2)} = 1.000$;

$\Theta_2^{(2)} = -0.8471$

PROBLEM 11.8 Inverting the matrices, determine the lowest natural frequency and mode shape for the system is Problem 11.7.

Answer: $\omega_1^2 = 0.917 \dfrac{g}{l}$; $\Theta_1^{(1)} = 1.000$; $\Theta_2^{(1)} = 1.1805$

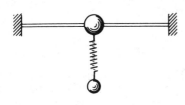

PROBLEM 11.9 The mass of an elastic beam, built in at each end, is negligible. A mass of 4 kg causes the beam to deflect 6 mm. A second mass of 2 kg is supported from the top by a spring with a constant of 4000 N/m. Estimate the lowest natural frequency of the system and mode shape, using the method of influence coefficients and matrix iteration (see also problem 11.12).

Answer: $\omega_1 = 29.46 \ s^{-1}$; $\chi_1 = 1.767$

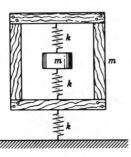

PROBLEM 11.10 The tension spring package of Problem 9.25 is repeated. The inner package is supported by tension springs from an outer case. The elasticity of the case is shown as a spring k. Using influence coefficients, set up a matrix equation for the natural frequencies of the system, and by matrix iteration determine its

lowest natural frequency and mode shape (see also Problem 11.11).

Answer: $\omega_1^2 = 0.4385\dfrac{k}{m}$; $X_1^{(1)} = 1.000$, $X_2^{(1)} = 1.281$

PROBLEM 11.11 Using equations of motion, determine the highest natural frequency and mode shape for the tension spring package of Problem 11.10.

Answer: $\omega_2^2 = 4.562\dfrac{k}{m}$; $X_1^{(2)} = 1.000$, $X_2^{(2)} = -0.781$

PROBLEM 11.12 Using the equations of motion, and matrix iteration, determine the highest natural frequency and mode shape for the beam, spring and two masses shown in Problem 11.9.

Answer: $\omega_2^2 = 3767 \ s^{-2}$; $X_1^{(2)} = 1.000$, $X_2^{(2)} = -1.132$

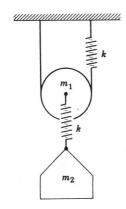

PROBLEM 11.13 A pulley of mass m is supported by a cable and a spring with a stiffness, k. A second and equal mass is supported by an identical spring from the pulley axle. Using equations of motion, determine the highest of the two natural frequencies and the corresponding mode shape for that frequency. The pulley is free to rotate about an axis through its geometric center (axle), and the cable does not slip. There is no horizontal motion of either mass (see Problem 11.16, also).

Answer: $\omega_2^2 = 3.59\dfrac{k}{m}$; $X_1^{(2)} = 1.000$; $X_2^{(2)} = -0.386$

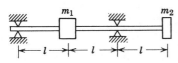

PROBLEM 11.14 Determine the first natural frequency and mode shape for the shaft rotating in bearings and carrying two concentrated masses m_1 and m_2, with $m_1 = 2m_2$ (see also Problem 11.17).

Answer: $\omega_1^2 = 0.868\dfrac{EI}{m_2 l^3}$; $X_1^{(1)} = 1.000$; $X_2^{(1)} = 3.277$

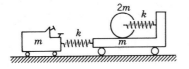

PROBLEM 11.15 A tractor and trailer system is used to haul large cylinders (paper, pipe, etc.) around an industrial plant. Using equations of motion, find the highest natural frequency and

mode shape of the system. Assume that the cyliner rolls without slipping on the trailer. The tractor and trailer each have a mass of 3000 kg, and the single large cylinder has a mass of 6000 kg. The springs k have an elasticity of 175 N/mm.

Answer: $\omega_3^2 = 1.743\dfrac{k}{m}$; $X_1^{(3)} = 1.000$; $X_2^{(3)} = 0.175$; $X_3^{(3)} = -1.350$

PROBLEM 11.16 Inverting the matrix equation of motion for Problem 11.13, determine the lowest of the two natural frequencies and the corresponding mode shape for that frequency. Again, there is no horizontal motion of either mass.

Answer: $\omega_1^2 = 0.741\dfrac{k}{m}$; $X_1^{(1)} = 1.000$; $X_2^{(1)} = 3.886$

PROBLEM 11.17 Inverting the matrix equation of Problem 11.14, obtained by using influence coefficients, determine the highest of the two natural frequencies and the corresponding mode shape for that frequency

Answer: $\omega_2^2 = \dfrac{5.532\ EI}{m_2 l^3}$; $X_1^{(2)} = 1.000$; $X_2^{(2)} = -0.6103$

PROBLEM 11.18 Show that the first and second modes of the following problems are orthogonal.

(a) Problem 10.2.
(b) Problem 10.3.
(c) Problems 11.5 and 11.6.
(d) Problems 11.10 and 11.11.
(e) Problems 11.13 and 11.16.
(f) Problems 11.14 and 11.17.

PROBLEM 11.19 Three railroad cars are coupled elastically in a three car train. If each car has an identical mass m, and each coupling has a stiffness k, what are the natural frequencies and mode shapes of free vibration?

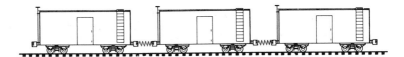

Answer:

$$\omega_1^2 = 0; \quad X_1^{(1)} = 1, \; X_2^{(1)} = \quad 1, \; X_3^{(1)} = \quad 1$$

$$\omega_2^2 = \frac{k}{m}; \quad X_1^{(2)} = 1, \; X_2^{(2)} = \quad 0, \; X_3^{(2)} = -1$$

$$\omega_3^2 = 3\frac{k}{m}; \quad X_1^{(3)} = 1, \; X_2^{(3)} = -2, \; X_3^{(3)} = \quad 1$$

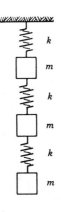

PROBLEM 11.20 Using matrix iteration and influence coefficients, determine the lowest natural frequency and mode shape of the three-mass system.

Answer: $\omega_1^2 = 0.198\dfrac{k}{m}$

$$X_1^{(1)} = 1.000, \; X_2^{(1)} = 1.802, \; X_3^{(1)} = 2.247$$

PROBLEM 11.21 Using matrix iteration and the equations of motion, determine the highest natural frequency and mode shape of Problem 11.20.

Answer: $\omega_3^2 = 3.247\dfrac{k}{m}$

$$X_1^{(3)} = 1.000, \; X_2^{(3)} = -1.247, \; X_2^{(3)} = 0.555$$

PROBLEM 11.22 Using matrix iteration and the principle of orthogonality of principal modes, determine the second lowest (or second highest) natural frequency, and the mode shape corresponding to that frequency, for the three-mass system of Problem 11.20.

Answer: $\omega_2^2 = 1.555\dfrac{k}{m}; \; X_1^{(2)} = 1.000, \; X_2^{(2)} = 0.445;$

$X_3^{(2)} = -0.802$

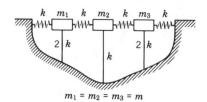

$m_1 = m_2 = m_3 = m$

PROBLEM 11.23 Using matrix iteration, determine the lowest natural frequency and its mode shape for the three section railroad bridge. Each section has a mass m and is supported on a column with a lateral stiffness of either k or $2k$, depending on the height of the column. Each bridge section is rigid and separated from the next section by an elastic member, which also has a stiffness. *Hint*: You may wish to start with the equations of motion for the bridge sections and invert the stiffness matrix to find the flexibility matrix. Note that there is symmetry.

Answer: $\omega_1^2 = 2\dfrac{k}{m}$; $X_1^{(1)} = 1$; $X_2^{(1)} = 2$; $X_3^{(1)} = 1$

PROBLEM 11.24 Using matrix iteration and the principle of the orthogonality of principal modes, determine the second highest natural frequency of the three section railroad bridge of Problem 11.23 and the mode shape corresponding to that frequency.

Answer: $\omega_2^2 = 4\dfrac{k}{m}$; $X_1^{(2)} = 1$; $X_2^{(2)} = 0$; $X_2^{(2)} = -1$

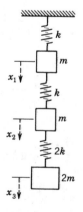

PROBLEM 11.25 Using matrix iteration and influence coefficients, determine the lowest natural frequency and mode shape of the three-mass system.

$$[M] = m \cdot \begin{bmatrix} 1 & 0 & 0 \\ 0 & 1 & 0 \\ 0 & 0 & 2 \end{bmatrix}$$

$$[K] = k \cdot \begin{bmatrix} 2 & -1 & 0 \\ -1 & 3 & -2 \\ 0 & -2 & 2 \end{bmatrix}$$

$$[K^{-1}] = \frac{1}{k} \cdot \begin{bmatrix} 1 & 1 & 1 \\ 1 & 2 & 2 \\ 1 & 2 & 2.5 \end{bmatrix}$$

Answer: $\omega_1^2 = 0.1392\dfrac{k}{m}$

$$X_1^{(1)} = 1.000$$

$$X_2^{(1)} = 1.861$$

$$X_3^{(1)} = 2.162$$

PROBLEM 11.26 Using matrix iteration and the equations of motion, determine the highest natural frequency, and the corresponding mode shape for that frequency, for the three-mass system of Problem 11.25

Answer: $\omega_3^2 = 4.115\dfrac{k}{m}; X_1^{(3)} = 1.000, X_2^{(3)} = -2.115,$ $X_3^{(3)} = 0.679$

PROBLEM 11.27 Using the principle of the orthogonality of principal modes and matrix iteration, determine the second lowest (or second highest) natural frequency, and the mode shape corresponding to that frequency, for the three-mass system of Problem 11.25.

Answer: $\omega_2^2 = 1.746\dfrac{k}{m}; X_1^{(2)} = 1.000; X_2^{(2)} = 0.2541;$ $X_3^{(2)} = -0.3407$

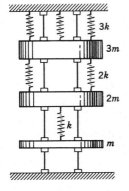

PROBLEM 11.28 Using matrix iteration, determine the lowest natural frequency and its corresponding mode shape for the three-mass system. Each mass moves without friction along the guide shafts.

Answer: $\omega_1^2 = 0.2991\dfrac{k}{m}; X_1^{(1)} = 1.0000,$ $X_2^{(1)} = 2.0513; X_3^{(1)} = 2.9268$

PROBLEM 11.29 Using the equations of motion and matrix iteration, determine the highest natural frequency and corresponding mode shape of Problem 11.28.

Answer: $\omega_3^2 = 2.563\dfrac{k}{m}$; $X_1^{(3)} = 1.000$

$X_2^{(3)} = -1.344$; $X_3^{(3)} = 0.859$

PROBLEM 11.30 Using matrix iteration and the principle of orthogonality of principal modes, determine the second lowest (or second highest) natural frequency, and the mode shape corresponding to that frequency, for the three-mass system of Problem 11.28.

Answer: $\omega_2^2 = 1.3042\dfrac{k}{m}$; $X_1^{(2)} = 1.000$; $X_2^{(2)} = 0.5436$;

$X_3^{(2)} = -1.787$

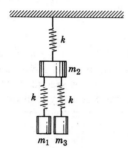

PROBLEM 11.31 Using influence coefficients and matrix iteration, determine the lowest natural frequency and mode shape for the three-mass system shown. All the springs have the same stiffness k, and $m_1 = m_3 = m$, $m_2 = 2m$, and in all modes, m_2 does not rotate.

Answer: $\omega_1^2 = 0.219\dfrac{k}{m}$; $X_1^{(1)} = 1.000$; $X_2^{(1)} = 0.781$;

$X_3^{(1)} = 1.000$

PROBLEM 11.32 Using matrix iteration and the principle of orthogonality of principal modes, determine the second lowest natural frequency, and the mode shape corresponding to that frequency, for the three-mass system of Problem 11.31.

Answer: $\omega_2^2 = \dfrac{k}{m}$; $X_1^{(2)} = 1$; $X_2^{(2)} = 0$; $X_3^{(2)} = -1$

PROBLEM 11.33 Using matrix iteration and the principle of the orthogonality of principal modes, determine the third lowest natural frequency and mode shape. In this case, it will be the highest natural frequency. There are two things to remember: it will be orthogonal to both the first and second modes (see Problems 11.31 and 11.32) and you cannot normalize on m_2. Why not?

Answer: $\omega_2^2 = 2.281\dfrac{k}{m}$; $X_1^{(3)} = 1$; $X_2^{(3)} = -1.281$;

$X_3^{(3)} = 1$

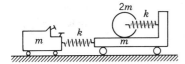

PROBLEM 11.34 The tractor and trailer system of Problem 11.15, used to haul large cylinders around an industrial plant, is repeated here. Beginning with the equations of motion and the mode shape for the highest natural frequency, found in Problem 11.15, and using matrix iteration and the principle of the orthogonality of principal modes, determine the second highest natural frequency and mode shape. In this case, the second highest will be the lowest natural frequency of any concern, since $\omega_1^2 = 0$. Remember to include both translation and rotation in any inertial matrix.

Answer: $\omega_2^2 = 0.459\dfrac{k}{m}$; $X_1^{(2)} = 1.000$; $X_2^{(2)} = -1.43$; $X_3^{(2)} = 1.85$

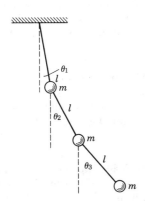

PROBLEM 11.35 Using matrix iteration, determine the lowest natural frequency and the corresponding mode shape for that frequency for the three-mass pendulum.

Answer: $\omega_1^2 = 0.4158\dfrac{g}{l}$; $\Theta_1^{(1)} = 1.0000$; $\Theta_2^{(1)} = 1.2921$; $\Theta_3^{(1)} = 1.6312$

PROBLEM 11.36 Using matrix iteration, determine the highest natural frequency and its corresponding mode shape for the three-mass pendulum of Problem 11.35.

Answer: $\omega_3^2 = 6.29\dfrac{g}{l}$, $\Theta_1^{(2)} = 1.000$, $\Theta_2^{(3)} = 1.645$, $\Theta_3^{(3)} = 0.767$

PROBLEM 11.37 For the three-mass pendulum of Problems 11.35 and 11.36, use the principle of the orthogonality of principal modes and determine the second natural frequency (not the highest) and mode shape. More on next page.

Hint: Think about the orthogonality of principal modes, or as an alternative, read the last paragraphs of p. 335.

$$\sum_{i=1}^{3} m_i \Theta_i^{(1)}\Theta_i^{(2)} \neq m_1\Theta_1^{(1)}\Theta_1^{(2)} + m_2\Theta_2^{(1)}\Theta_2^{(2)} + m_3\Theta_3^{(1)}\Theta_3^{(2)}$$

but why not? If you know why not, you'll have no difficulty with this problem.

Answer: $\omega_2^2 = 2.2934\dfrac{g}{l}$

$$\Theta_1^{(2)} = \quad 1.0000$$

$$\Theta_2^{(2)} = \quad 0.3528$$

$$\Theta_3^{(2)} = -2.3980$$

PROBLEM 11.38 Determine the principal coordinates of Problem 11.28, in terms of the generalized coordinates x_1, x_2, x_3. *Hint:* Use the answers of Problem 11.28 11.29, and 11.30.

PROBLEM 11.39 Determine the response for free vibration of each mass of Problem 11.28 if the bottom mass is statically deflected 10 mm and released from the rest position.

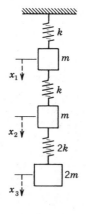

PROBLEM 11.40 The three mass, three spring mechanical system (Problem 11.25) has three natural frequencies. Refer to Problems 11.25 to 11.27 for these frequencies and their corresponding mode shapes. From a rest position, $X_1(0) = X_2(0) = \dot{X}_1(0) = \dot{X}_2(0) = \dot{X}_3(0) = 0$, the last mass is given a displacement of $X_3(0) = 10$ mm, downward from its equilibrium position. What are the equations of motion $x_1(t)$, $x_2(t)$, $x_3(t)$?

Answer:

$$x_1(t) = 3.13 \cos \sqrt{0.1392\frac{k}{m}\cdot t}$$

$$-5.25 \cos \sqrt{1.7459\frac{k}{m}\cdot t}$$

$$+2.12 \cos \sqrt{4.1149\frac{k}{m}\cdot t}$$

$$x_2(t) = 5.82 \cos \sqrt{0.1392 \frac{k}{m} \cdot t}$$

$$- 1.33 \cos \sqrt{1.7459 \frac{k}{m} \cdot t}$$

$$- 4.48 \cos \sqrt{4.1149 \frac{k}{m} \cdot t}$$

$$x_3(t) = 6.77 \cos \sqrt{0.1392 \frac{k}{m} \cdot t}$$

$$+ 1.79 \cos \sqrt{1.7459 \frac{k}{m} \cdot t}$$

$$+ 1.44 \cos \sqrt{4.1149 \frac{k}{m} \cdot t}$$

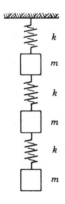

PROBLEM 11.41 The natural frequencies and mode shapes for Problems 11.20 to 11.22 are

$$\omega_1^2 = 1.198 \frac{k}{m}, \omega_2^2 = 1.555 \frac{k}{m}, \omega_3^2 = 3.247 \frac{k}{m},$$

The normal modes are

$$X_1^{(1)} = 1.000 \quad X_1^{(2)} = 1.000 \quad X_1^{(3)} = 1.000$$

$$X_2^{(1)} = 1.802 \quad X_2^{(2)} = 0.445 \quad X_2^{(3)} = -1.247$$

$$X_3^{(1)} = 2.247 \quad X_3^{(2)} = -0.802 \quad X_3^{(3)} = 0.555$$

All springs have a modulus of 4000 N/m and each mass is 5 kg. When the system is at rest, the bottom mass is displaced 0.1 m and released (100 mm). The other two masses are not permitted to be displaced until mass 3 is released. Express $x_1(t)$, $x_2(t)$, $x_3(t)$ when all three mass are released.

$$X_1(0) = 0 \qquad \dot{X}_1(0) = 0$$
$$X_2(0) = 0 \qquad \dot{X}_2(0) = 0$$
$$X_3(0) = 100 \qquad \dot{X}_3(0) = 0$$

Answer: in mm,

$$x_1(t) = 24.17 \cos 12.59\ t - 43.55 \cos 35.27\ t$$
$$+ 19.385 \cos 50.97\ t$$

$$x_2(t) = 43.55 \cos 12.59\ t - 19.382 \cos 35.27\ t$$
$$- 24.17 \cos 50.97\ t$$

$$x_3(t) = 54.31 \cos 12.59\ t + 34.93 \cos 35.27\ t$$
$$+ 10.76 \cos 50.97\ t$$

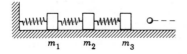

m_1 m_2 m_3

PROBLEM 11.42 A structure is modeled by the same three-mass system. All springs have a modulus of 4000 N/m, and each mass is 5 kg. With the system at rest, the end mass is now struck with a 0.5-kg ball, which imparts a velocity of 100 mm/s to the end mass and then rebounds away. Determine the equations of motion for each of the three masses. Be careful of signs!

Answer: in mm,

$$x_1(t) = -2.447 \sin 12.59\ t + 1.5736 \sin 35.27\ t$$
$$-0.4847 \sin 50.97\ t$$

$$x_2(t) = -4.409 \sin 12.59\ t + 0.700 \sin 35.27\ t$$
$$+0.6044 \sin 50.97\ t$$

$$x_3(t) = -5.498 \sin 12.59\ t - 0.790 \sin 35.27\ t$$
$$-0.269 \sin 50.97\ t$$

PROBLEM 11.43 The three car railroad train of Problem 11.19 is at rest when the first railroad car suddenly is given a velocity v. Determine the equation of motion for each of the three railroad cars.

Hint: The principal coordinate for m_1 will be $p_1 = Ct$. This is a rigid-body mode.

Answer: $x_1(t) = v\dfrac{t}{3} + \dfrac{v}{2\omega_2} \sin \omega_2 t + \dfrac{v}{6\omega_3} \sin \omega_3 t$

$$x_2(t) = v\dfrac{t}{3} - \dfrac{v}{3\omega_3} \sin \omega_3 t$$

$$x_3(t) = \dfrac{vt}{3} - \dfrac{v}{2\omega_2} \sin \omega_2 t + \dfrac{v}{6\omega_3} \sin \omega_3 t$$

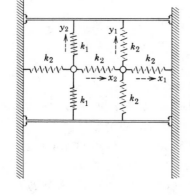

PROBLEM 11.44 Two identical masses, each of 2 kg, are suspended in a vertical frame as shown. The right-hand mass is displaced 20 mm to the right and 10 mm down and the left-hand mass is held in its original position. Simultaneously, both masses are released from rest. Describe the motion of both masses. Displacement of 10 and 20 mm can be considered to be small ($k_1 = 4$ kM/m, $k_2 = 2$ kN/m).

Answer: $x_1(t) = 10 \cos \omega_1 t + 10 \cos \omega_3 t$

$$y_1(t) = -10 \cos \omega_2 t$$

$$x_2(t) = 10 \cos \omega_1 t - 10 \cos \omega_3 t$$

$$y_2(t) = 0$$

PROBLEM 11.45 Using matrix iteration, determine the lowest natural frequency of the four section railroad bridge. Each section has a mass m and is supported on a column with a lateral stiffness of either k or $2k$ depending on the height of the column. Each bridge section is rigid and separated from the next section by an elastic member, which also has a stiffness k.

Answer: $\omega_1^2 = 1.586\dfrac{k}{m}$

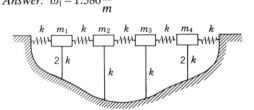

PROBLEM 11.46 A four story building can be modeled as a system of four masses supported on an elastic structure capable of deflecting only in shear. Estimate the lowest natural frequency and its corresponding mode shape.

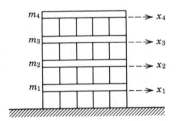

$m_1 = m_2 = m_3 = m_4 = 10^6$ kg

$k_{34} = 1 \times 10^8$ N/m

$k_{23} = 2 \times 10^8$ N/m

$k_{12} = 3 \times 10^8$ N/m

$k_1 = 4 \times 10^8$ N/m

Answer: $f_1 = 0.91$ Hz; 1, 2.22, 3.70, 5.46

PROBLEM 11.47 The four-mass system has the following constants. What is the frequency of the first mode, and what is the first mode shape? See figure on next page.

$$[K^{-1}] = \frac{1}{k}\begin{bmatrix} 1 & 1 & 1 & 1 \\ 1 & 2 & 2 & 2 \\ 1 & 2 & 3 & 3 \\ 1 & 2 & 3 & 4 \end{bmatrix}$$

$$[K] = k\begin{bmatrix} 2 & -1 & 0 & 0 \\ -1 & 2 & -1 & 0 \\ 0 & -1 & 2 & -1 \\ 0 & 0 & -1 & 1 \end{bmatrix}$$

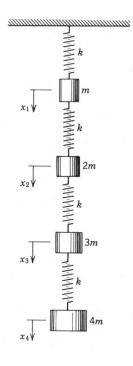

$$[M] = m \begin{bmatrix} 1 & 0 & 0 & 0 \\ 0 & 2 & 0 & 0 \\ 0 & 0 & 3 & 0 \\ 0 & 0 & 0 & 4 \end{bmatrix}$$

Answer: $\omega_1^2 = 0.038\dfrac{k}{m}$; 1, 1.962, 2.775, 3.272

PROBLEM 11.48 What is the frequency of the second mode, and what is the second mode shape for Problem 11.47?

Answer: $\omega_2^2 = 0.435\dfrac{k}{m}$; 1, 1.566, 0.766, -1.033

11.6. THE DUNKERLEY–SOUTHWELL EQUATIONS

An approximate method that bears some similarity to the techniques of matrix iteration is attributed to S. Dunkerley and R. V. Southwell. Empirically, S. Dunkerley developed a way to find the natural frequency of a flexural beam carrying concentrated masses by separately finding a series of individual natural frequencies for the beam carrying the individual concentrated masses one at a time. For this reason, it is related to the method of influence coefficients.

Going back to equation 11.8, another way of expressing the frequency equation in matrix form would be

$$\text{Det}(\omega^2) = \begin{vmatrix} \left(a_{11}m_1 - \dfrac{1}{\omega^2}\right) & a_{12}m_2 & a_{13}m_3 \\[2mm] a_{21}m_1 & \left(a_{22}m_2 - \dfrac{1}{\omega^2}\right) & a_{23}m_3 \\[2mm] a_{31}m_1 & a_{32}m_2 & \left(a_{33}m_3 - \dfrac{1}{\omega^2}\right) \end{vmatrix} = 0 \quad (11.21)$$

Expanding the matrix,

$$\begin{aligned}
\text{Det}(\omega^2) = {} & \omega^6(m_1 m_2 m_3) \\
& \times (a_{11}a_{22}a_{33} + 2a_{21}a_{13}a_{23} - a_{22}a_{13}^2 - a_{33}a_{21}^2 - a_{11}a_{23}^2) \\
& - \omega^4(m_1 m_2(a_{22}a_{11} - a_{21}^2) + m_1 m_3(a_{33}a_{11} - a_{13}^2) \\
& + m_2 m_3(a_{22}a_{33} - a_{23}^2)) + \omega^2(a_{11}m_1 + a_{22}m_2 + a_{33}m_3) - 1 = 0
\end{aligned}$$

This is the frequency equation with roots ω_1^2, ω_2^2, and ω_3^2 or

$$\text{Det}(\omega^2) = (\omega^2 - \omega_1^2)(\omega^2 - \omega_2^2)(\omega^2 - \omega_3^2) = 0 \quad (11.22)$$

Expanding this second equation,

$$\text{Det}(\omega^2) = \omega^6 - \omega^4(\omega_1^2 + \omega_2^2 + \omega_3^2) + \omega^2(\omega_1^2\omega_2^2 + \omega_2^2\omega_3^2 + \omega_1^2\omega_3^2) - \omega_1^2\omega_2^2\omega_3^2 = 0$$

Since these two equations are identical, Dunkerley reasoned that the coefficients of the ω^2, ω^4 and ω^6 terms were identical. In particular, for the ω^2 term,

$$\frac{1}{\omega_1^2} + \frac{1}{\omega_2^2} + \frac{1}{\omega_3^2} = a_{11}m_1 + a_{22}m_2 + a_{33}m_3 \quad (11.23)$$

The example could be expanded to include a larger number of masses.

For the instance where $\omega_3^2 \gg \omega_2^2 \gg \omega_1^2$, the two terms involving the higher frequencies can be neglected. The result is an approximate expression for ω_1^2, which will be low. This is Dunkerley's equation,

$$1/\omega_1^2 \approx a_{11}m_1 + a_{22}m_2 + a_{33}m_3 \quad (11.24)$$

in terms of $a_{11}m_1$, $a_{22}m_2$, $a_{33}m_3$, which are the natural frequencies of each mass acting separately. Its accuracy is very dependent on ω_1^2 being much lower than ω_2^2 or ω_3^2.

R. V. Southwell showed that there was a complementary equation using spring constants instead of influence coefficients. Starting with equation

11.2 and following a similar development

$$\omega_3^2 \approx \frac{1}{a_{11}m_1} + \frac{1}{a_{22}m_2} + \frac{1}{a_{33}m_3} \tag{11.25}$$

Dunkerley's equation is an approximate method of finding the lowest natural frequency, and Southwell's equation is an approximate method of finding the highest. Again, accuracy depends on the separation of the natural frequencies. These equations are of limited use, but they are simple, and do permit a quick approximation of the lowest and highest natural frequencies, with the same data.

11.7. FORCED VIBRATION OF SYSTEMS WITH MANY DEGREES OF FREEDOM

The vibration of systems with many degrees of freedom forced by harmonic forcing functions can be treated quite simply as an extension of our matrix methods. Consider forcing functions $F_1(t) = Q_1 e^{i\omega t}$, $F_2(t) = Q_2 e^{i\omega t}$, and $F_3(t) = Q_3 e^{i\omega t}$ being applied in the direction of the generalized coordinates q_1, q_2, and q_3. The symbols Q_1, Q_2 and Q_3 have been used to emphasize the subtle point that they are generalized forces in the direction of the generalized coordinates.

Each of the three equations of motion can be rewritten

$$m_1 \ddot{x}_1 = -k_{12}(x_1 - x_2) - k_1 x_1 + Q_1 e^{i\omega t}$$
$$m_2 \ddot{x}_2 = -k_{21}(x_2 - x_1) - k_{32}(x_2 - x_3) + Q_2 e^{i\omega t}$$
$$m_3 \ddot{x}_3 = -k_{32}(x_3 - x_2) + Q_3 e^{i\omega t}$$

For harmonic motion at a frequency ω, the equations of motion in matrix form become

$$\begin{bmatrix} (k_{12} + k_1 - m_1 \omega^2) & -k_{12} & 0 \\ -k_{21} & (k_{21} + k_{23} - m_2 \omega^2) & -k_{23} \\ 0 & -k_{32} & (k_{32} - m_3 \omega^2) \end{bmatrix} \begin{Bmatrix} X_1 \\ X_2 \\ X_3 \end{Bmatrix} = \begin{Bmatrix} Q_1 \\ Q_2 \\ Q_3 \end{Bmatrix} \tag{11.26}$$

This is the same set as equation 11.2, except that the set is equal to the force vector Q_1, Q_2, Q_3 and is not equal to zero.

This set of equations can be solved explicitly for X_1, X_2, and X_3. Using Cramer's rule and substituting the column vector of forces succussively in the first, second, and then the third column,

$$X_1 = \frac{\begin{bmatrix} Q_1 & -k_{12} & 0 \\ Q_2 & (k_{21}+k_{23}-m_2\omega^2) & -k_{23} \\ Q_3 & -k_{32} & (k_{32}-m_3\omega^2) \end{bmatrix}}{\begin{bmatrix} (k_{12}+k_1-m_1\omega^2) & -k_{12} & 0 \\ -k_{12} & (k_{21}+k_{23}-m_2\omega^2) & -k_{23} \\ 0 & -k_{32} & (k_{32}-m_3\omega^2) \end{bmatrix}}$$

$$= \frac{\begin{bmatrix} Q_1 & -k_{12} & 0 \\ Q_2 & (k_{21}+k_{23}-m_2\omega^2) & -k_{23} \\ Q_3 & -k_{32} & (k_{32}-m_3\omega^2) \end{bmatrix}}{\text{Det}(\omega^2)} \tag{11.27a}$$

$$X_2 = \frac{\begin{bmatrix} (k_{12}+k_1-m_1\omega^2) & Q_1 & 0 \\ -k_{21} & Q_2 & -k_{23} \\ 0 & Q_3 & (k_{32}-m_3\omega^2) \end{bmatrix}}{\begin{bmatrix} (k_{12}+k_1-m_1\omega^2) & -k_{12} & 0 \\ -k_{21} & (k_{21}+k_{23}-m_2\omega^2) & -k_{23} \\ 0 & -k_{32} & (k_{32}-m_3\omega^2) \end{bmatrix}}$$

$$= \frac{\begin{bmatrix} (k_{12}+k_1-m_1\omega^2) & Q_1 & 0 \\ -k_{21} & Q_2 & -k_{23} \\ 0 & Q_3 & (k_{32}-m_3\omega^2) \end{bmatrix}}{\text{Det}(\omega^2)} \tag{11.27b}$$

$$X_3 = \frac{\begin{bmatrix} (k_{12}+k_1-m_1\omega^2) & -k_{12} & Q_1 \\ -k_{21} & (k_{21}+k_{23}-m_2\omega^2) & Q_2 \\ 0 & -k_{32} & Q_3 \end{bmatrix}}{\begin{bmatrix} (k_{12}+k_1-m-_1\omega^2) & -k_{12} & 0 \\ -k_{23} & (k_{21}+k_{23}-m_2\omega^2) & -k_{23} \\ 0 & -k_{32} & (k_{32}-m_3\omega^2) \end{bmatrix}}$$

$$= \frac{\begin{bmatrix} (k_{12}+k_1-m_1\omega^2) & -k_{12} & Q_1 \\ -k_{21} & (k_{21}+k_{23}-m_2\omega^2) & Q_2 \\ 0 & -k_{32} & Q_3 \end{bmatrix}}{\text{Det}(\omega^2)} \tag{11.27c}$$

In each case, the denominator is the value of the frequency determinant at the forcing frequency ω^2, Det (ω^2).

One particular variation of this analysis is worth special mention. If only one force acts, and all the others are zero, the displacements X_1, X_2, and X_3 are the products of the force Q_1 and the respective cofactors of the frequency determinant, divided by the value of the frequency determinant.

$$X_1 = \frac{\begin{bmatrix} Q_1 & -k_{12} & 0 \\ 0 & (k_{21}+k_{23}-m_2\omega^2) & -k_{23} \\ 0 & -k_{32} & (k_{32}-m_3\omega^2) \end{bmatrix}}{\begin{bmatrix} (k_{12}+k_1-m_1\omega^2) & -k_{12} & 0 \\ -k_{21} & (k_{21}+k_{23}-m_2\omega_2) & -k_{23} \\ 0 & -k_{32} & (k_{32}-m_3\omega^2) \end{bmatrix}}$$

$$= \frac{\begin{bmatrix} (k_{12}+k_{23}-m_2\omega^2) & -k_{32} \\ -k_{32} & (k_{32}-m_3\omega^2) \end{bmatrix}}{\mathrm{Det}(\omega^2)} Q_1 = \alpha_{11}Q_1 \qquad (11.28)$$

The fraction is called the *receptance*, symbolized as α_{11}. For the displacement X_2

$$X_2 = \frac{\begin{bmatrix} (k_{12}+k_1-m_1\omega^2) & Q_1 & 0 \\ -k_{21} & 0 & -k_{23} \\ 0 & 0 & (k_{32}-m_3\omega^2) \end{bmatrix}}{\begin{bmatrix} (k_{12}+k_1-m_1\omega^2) & -k_{12} & 0 \\ -k_{21} & (k_{21}+k_{23}-m_2\omega^2) & -k_{23} \\ 0 & -k_{32} & (k_{32}-m_3\omega^2) \end{bmatrix}}$$

$$= -\frac{\begin{bmatrix} -k_{12} & -k_{32} \\ 0 & (k_{32}-m_3\omega^2) \end{bmatrix}}{\mathrm{Det}(\omega^2)} Q_1 = a_{21}Q_1 \qquad (11.29)$$

The parallel with influence coefficients is obvious.

$$X_1 = \alpha_{11}Q_1 + \alpha_{12}Q_2 + \alpha_{13}Q_3 \qquad (11.30)$$

$\alpha_{12}, \alpha_{13}, \ldots, \alpha_{21}, \alpha_{23}, \ldots$ are cross-receptances as differentiated from the *direct receptances* $\alpha_{11}, \alpha_{22}, \alpha_{33}$.

In some literature, much is made of receptance. It is effectively mechanical admittance, and it can be evaluated for a single mass, a spring or damper, or any combination of discrete elements. The frequency at which a receptance becomes infinite is a natural frequency of the system. The concept is simply another way of setting up equations of motion that some engineers find to be convenient for them.

PROBLEM 11.49 A cantilever beam deflects 2 mm under a load of 500 N applied at its tip and has an observed natural frequency of 10.2 Hz. If the beam is used to support machinery with a mass of 75 kg, approximate the new natural frequency.

Answer: 6.83 Hz

PROBLEM 11.50 A bridge trestle has a natural frequency of 6.1 Hz, as determined by test, and deflects 3 mm at midspan under a 10,000-kg vehicle. Approximate the natural frequency of the bridge and vehicle.

Answer: 5.07 Hz

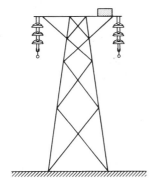

PROBLEM 11.51 A 20-m steel transmission tower shows a resonant frequency of 5 Hz when vibrated by an eccentric mass shaker placed at the cross arm. The mass of the shaker is 25 kg. With an additional weight of 30 kg at the cross arm, the resonant frequency is decreased to 4 Hz. What is the fundamental natural frequency of the steel transmission tower?

Answer: 6.86 Hz

PROBLEM 11.52 The first car of the three car railroad train of Problem 11.19 pulls the other two cars. If it is subjected to a cyclical driving force, $F(t) = 2 \times 10^6 e^{i30t}$, $m = 100,000$ kg and $k = 50 \times 10^6$ N/m, what is the maximum amplitude of motion of the third car?

Answer: 26.9 mm

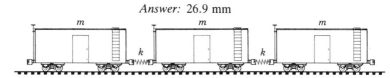

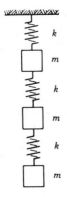

PROBLEM 11.53 For Problem 11.20, $k = 4000$ N/m and $m = 5$ kg. If the middle mass is forced by the harmonic function $F(t) = 10e^{i50t}$, determine the steady state amplitude of each of the three masses.

$$\omega_1^2 = 158.5 \text{ s}^{-2}; \quad \omega_2^2 = 1247 \text{ s}^{-2}; \quad \omega_3^2 = 2598 \text{ s}^{-2}$$

Answer: $X_1 = -9.5$ mm; $X_2 = 10.7$ mm; $X_3 = -5.0$ mm

PROBLEM 11.54 For Problem 11.53, what would the amplitude of each mass be, if the forcing function had been applied to the top mass?

PROBLEM 11.55 For Problem 11.25, $k = 2000$ N/m and $m = 2$ kg. If the middle mass is forced by the harmonic function $F(t) = 20\,e^{i20t}$, determine the steady state amplitude of each of the three masses.

Answer: $X_1 = 4.6$ mm; $X_2 = -7.4$ mm; $X_3 = 12.3$ mm

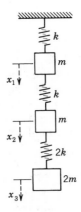

PROBLEM 11.56 For Problem 11.55, what would the amplitude of each mass be, if the forcing function had been applied to the bottom mass?

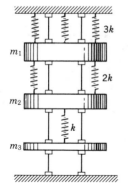

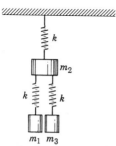

PROBLEM 11.57 For Problem 11.28, $k = 1000$ N/m and $m = 2$ kg. If mass 3 is forced by the harmonic function $F(t) = Fe^{i\omega t}$, where $\omega = 5$ Hz and $F = 10$ N, determine the steady state amplitude of m_1, m_2, and m_3.

$$m_1 = m, \ m_2 = 2m, \ m_3 = 3m$$

Answer: $X_1 = 5.05$ mm; $X_2 = 14.7$ mm; $X_3 = 3.83$ mm

PROBLEM 11.58 For Problem 11.31, assume that an oscillating force, $F(t) = 5 \sin \omega t$ is applied at one of the two smaller masses. For one value of ω^2, choosing $\omega^2 = 100$, 400, 900, 1600, 2500, or 3600 s^{-2}, determine the amplitudes of motion at m_1 and m_2, for $k = 1500$ N/m, $m = 1$ kg.

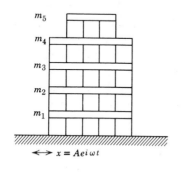

PROBLEM 11.59 A five story building can be visualized as a system of five masses supported on a structure capable of deflecting only in shear. For a horizontal displacement of 2 mm with a period of $\frac{1}{4}$ s, determine the maximum amplitude at each floor.

$k_5 = \ 5 \times 10^8$ N/m	$m_5 = 0.5 \times 10^6$ kg
$k_4 = \ 9 \times 10^8$ N/m	$m_4 = 0.6 \times 10^6$ kg
$k_3 = 14 \times 10^8$ N/m	$m_3 = 0.7 \times 10^6$ kg
$k_2 = 14 \times 10^8$ N/m	$m_2 = 0.8 \times 10^6$ kg
$k_1 = 18 \times 10^8$ N/m	$m_1 = 1.0 \times 10^6$ kg

Answer:

$X_1 = +3$ mm	$X_2 = -4.5$ mm
$X_3 = -1.5$ mm	$X_4 = +4$ mm
$X_5 = +11$ mm	

CHAPTER TWELVE

DISCRETE SYSTEMS: FINITE ELEMENTS

12.1. INTRODUCTION

Matrix iteration uses the behavioral properties of a square matrix and is an excellent technique for solving the characteristic value problem, but the technique does have limitations. For one, errors are cumulative. If there is a residual error in the first characteristic vector when iteration is stopped, that error will be entered into the matrix, if the orthogonality principle is used. The second characteristic vector will have an error inherited from the first characteristic vector. After two, three, or more successive iterations, the total error can become quite large. Of course, the use of a modern computer and a large number of significant figures can reduce errors in the higher modes. Second, although computational errors only slow convergence of the iteration process, errors in the original matrix will bar a correct solution, and these errors may be very difficult to detect. Finally, the size of the matrix increases as n^2, the square of the number of degrees of freedom. The problems listed here are quite simple, with two, three, and four degrees of freedom, requiring two by two, three by three, and four by four matrices. A thousand degrees of freedom is a large number, but some complex problems could require a thousand generalized coordinates to describe motion. This would require 10^6 storage terminals in a memory bank. Even for a large, modern computer, the technique of matrix iteration can be slow and cumbersome. As an alternate, a finite number of elements may be used to describe the system.

It will also limit the size of the matrices to the number of variables necessary to completely describe the motion of the system. This technique is called the *finite element* method. It can be used to solve many differential equations involving physics and engineering systems.

The fundamental concept of the finite element method is that a model of a system with many degrees of freedom can be constructed so that the characteristic quantities, such as force, displacement, velocity, and acceleration can be approximated piecewise over a finite number of elements. These elements transfer the value or state of these characteristic quantities from one point to another. These points are called *nodes*. This terminology is not to be interpreted as zero vibration, which is another use of the same word.

The discretized model consists of nodes and elements. It is constructed as follows:

1. The system is divided into a finite number of elements, which collectively approximate the physical description of the system.
2. These elements are connected at nodes.
3. The characteristic quantities of force, displacement, velocity, acceleration, and so on, are called the state variables. These describe the physical *state variables* of the system.
4. The value of these state variables are equated at the nodes. This provides the continuity required for these characteristic functions.

For example, in Figure 12.1a the axial force F_i causes an axial displacement x_i. The state variable $\{z_i\}$ is simply

$$\{z_i\} = \begin{Bmatrix} x_i \\ F_i \end{Bmatrix} \tag{12.1a}$$

Fig. 12.1

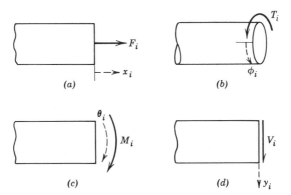

(a)   (b)   (c)   (d)

The axial displacement and axial force describe the state of displacement and force. In Figure 12.1b, the torque T_i and angular displacement ϕ_i describe the state of force and displacement. In Figure 12.1c, it is the bending moment M_i and angular displacement θ_i. In Figure 12.1d, it is the shear V_i and vertical displacement y_i.

$$\{z_i\} = \begin{Bmatrix} \phi_i \\ T_i \end{Bmatrix} \tag{12.1b}$$

$$\{z_i\} = \begin{Bmatrix} \theta_i \\ M_i \end{Bmatrix} \tag{12.1c}$$

$$\{z_i\} = \begin{Bmatrix} y_i \\ V_i \end{Bmatrix} \tag{12.1d}$$

Any combination of two, three, or four quantities can occur, requiring four, six, or eight unknowns in the state variables. For example, if shear and bending moment occur together

$$\{z_i\} = \begin{Bmatrix} y_i \\ \theta_i \\ M_i \\ V_i \end{Bmatrix} \tag{12.1e}$$

One more component needs to be defined, and that is the transfer matrix for an element. A transfer matrix transfers the state variables across an element, connecting one nodal point to another. This can be illustrated best by referring to examples from systems that we have already studied.

12.2. SPRING AND MASS SYSTEMS

In many vibration problems, particularly the problems with which we have been working, the physical systems are divided into elements that are masses and elements that are springs. All of the inertial properties are concentrated at the discrete masses, and all elasticity is concentrated at discrete but massless springs. The separation of these two properties results in two distinct and different transfer matrices, one for a point mass, and another for a spring.

The state variables are force and acceleration, which in harmonic motion is the second derivative of displacement. We can and will use force and displacement as the state variables, rather than using force and acceleration. The transfer matrix for the state variables transfers the state of the variables at node $i-1$ to the state of the variables at node i, where $i-1$ and i are the nodes on either side of the element of mass.

Fig. 12.2

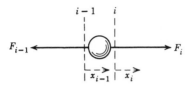

Referring to Figure 12.2, there are two equations that describe the displacement and force, one kinematic and one kinetic. Since the mass m is concentrated at a point, the displacement at station $i-1$ and station i are the same.

$$x_{i-1} = x_i$$

From the equation of motion,

$$\sum \mathbf{F}_i = m\ddot{x}_i$$

$$F_i - F_{i-1} = m\ddot{x}_i$$

For harmonic motion in a principal mode, $x_i = X_i e^{i\omega t}$ and these two equations can be restated as

$$X_i = X_{i-1}$$

and,

$$F_i = F_{i-1} - m\omega^2 X_i$$

In matrix form,

$$\begin{Bmatrix} X_i \\ F_i \end{Bmatrix} = \begin{bmatrix} 1 & 0 \\ -m\omega^2 & 1 \end{bmatrix} \begin{Bmatrix} X_{i-1} \\ F_{i-1} \end{Bmatrix}$$

$$\begin{Bmatrix} X_i \\ F_i \end{Bmatrix} = [M_{i(i-1)}] \begin{Bmatrix} X_{i-1} \\ F_{i-1} \end{Bmatrix}$$

(12.2)

The transfer matrix for a spring transfers the state variables at node $i-1$ to the state variables at node i across an elastic element. Since an elastic element deflects, where a mass does not, the displacement at node i is not the same as at node $i-1$.

Fig. 12.3

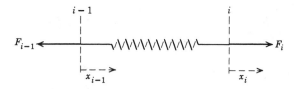

Referring to Figure 12.3, for a massless spring,

$$\sum \mathbf{F} = 0$$

$$F_i = F_{i-1} = k(X_i - X_{i-1})$$

again, stating these as two equations, for harmonic motion

$$\begin{Bmatrix} X_i \\ F_i \end{Bmatrix} = \begin{bmatrix} 1 & \dfrac{1}{k} \\ 0 & 1 \end{bmatrix} \begin{Bmatrix} X_{i-1} \\ F_{i-1} \end{Bmatrix}$$

$$\begin{Bmatrix} X_i \\ F_i \end{Bmatrix} = [F_{i(i-1)}] \begin{Bmatrix} X_{i-1} \\ F_{i-1} \end{Bmatrix}$$

(12.3)

One advantage of using transfer matrices is that once they are determined, they can be used over again. All massless springs have the same flexibility transfer matrix, except for the value of the modulus, and all concentrated masses have the same mass transfer matrix, except for the inertial properties.

State variables and transfer matrices can be used to advantage in large systems. A system can be modeled by a succession of springs and masses without increasing the number of terms in the matrix beyond those necessary to describe the state variable.

Since each generalized coordinate is accompanied by a generalized force, the computer storage necessary is limited to $2n$ terms. The state vector at one point is related to the state variable at another by multiplying successive matrices.

$$\{z_{i+1}\} = [A]\{z_i\}$$

$$\{z_i\} = [B]\{z_{i-1}\}$$

$$\{z_{i+1}\} = [A][B]\{z_{i-1}\}$$

There is no limit to the number of matrix multiplications that can be made, and theoretically, no limit to the number of degrees of freedom that can be handled. Each degree of freedom would require one additional multiplication step. Boundary conditions determine the value of the state vectors at each end of the chain, and with the boundary conditions established, we can determine equations for force and displacement. Remember, each row of the matrix equation is an equation of force or displacement. The equations of force determine the frequency equation; the equations of displacement will determine mode shape.

EXAMPLE PROBLEM 12.1

Solve for the natural frequency of a simple spring and mass system using state variables and transfer matrices.

Solution:

All of the inertial properties are contained in the suspended mass m. The transfer matrix is

$$[M_{2-1}] = \begin{bmatrix} 1 & 0 \\ -m\omega^2 & 1 \end{bmatrix}$$

All of the elastic properties are contained in the spring. The flexibility matrix is

$$[F_{1-0}] = \begin{bmatrix} 1 & \dfrac{1}{k} \\ 0 & 1 \end{bmatrix}$$

The state variable at node 1 can be found from the state variable at node 0

$$\{z_1\} = [F_{1-0}]\{z_0\}$$

and,

$$\{z_2\} = [M_{2-1}]\{z_1\} = [M_{2-1}][F_{1-0}]\{z_0\}$$

If we multiply the two matrices we can determine the force and displacement at station 2 directly from station 0

$$\begin{Bmatrix} X_2 \\ F_2 \end{Bmatrix} = \begin{bmatrix} 1 & 0 \\ -m\omega^2 & 1 \end{bmatrix} \begin{bmatrix} 1 & \dfrac{1}{k} \\ 0 & 1 \end{bmatrix} \begin{Bmatrix} X_0 \\ F_0 \end{Bmatrix}$$

$$\begin{Bmatrix} X_2 \\ F_2 \end{Bmatrix} = \begin{bmatrix} 1 & \dfrac{1}{k} \\ -m\omega^2 & \left(1 - \dfrac{m\omega^2}{k}\right) \end{bmatrix} \begin{Bmatrix} X_0 \\ F_0 \end{Bmatrix}$$

For free vibration, the boundary conditions $F_2 = 0$ and $X_0 = 0$ are established. These lead to two equations,

$$X_2 = \frac{F_0}{k}$$

and

$$0 = \left(1 - \frac{m\omega^2}{k}\right) F_0$$

The first is an expression for the displacement of the mass in terms of the force in the spring. The second is the frequency equation,

$$\omega^2 = \frac{k}{m}$$

EXAMPLE PROBLEM 12.2

Solve for the natural frequencies and mode shapes of a two-mass system consisting of two equal springs and two equal masses using state variables and transfer matrices.

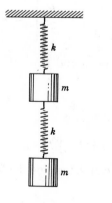
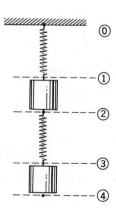

Solution:

This is the same problem as Figure 9.8, for which

$$\omega_1^2 = \frac{(3-\sqrt{5})}{2}\frac{k}{m} \quad \text{and} \quad \omega_2^2 = \frac{(3+\sqrt{5})}{2}\frac{k}{m}$$

For the two equal masses, the mass transfer matrices are

$$[M_{4-3}] = [M_{2-1}] = \begin{bmatrix} 1 & 0 \\ -m\omega^2 & 1 \end{bmatrix}$$

For the two elastic springs, the flexibility transfer matrices are

$$[F_{3-2}] = [F_{1-0}] = \begin{bmatrix} 1 & \dfrac{1}{k} \\ 0 & 1 \end{bmatrix}$$

The state variable at node 4 can be found from the state variable at station 0, if we follow the successive multiplication

$$\{z_1\} = [F_{1-0}]\{z_0\}$$

$$\{z_2\} = [M_{2-1}]\{z_1\}$$

$$\{z_3\} = [F_{3-2}]\{z_2\}$$

$$\{z_4\} = [M_{4-3}]\{z_3\}$$

or,

$$\{z_4\} = [M_{4-3}][F_{3-2}][M_{2-1}][F_{1-0}]\{z_0\}$$

To form this successive multiplication is a task that requires a technique. One such technique is to place the matrix that is to be multiplied at the top, and the multiplier at the left side of the matrix that is to be the product of the two. Thus,

$$[F_{1-0} \qquad]$$

$$[M_{2-1}][M_{2-1} \cdot F_{1-0}]$$

The intersection of columns in the matrix $[F_{2-1}]$ and rows in the matrix $[F_{1-0}]$ locate terms in the matrix that is the product of $[M_{2-1}][F_{1-0}]$. For example,

$$\begin{bmatrix} a_{11} & a_{12} \\ a_{21} & a_{22} \end{bmatrix}$$

$$\begin{bmatrix} b_{11} & b_{12} \\ b_{21} & b_{22} \end{bmatrix}\begin{bmatrix} c_{11} & c_{12} \\ c_{21} & c_{22} \end{bmatrix}$$

Here,

$$c_{11} = b_{11} \cdot a_{11} + b_{12} \cdot a_{21}$$
$$c_{12} = b_{11} \cdot a_{12} + b_{12} \cdot a_{22}$$
$$c_{21} = b_{21} \cdot a_{11} + b_{22} \cdot a_{21}$$
$$c_{22} = b_{21} \cdot a_{12} + b_{22} \cdot a_{22}$$

or,

$$
\begin{bmatrix} 1 & \dfrac{1}{k} \\ 0 & 1 \end{bmatrix}
$$

$$
\begin{bmatrix} 1 & 0 \\ -m\omega^2 & 1 \end{bmatrix}
\begin{bmatrix} 1 & \dfrac{1}{k} \\ -m\omega^2 & 1-\dfrac{m\omega^2}{k} \end{bmatrix}
$$

and,

$$
\begin{bmatrix} 1 & \dfrac{1}{k} \\ -m\omega^2 & 1-\dfrac{m\omega^2}{k} \end{bmatrix}
\{z_0\} = \{z_2\}
$$

This scheme places the product matrix in position to be multiplied by the next matrix in the chain.

$$
\begin{bmatrix} 1 & \dfrac{1}{k} \\ 0 & 1 \end{bmatrix}
\{z_0\} = \{z_1\}
$$

$$
\begin{bmatrix} 1 & 0 \\ -m\omega^2 & 1 \end{bmatrix}
\begin{bmatrix} 1 & \dfrac{1}{k} \\ -m\omega^2 & 1-\dfrac{m\omega^2}{k} \end{bmatrix}
\{z_0\} = \{z_2\}
$$

$$[M_{2-1}][K_{1-0}]\{z_0\} = \{z_2\}$$

$$\begin{bmatrix} 1 & \dfrac{1}{k} \\ 0 & 1 \end{bmatrix} \begin{bmatrix} 1-\dfrac{m\omega^2}{k} & \dfrac{2}{k}-\dfrac{m\omega^2}{k^2} \\ -m\omega^2 & 1-\dfrac{m\omega^2}{k} \end{bmatrix} \{z_0\}=\{z_3\}$$

$$\begin{bmatrix} 1 & 0 \\ -m\omega^2 & 1 \end{bmatrix} \begin{bmatrix} 1-\dfrac{m\omega^2}{k} & \dfrac{2}{k}-\dfrac{m\omega^2}{k^2} \\ \dfrac{m^2\omega^4}{k}-2m\omega^2 & -m\omega^2\left(\dfrac{2}{k}-\dfrac{m\omega^2}{k^2}\right)+1-\dfrac{m\omega^2}{k} \end{bmatrix} \{z_0\}=\{z_4\}$$

Free Vibration. For free vibration, the boundary conditions are $X_0=0$ and $F_4=0$. That is,

$$\begin{bmatrix} 1-\dfrac{m\omega^2}{k} & \dfrac{2}{k}-\dfrac{m\omega^2}{k^2} \\ \dfrac{m^2\omega^4}{k}-2m\omega^2 & -m\omega^2\left(\dfrac{2}{k}-\dfrac{m\omega^2}{k^2}\right)+1-\dfrac{m\omega^2}{k} \end{bmatrix} \begin{Bmatrix} 0 \\ F_0 \end{Bmatrix}=\begin{Bmatrix} X_4 \\ 0 \end{Bmatrix}$$

This matrix statement is comprised of two equations,

$$X_4=\left(\frac{2}{k}-\frac{m\omega^2}{k^2}\right)F_0$$

and

$$0=F_0\left[-m\omega^2\left(\frac{2}{k}-\frac{m\omega^2}{k^2}\right)+1-\frac{m\omega^2}{k}\right]$$

The last is the frequency equation.

$$-m\omega^2\left(\frac{2}{k}-\frac{m\omega^2}{k^2}\right)+1-\frac{m\omega^2}{k}=0$$

$$\omega^4-3\frac{k}{m}\omega^2+\frac{k^2}{m^2}=0$$

from which,

$$\omega_1^2=\frac{(3-\sqrt{5})}{2}\frac{k}{m} \quad \text{and} \quad \omega_2^2=\frac{(3+\sqrt{5})}{2}\frac{k}{m}$$

Although this appears to be a cumbersome way to find the frequency equation, it can be shortened. For example, if we note that the first column is multiplied by the boundary condition $X_0=0$, the first column need not be evaluated. It does not enter the calculation.

Normal Modes. To establish normal modes, the amplitude at one of the masses must be normalized, setting $F_0/k = 1.000$. For $\omega_1^2 = [(3 - \sqrt{5})/2](k/m)$

$$\{z_1\} = \begin{bmatrix} 1 & \dfrac{1}{k} \\ 0 & 1 \end{bmatrix} \begin{Bmatrix} 0 \\ F_0 \end{Bmatrix} = \begin{Bmatrix} \dfrac{F_0}{k} \\ F_0 \end{Bmatrix} = \begin{Bmatrix} 1.000 \\ k \end{Bmatrix}$$

$$\{z_2\} = \begin{bmatrix} 1 & \dfrac{1}{k} \\ -m\omega^2 & 1 - \dfrac{m\omega^2}{k} \end{bmatrix} \begin{Bmatrix} 0 \\ F_0 \end{Bmatrix} = \begin{Bmatrix} \dfrac{F_0}{k} \\ \dfrac{(-1+\sqrt{5})F_0}{2} \end{Bmatrix} = \begin{Bmatrix} 1.000 \\ 0.618k \end{Bmatrix}$$

$$\{z_3\} = \begin{bmatrix} 1 - \dfrac{m\omega^2}{k} & \dfrac{2}{k} - \dfrac{m\omega^2}{k^2} \\ -m\omega^2 & 1 - \dfrac{m\omega^2}{k} \end{bmatrix} \begin{Bmatrix} 0 \\ F_0 \end{Bmatrix} = \begin{Bmatrix} \dfrac{(1+\sqrt{5})F_0}{2k} \\ \dfrac{(-1+\sqrt{5})F_0}{2} \end{Bmatrix} = \begin{Bmatrix} 1.618 \\ 0.618k \end{Bmatrix}$$

$$\{z_4\} = \begin{bmatrix} 1 - \dfrac{m\omega^2}{k} & \dfrac{2}{k} - \dfrac{m\omega^2}{k^2} \\ \dfrac{m^2\omega^4}{k} - 2 - m\omega^2 & -m\omega^2\left(\dfrac{2}{k} - \dfrac{m\omega^2}{k^2}\right) + 1 - \dfrac{m\omega^2}{k} \end{bmatrix} \begin{Bmatrix} 0 \\ F_0 \end{Bmatrix} = \begin{Bmatrix} \dfrac{(1+\sqrt{5})F_0}{2k} \\ 0 \end{Bmatrix} = \begin{Bmatrix} 1.618 \\ 0 \end{Bmatrix}$$

and for $\omega_2^2 = [(3 + \sqrt{5})/2]k/m$

$$\{z_1\} = \begin{bmatrix} 1 & \dfrac{1}{k} \\ 0 & 1 \end{bmatrix} \begin{Bmatrix} 0 \\ F_0 \end{Bmatrix} = \begin{Bmatrix} \dfrac{F_0}{k} \\ F_0 \end{Bmatrix} = \begin{Bmatrix} 1.000 \\ k \end{Bmatrix}$$

$$\{z_2\} = \begin{bmatrix} 1 & \dfrac{1}{k} \\ -m\omega^2 & 1 - \dfrac{m\omega^2}{k} \end{bmatrix} \begin{Bmatrix} 0 \\ F_0 \end{Bmatrix} = \begin{Bmatrix} \dfrac{F_0}{k} \\ \dfrac{(-1-\sqrt{5})F_0}{2} \end{Bmatrix} = \begin{Bmatrix} 1.000 \\ -1.618k \end{Bmatrix}$$

$$\{z_3\} = \begin{bmatrix} 1 - \dfrac{m\omega^2}{k} & \dfrac{2}{k} - \dfrac{m\omega^2}{k^2} \\ -m\omega^2 & 1 - \dfrac{m\omega^2}{k} \end{bmatrix} \begin{Bmatrix} 0 \\ F_0 \end{Bmatrix} = \begin{Bmatrix} \dfrac{(1-\sqrt{5})F_0}{2}\dfrac{}{k} \\ \dfrac{(-1-\sqrt{5})F_0}{2} \end{Bmatrix} = \begin{Bmatrix} -0.618 \\ -1.618k \end{Bmatrix}$$

$$\{z_4\} = \begin{bmatrix} 1 - \dfrac{m\omega^2}{k} & \dfrac{2}{k} - \dfrac{m\omega^2}{k^2} \\ \dfrac{m^2\omega^4}{k} - 2m\omega^2 & -m\omega^2\left(\dfrac{2}{k} - \dfrac{m\omega^2}{k^2}\right) + 1 - \dfrac{m\omega^2}{k} \end{bmatrix} \begin{Bmatrix} 0 \\ F_0 \end{Bmatrix} = \begin{Bmatrix} \dfrac{(1-\sqrt{5})F_0}{2}\dfrac{}{k} \\ 0 \end{Bmatrix} = \begin{Bmatrix} -0.618 \\ 0 \end{Bmatrix}$$

The normal modes can be picked from the state variables. The first mode is 1.000 and 1.618; the second mode is 1.000 and -0.618.

This appears to be much more work than it is. You should bear in mind that this technique is to be used with computers and numerical values. The method is not designed to determine natural frequencies in terms of abstract expressions for m and k. For anything over two or three degrees of freedom, the multiplication of matrices containing symbols for mass, elasticity, or displacement instead of numbers is an impossible task. The next example problem illustrates this.

EXAMPLE PROBLEM 12.3

Solve Example Problem 12.2 for $\omega^2 = 1000 \ s^{-2}$, if $k = 1000$ N/m and $m = 2$ kg.

Solution:
The most common use of state variables and transfer matrices to solve vibration problems is through numerical methods and using computers.

In the last problem, $k = 1000$ N/m. The flexibility transfer matrix is

$$\begin{bmatrix} 1 & \dfrac{1}{k} \\ 0 & 1 \end{bmatrix} = \begin{bmatrix} 1 & 0.001 \\ 0 & 1 \end{bmatrix}$$

The mass transfer matrix, for $m\omega^2 = 2(1000) = 2000$ N/m, is

$$\begin{bmatrix} 1 & 0 \\ -m\omega^2 & 1 \end{bmatrix} = \begin{bmatrix} 1 & 0 \\ -2000 & 1 \end{bmatrix}$$

Going through the same multiplication scheme as before,

$$\begin{bmatrix} 1 & 0.001 \\ 0 & 1 \end{bmatrix} \{z_0\} = \{z_1\}$$

$$\begin{bmatrix} 1 & 0 \\ -2000 & 1 \end{bmatrix} \begin{bmatrix} 1 & 0.001 \\ -2000 & -1 \end{bmatrix} \{z_0\} = \{z_2\}$$

$$\begin{bmatrix} 1 & 0.001 \\ 0 & 1 \end{bmatrix} \begin{bmatrix} -2 & 0 \\ -2000 & -1 \end{bmatrix} \{z_0\} = \{z_3\}$$

$$\begin{bmatrix} 1 & 0 \\ -2000 & 1 \end{bmatrix} \begin{bmatrix} -2 & 0 \\ -6000 & -1 \end{bmatrix} \{z_0\} = \{z_4\}$$

The end result is a matrix equation in which the state variable at node 4 is stated in terms of the state variable at node 0 and a matrix containing numbers, u_{11}, u_{12}, u_{21}, and u_{22}

$$\{z_4\} = \begin{bmatrix} u_{11} & u_{12} \\ u_{21} & u_{22} \end{bmatrix} \{z_0\}$$

or,

$$\begin{Bmatrix} X_4 \\ F_4 \end{Bmatrix} = \begin{bmatrix} u_{11} & u_{12} \\ u_{21} & u_{22} \end{bmatrix} \begin{Bmatrix} X_0 \\ F_0 \end{Bmatrix}$$

This is really two statements,

$$X_4 = u_{11}X_0 + u_{12}F_0$$
$$F_4 = u_{21}X_0 + u_{22}F_0$$

For boundary conditions, $X_0 = 0$ and $F_4 = 0$, u_{22} must be zero. Since it is not, $\omega^2 = 1000$ is not a natural frequency.

Suppose we had used $\omega^2 = [(3 - \sqrt{5})/2]k/m = 190.98$ s^{-2},

$$\begin{bmatrix} 1 & 0.001 \\ 0 & 1 \end{bmatrix} \{z_0\} = \{z_1\}$$

$$\begin{bmatrix} 1 & 0 \\ -381.97 & 1 \end{bmatrix} \begin{bmatrix} 1 & 0.001 \\ -381.97 & 0.618 \end{bmatrix} \{z_0\} = \{z_2\}$$

$$\begin{bmatrix} 1 & 0.001 \\ 0 & 1 \end{bmatrix} \begin{bmatrix} 0.6180 & 0.00162 \\ -381.97 & 0.618 \end{bmatrix} \{z_0\} = \{z_3\}$$

$$\begin{bmatrix} 1 & 0 \\ -381.97 & 1 \end{bmatrix} \begin{bmatrix} 0.6180 & 0.00162 \\ -236.1 & 0 \end{bmatrix} \{z_0\} = \{z_4\}$$

here $u_{22} = 0$ and $\omega^2 = 190.98$ s^{-2} is a natural frequency.

It follows logically to plot u_{22} as a function of ω^2. The roots of this function will be the natural frequencies.

This is quite reminiscent of Holzer's method, and indeed it is the same. The difference is that state variables and transfer matrices are adapted to the use of a computer. In effect, it is a modernized Holzer method.

To determine the state variable at any given node, simply normalize on the first mass, setting $F_0/k = 1.000$, or use $F_0 = 0.0005$ N. The state variable will all be in terms of 1 m of displacement at the first mass. Of

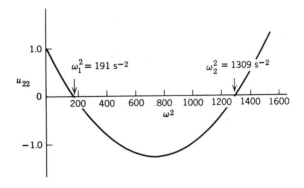

course, we could have normalized on the second mass, or any mass, if there were more than these two.

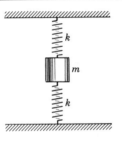

PROBLEM 12.4 Determine the natural frequency of the spring and mass system, using mass and flexibility transfer matrices.

Answer: $\omega_n^2 = 2k/m$

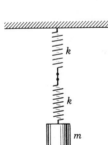

PROBLEM 12.5 Determine the natural frequency of the spring and mass system, using mass and flexibility transfer matrices.

Answer: $\omega_n^2 = k/2m$

PROBLEM 12.6 Repeat Problem 9.4, using mass and flexibility transfer matrices, and solve for the two natural frequencies and mode shapes of the system.

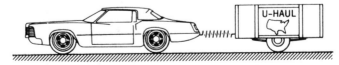

PROBLEM 12.7   Repeat Problem 9.9, using mass and flexibility transfer matrices, and solve for the two natural frequencies and mode shapes of the system.

PROBLEM 12.8   Repeat Problem 9.11, using mass and flexibility transfer matrices, and solve for the two natural frequencies and mode shapes of the system.

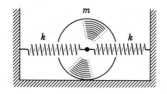

PROBLEM 12.9

(a) Determine the mass transfer matrix for a cylinder rolling without slipping on a horizontal plane.

(b) Use (a) to find the natural frequency of this system.

Answer: (a) $[M] = \begin{bmatrix} 1 & 0 \\ -\frac{3}{2}m\omega^2 & 1 \end{bmatrix}$

(b) $\omega_n^2 = \dfrac{4k}{3m}$

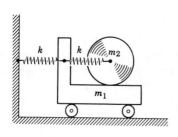

PROBLEM 12.10   Set up the mass and flexibility transfer matrices and solve for the two natural frequencies and mode shapes of the system ($m_2 = m$, $m_1 = 2m$). Note that you must first find the point transfer matrix for a rolling cylinder (refer to Problem 12.9).

Answer: $\omega_1^2 = \dfrac{2k}{7m}$; $\omega_2^2 = \dfrac{k}{m}$

PROBLEM 12.11   For Problem 11.20, $k = 4000$ N/m and $m = 5$ kg. Using mass and flexibility transfer matrices, determine the first natural frequency and the mode shape for that frequency.

Answer: $\omega_1^2 = 158.5$ s^{-2}; $X_1^{(1)} = 1.000$; $X_2^{(1)} = 1.802$; $X_3^{(1)} = 2.247$

PROBLEM 12.12   For Problem 12.11, using mass and flexibility transfer matrices, determine the second and the third natural frequencies and the mode shapes for those frequencies.

Answer:

$$\omega_2^2 = 1247 \text{ s}^{-2}; \ X_1^{(2)} = 1.000,$$
$$X_2^{(2)} = 0.455; \quad X_3^{(2)} = -0.802$$
$$\omega_3^2 = 2598 \text{ s}^{-2}; \ X_1^{(3)} = 1.000,$$
$$X_2^{(3)} = -1.247; \ X_3^{(3)} = 0.555$$

PROBLEM 12.13 For Problem 11.25, $k = 2000$ N/m and $m = 2$ kg. Using mass and flexibility transfer matrices, determine the first natural frequency and the mode shape for that frequency.

Answer: $\omega_1^2 = 139.2 \text{ s}^{-2}; \ X_1^{(1)} = 1.000; \ X_2^{(1)} = 1.861;$ $X_3^{(1)} = 2.162$

PROBLEM 12.14 For Problem 12.13, using mass and flexibility transfer matrices, determine the second and the third natural frequencies and the mode shapes for those frequencies.

Answer:

$$\omega_2^2 = 1746 \text{ s}^{-2}; \ X_1^{(2)} = 1.000,$$
$$X_2^{(2)} = 0.254; \quad X_3^{(2)} = -0.341$$
$$\omega_3^2 = 4114 \text{ s}^{-2}; \ X_1^{(3)} = 1.000,$$
$$X_2^{(3)} = -2.115; \ X_3^{(3)} = 0.679$$

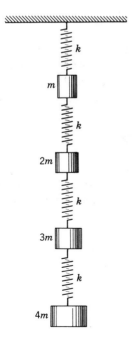

PROBLEM 12.15 For Problem 11.47, $k = 4000$ N/m and $m = 5$ kg. Using mass and flexibility transfer matrices, determine the lowest natural frequency and the mode shape for that frequency.

Answer: $\omega_1^2 = 30.38 \text{ s}^{-2}; \ X_1^{(1)} = 1.000; \ X_2^{(1)} = 1.962;$
$X_3^{(1)} = 2.775; \ X_4^{(1)} = 3.272$

PROBLEM 12.16   For Problem 12.15, determine the frequency of the second mode and the second mode shape.

Answer: $\omega_2^2 = 348 \text{ s}^{-2}$; $X_1^{(2)} = 1.000$; $X_2^{(2)} = 1.566$
$X_3^{(2)} = 0.766$; $X_4^{(2)} = -1.033$

PROBLEM 12.17   For Problem 11.46, using mass and flexibility transfer matrices, determine the lowest natural frequency and the mode shape for that frequency.

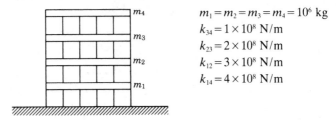

$m_1 = m_2 = m_3 = m_4 = 10^6 \text{ kg}$
$k_{34} = 1 \times 10^8 \text{ N/m}$
$k_{23} = 2 \times 10^8 \text{ N/m}$
$k_{12} = 3 \times 10^8 \text{ N/m}$
$k_{14} = 4 \times 10^8 \text{ N/m}$

PROBLEM 12.18   For Problem 12.17, determine the frequency of the second mode and the second mode shape.

12.3. FORCED VIBRATION AND EXTENDED MATRICES

An alternate means of handling forced vibrations, using state variable and transfer matrices, is to extend the matrices by one row and one column to accommodate applied forces. For example, the discrete mass of Figure 12.4 also carries the applied harmonic force $F(t) = Fe^{i\omega t}$. With the addition of this force, the equation of motion is

$$\sum F_i = m\ddot{x}_i$$
$$F_i - F_{i-1} + Fe^{i\omega t} = m\ddot{x}_i$$

This is the first of the two equations necessary to describe the state of the vectors of force and displacement at station i and station $i-1$. The

Fig. 12.4

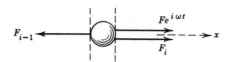

second is the kinematic statement that the displacements at station i and station $i-1$ are identical, the mass being concentrated at a point

$$X_i = X_{i-1}$$

These two equations lead to the matrix equation.

$$\begin{Bmatrix} X_i \\ F_i \end{Bmatrix} + \begin{Bmatrix} 0 \\ F \end{Bmatrix} = \begin{bmatrix} 1 & 0 \\ -m\omega^2 & 1 \end{bmatrix} \begin{Bmatrix} X_{i-1} \\ F_{i-1} \end{Bmatrix} \tag{12.4}$$

This is the same equation as equation 12.2, with the additional term $\begin{Bmatrix} 0 \\ F \end{Bmatrix}$. To accommodate the additional term in one matrix, we can add a third equation to the two we already have.

$$1 = \begin{bmatrix} 0 & 0 \end{bmatrix} \begin{Bmatrix} X_{i-1} \\ F_{i-1} \end{Bmatrix} + 1 \tag{12.5}$$

This is simply the identity, $1 \equiv 1$. Combining, the extended matrix for the discrete mass is

$$\begin{Bmatrix} X_i \\ F_i \\ 1 \end{Bmatrix} = \begin{bmatrix} 1 & 0 & 0 \\ -m\omega^2 & 1 & -F \\ 0 & 0 & 1 \end{bmatrix} \begin{Bmatrix} X_{i-1} \\ F_{i-1} \\ 1 \end{Bmatrix}$$

Similarly, the extended matrix equation for a spring is

$$\begin{Bmatrix} X_i \\ F_i \\ 1 \end{Bmatrix} = \begin{bmatrix} 1 & \frac{1}{k} & 0 \\ 0 & 1 & 0 \\ 0 & 0 & 1 \end{bmatrix} \begin{Bmatrix} X_{i-1} \\ F_{i-1} \\ 1 \end{Bmatrix} \tag{12.6}$$

Problems arising from the forced vibration of discrete systems can be solved using state variables and transfer matrices in the same way as free vibrations. The size of the matrix has been extended to include the addition of an applied force. These equations are nonhomogeneous, however, and can be solved explicitly for amplitude and force. We could, of course, extend these equations still further.

EXAMPLE PROBLEM 12.19

Determine an expression for the forced vibration of the spring and mass system.

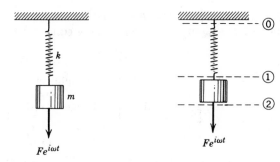

Solution:

For vibration forced by $F_2(t) = Fe^{i\omega t}$, the maximum value of the harmonic forcing function is F. In this case the boundary conditions are $F_2 = F$ and $X_0 = 0$. Using an extended matrix,

$$
\begin{Bmatrix} X_2 \\ F_2 \\ 1 \end{Bmatrix} = \begin{bmatrix} 1 & 0 & 0 \\ -m\omega^2 & 1 & -F \\ 0 & 0 & 1 \end{bmatrix} \begin{bmatrix} 1 & \dfrac{1}{k} & 0 \\ 0 & 1 & 0 \\ 0 & 0 & 1 \end{bmatrix} \begin{Bmatrix} X_0 \\ F_0 \\ 1 \end{Bmatrix}
$$

$$
\begin{Bmatrix} X_2 \\ F \\ 1 \end{Bmatrix} = \begin{bmatrix} 1 & \dfrac{1}{k} & 0 \\ -m\omega^2 & \left(1 - \dfrac{m\omega^2}{k}\right) & 0 \\ 0 & 0 & 1 \end{bmatrix} \begin{Bmatrix} 0 \\ F_0 \\ 1 \end{Bmatrix}
$$

Using the boundary conditions, which have been set,

$$
X_2 = \frac{F_0}{k}
$$

$$
F = \left(1 - \frac{m\omega^2}{k}\right) F_0
$$

and,

$$
1 = 1
$$

rearranging terms,

$$
X_2 = \frac{F}{k\left(1 - \dfrac{m\omega^2}{k}\right)}
$$

which is the familiar response of a simple system with a single degree of freedom to a harmonic forcing function.

PROBLEM 12.20 Repeat Problem 9.9, with $k = 800$ N/m and $m = 2$ kg. Using extended matrices, solve for the forced amplitude of each mass if the lower mass is forced by $F(t) = 50 \sin 30t$. Note that this is identical with Problem 9.73.

PROBLEM 12.21 Repeat Problem 9.11, with $k = 2000$ N/m and $m = 2$ kg. Using extended matrices, solve for the forced amplitude of each mass if the lower mass is forced by $F(t) = 20 \sin 50t$. Note that this is identical with Problem 9.75.

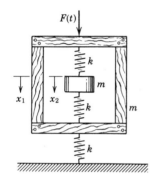

PROBLEM 12.22 Repeat Problem 9.74, with $k = 1930$ N/m and $m = 2.5$ kg. Using extended matrices, solve for the maximum amplitude of the center mass, if $F(t) = 10 \sin 30t$

$$\omega_1^2 = 0.4384 \frac{k}{m}; \ X_1^{(1)} = 1.0000; \ X_1^{(2)} = 1.0000$$

$$\omega_2^2 = 4.5616 \frac{k}{m}; \ X_2^{(1)} = 1.2808; \ X_2^{(2)} = -0.7808$$

PROBLEM 12.23 Do Problem 11.53 using extended matrices.

PROBLEM 12.24 Do problem 11.55 using extended matrices.

12.4. TRANSFER MATRICES FOR BEAMS

The lateral vibration of beams is a vibration problem that can be solved successfully using state variables and transfer matrices. The inertial properties of mass are concentrated at one point in a mass transfer matrix, and the elastic properties of the beam are distributed in a elasticity transfer matrix.

For the massless beam of Figure 12.5a, the equations for equilibrium

Fig. 12.5

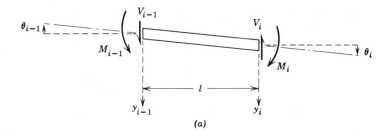

(a)

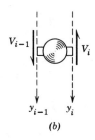

(b)

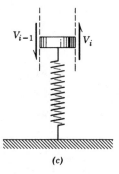

(c)

of an element with a length l are

$$\sum \mathbf{F} = 0$$
$$V_i = V_{i-1} \tag{12.7a}$$

and

$$\sum \mathbf{M} = 0$$
$$M_{i-1} = M_i - V_i l \tag{12.7b}$$
$$M_i = M_{i-1} + V_{i-1} l$$

The length l is a uniform section of the beam and is finite. The change in angular curvature of the beam is

$$\Theta_i - \Theta_{i-1} = \frac{M_i l}{EI} - \frac{V_i l^2}{2EI}$$

The beam is considered to be a short cantilever loaded at its ends. The sign of the angular deflection for the bending moment is positive since concave curvature downward is conventionally accepted as positive curvature. Since shear causes convex curvature, it is negative. Substituting for M_i and V_i,

$$\Theta_i = \Theta_{i-1} + \frac{(M_{i-1} + V_{i-1}l)l}{EI} - \frac{V_{i-1}l^2}{2EI}$$

$$\Theta_i = \Theta_{i-1} + \frac{M_{i-1}l}{EI} + \frac{V_{i-1}l^2}{2EI}$$

(12.7c)

The deflection is found similarly,

$$Y_i - Y_{i-1} = \Theta_{i-1}l - \frac{V_i l^3}{3EI} + \frac{M_i l^2}{2EI}$$

Substituting for V_i and M_i, we have a fourth equation,

$$Y_i = Y_{i-1} + \Theta_{i-1}l + \frac{M_{i-1}l^2}{2EI} + \frac{V_{i-1}l^3}{6EI}$$

(12.7d)

Equations 12.7c and 12.7d state θ_i and Y_i in terms of the lateral displacement, angular displacement, bending moment, and shear at station $i-1$. These four equations are all that we need to construct a transfer matrix that will describe the elastic properties of a section of beam with a length l and flexural rigidity EI.

$$\begin{Bmatrix} Y_i \\ \Theta_i \\ M_i \\ V_i \end{Bmatrix} = \begin{bmatrix} 1 & l & \dfrac{l^2}{2EI} & \dfrac{l^3}{6EI} \\ 0 & 1 & \dfrac{l}{EI} & \dfrac{l^2}{2EI} \\ 0 & 0 & 1 & l \\ 0 & 0 & 0 & 1 \end{bmatrix} \begin{Bmatrix} Y_{i-1} \\ \Theta_{i-1} \\ M_{i-1} \\ V_{i-1} \end{Bmatrix}$$

(12.8)

There are four terms in the state variable, which makes it larger than the previous set, but, if we once determine the transfer matrix for a beam, it can be used for all sections of the same length and same flexural rigidity.

The corresponding transfer matrix for a mass must also have four terms in the state variable. In Figure 12.5b from the equation of motion,

$$\sum F = m\ddot{y}$$

(12.9a)

$$V_{i-1} - V_i = m\ddot{y} = -mY_i\omega^2$$

and for a point,

$$Y_i = Y_{i-1}$$

(12.9b)

from which,

$$V_i = V_{i-1} + m\omega^2 Y_{i-1}$$

The transfer matrix is

$$\begin{Bmatrix} Y_i \\ \Theta_i \\ M_i \\ V_i \end{Bmatrix} = \begin{bmatrix} 1 & 0 & 0 & 0 \\ 0 & 1 & 0 & 0 \\ 0 & 0 & 1 & 0 \\ m\omega^2 & 0 & 0 & 1 \end{bmatrix} \begin{Bmatrix} Y_{i-1} \\ \Theta_{i-1} \\ M_{i-1} \\ V_{i-1} \end{Bmatrix} \qquad (12.10)$$

The first three rows are simply identities. The equation of motion is the fourth. Note that in writing this transfer matrix $m\omega^2$ has a positive sign, where in the last section, $m\omega^2$ was negative. In the most used sign convention for a beam, concave curvature and positive deflection are downward, and the shear $V_{i-1} - V_i$ is positive. If we had used concave curvature upward and positive deflection upward as a positive convention, our signs would be reversed, since $V_i - V_{i-1}$ would be positive. We must remember that signs are only a convention, and should be used as a convenience. It doesn't make any difference which we use so long as we are constant in any given problem.

For a spring, in contact with the beam at one point, using the sign convention above, and referring to Figure 12.5c

$$\sum \mathbf{F} = 0 \qquad (12.11)$$
$$V_{i-1} = V_i + k Y_i$$

from which the transfer matrix is

$$\begin{Bmatrix} Y_i \\ \theta_i \\ M_i \\ V_i \end{Bmatrix} = \begin{bmatrix} 1 & 0 & 0 & 0 \\ 0 & 1 & 0 & 0 \\ 0 & 0 & 1 & 0 \\ -k & 0 & 0 & 1 \end{bmatrix} \begin{Bmatrix} Y_{i-1} \\ \theta_{i-1} \\ M_{i-1} \\ V_{i-1} \end{Bmatrix} \qquad (12.12)$$

We now have three components, a beam section, a spring, and a concentrated mass with which to model our problems. Success will depend heavily on how good our model is, as well as our mathematics.

12.5. DIMENSIONLESS TRANSFER MATRICES

In several preceding sections, mass and elasticity have been considered to be discrete quantities that are distinct properties of a point or short section, depending on whether the quantities are stated in a point matrix or field matrix. Mass has always been symbolized by m, a spring has an elastic modulus k, and a short section of beam has a flexural rigidity EI. It is not convenient to keep such symbols when working with transfer matrices. Transfer matrices are only convenient when we use numbers and numerical techniques. Otherwise, even the most simple problem becomes a monumental experience.

Each term of the transfer matrix has a different set of units. For example, in the transfer matrix for a beam, in the first row, the first term is a pure number, the second has the units in metres, the third has reciprocal Newtons (N^{-1}), and the fourth metres per Newton (m/N). To avoid the likelihood of transcription errors, dimensionless transfer matrices can be used. In a dimensionless matrix, each term is a pure number.

For the elastic transfer matrix for a beam, the first equation, or first row, can be made dimensionless by dividing by l. $\overline{Y}_i$ is a dimensionless number, Y_i/l

$$\overline{Y}_i = \frac{Y_i}{l} = \frac{Y_{i-1}}{l} + \Theta_{i-1} + \frac{M_{i-1}l}{2EI} + \frac{V_{i-1}l^2}{6EI}$$

The second equation is dimensionless, but the third and fourth are not. The third can be made dimensionless by multiplying by l/EI, and the fourth can be made dimensionless by multiplying by l^2/EI. These quantities are arbitrary combinations of physical dimensions, but a little experimentation with dimensionless numbers will show that they are unique. That is,

$$\overline{\Theta}_i = \Theta_{i-1} + \frac{M_{i-1}l}{EI} + \frac{V_{i-1}l^2}{2EI}$$

$$\overline{M}_i = \frac{M_i l}{EI} = \frac{M_{i-1}l}{EI} + \frac{V_{i-1}l^2}{EI}$$

$$\overline{V}_i = \frac{V_i l^2}{EI} = \frac{V_{i-1}l^2}{EI}$$

Writing these equations in matrix form

$$\begin{Bmatrix} \overline{Y}_i \\ \overline{\Theta}_i \\ \overline{M}_i \\ \overline{V}_i \end{Bmatrix} = \begin{bmatrix} 1 & 1 & \frac{1}{2} & \frac{1}{6} \\ 0 & 1 & 1 & \frac{1}{2} \\ 0 & 0 & 1 & 1 \\ 0 & 0 & 0 & 1 \end{bmatrix} \begin{Bmatrix} \overline{Y}_{i-1} \\ \overline{\Theta}_{i-1} \\ \overline{M}_{i-1} \\ \overline{V}_{i-1} \end{Bmatrix} \tag{12.13}$$

The transfer matrix for a short section of a beam is now a set of pure numbers.

There is a corresponding dimensionless transfer matrix for a mass. From the equation of motion,

$$\sum \mathbf{F} = m\ddot{\mathbf{y}}$$
$$V_{i-1} - V_i = -m\omega^2 Y_i$$

To make the shear force dimensionless, each term must be multiplied by l^2/EI.

$$\overline{V}_i = \overline{V}_{i-1} + \frac{m\omega^2 l^3}{EI}\overline{Y}_{i-1}$$

$$\begin{Bmatrix} \overline{Y}_i \\ \overline{\Theta}_i \\ \overline{M}_i \\ \overline{V}_i \end{Bmatrix} = \begin{bmatrix} 1 & 0 & 0 & 0 \\ 0 & 1 & 0 & 0 \\ 0 & 0 & 1 & 0 \\ \dfrac{m\omega^2 l^3}{EI} & 0 & 0 & 1 \end{bmatrix} \begin{Bmatrix} \overline{Y}_{i-1} \\ \overline{\Theta}_{i-1} \\ \overline{M}_{i-1} \\ \overline{V}_{i-1} \end{Bmatrix} \tag{12.14}$$

For a spring, which acts as a support, we must also multiply each term by l^2/EI.

$$\sum \mathbf{F} = 0$$
$$V_{i-1} = V_i + kY_i$$

$$\overline{V}_i = \overline{V}_{i-1} - \frac{kl^3}{EI}\overline{Y}_i$$

$$\begin{Bmatrix} \overline{Y}_i \\ \overline{\Theta}_i \\ \overline{M}_i \\ \overline{V}_i \end{Bmatrix} = \begin{bmatrix} 1 & 0 & 0 & 0 \\ 0 & 1 & 0 & 0 \\ 0 & 0 & 1 & 0 \\ -\dfrac{kl^3}{EI} & 0 & 0 & 1 \end{bmatrix} \begin{Bmatrix} \overline{Y}_{i-1} \\ \overline{\Theta}_{i-1} \\ \overline{M}_{i-1} \\ \overline{V}_{i-1} \end{Bmatrix} \tag{12.15}$$

EXAMPLE PROBLEM 12.25

Determine the natural frequency of a cantilevered beam that supports a mass m at one end and is built into a rigid wall at the other.

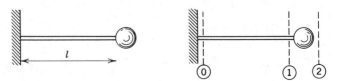

Solution:

The state vector at node 1 can be found from the state vector at node 0 by using the transfer matrix for a beam

$$\{z_1\} = [F_{1-0}]\{z_0\}$$

$$
\begin{Bmatrix} Y_i \\ \Theta_i \\ M_1 \\ V_1 \end{Bmatrix} =
\begin{bmatrix}
1 & l & \dfrac{l^2}{2EI} & \dfrac{l^3}{3EI} \\
0 & 1 & \dfrac{l}{EI} & \dfrac{l^2}{2EI} \\
0 & 0 & 1 & l \\
0 & 0 & 0 & 1
\end{bmatrix}
\begin{Bmatrix} Y_0 \\ \Theta_0 \\ M_0 \\ V_0 \end{Bmatrix}
$$

The state vector at node 2 can be found from the state vector at node 1.

$$\{z_2\} = [P_{2-1}]\{z_0\}$$

$$
\begin{Bmatrix} Y_2 \\ \Theta_2 \\ M_2 \\ V_2 \end{Bmatrix} =
\begin{bmatrix}
1 & 0 & 0 & 0 \\
0 & 1 & 0 & 0 \\
0 & 0 & 1 & 0 \\
m\omega^2 & 0 & 0 & 1
\end{bmatrix}
\begin{Bmatrix} Y_1 \\ \Theta_1 \\ M_1 \\ V_1 \end{Bmatrix}
$$

or combining,

$$
\begin{Bmatrix} Y_2 \\ \Theta_2 \\ M_2 \\ V_2 \end{Bmatrix} =
\begin{bmatrix}
1 & 0 & 0 & 0 \\
0 & 1 & 0 & 0 \\
0 & 0 & 1 & 0 \\
m\omega^2 & 0 & 0 & 1
\end{bmatrix}
\begin{bmatrix}
1 & l & \dfrac{l^2}{2EI} & \dfrac{l^3}{3EI} \\
0 & 1 & \dfrac{l}{EI} & \dfrac{l^2}{2EI} \\
0 & 0 & 1 & l \\
0 & 0 & 0 & 1
\end{bmatrix}
\begin{Bmatrix} Y_0 \\ \Theta_0 \\ M_0 \\ V_0 \end{Bmatrix}
$$

Multiplying the two transfer matrices leads to the matrix equation,

$$
\begin{Bmatrix} Y_2 \\ \Theta_2 \\ M_2 \\ V_2 \end{Bmatrix} =
\begin{bmatrix}
1 & \dfrac{l^2}{2EI} & \dfrac{l^2}{2EI} & \dfrac{l^3}{6EI} \\
0 & 1 & \dfrac{l}{EI} & \dfrac{l^2}{2EI} \\
0 & 0 & 1 & l \\
m\omega^2 & m\omega^2 l & \dfrac{m\omega^2 l^2}{2EI} & \left(1 + \dfrac{m\omega^2 l^3}{6EI}\right)
\end{bmatrix}
\begin{Bmatrix} Y_0 \\ \Theta_0 \\ M_0 \\ V_0 \end{Bmatrix}
$$

There are four boundary conditions, $Y_0 = 0$ and $\Theta_0 = 0$ at the wall, and $M_2 = 0$ and $V_2 = 0$ at the end of the beam. This leads to a matrix statement

$$\begin{Bmatrix} 0 \\ 0 \end{Bmatrix} = \begin{bmatrix} 1 & l \\ \dfrac{m\omega^2 l^2}{2EI} & \left(1 + \dfrac{m\omega^2 l^3}{6EI}\right) \end{bmatrix} \begin{Bmatrix} M_0 \\ V_0 \end{Bmatrix}$$

For this to be true the value of the determinant must be zero,

$$\mathrm{Det}(\omega^2) = \left(1 + \frac{m\omega^2 l^3}{6EI}\right) - \frac{m\omega^2 l^3}{2EI} = 0$$

or

$$\omega^2 = \frac{3EI}{ml^3}$$

which is the natural frequency of a cantilevered beam supporting a concentrated mass at one end.

We should make a careful comparison of this problem and Example Problem 12.2. Since we have four terms in the state variable instead of two, our frequency equation will be determined by a matrix of four terms instead of just one. It follows that if we had had a state variable of six terms, our frequency equation would have been determined by a matrix of twelve terms.

EXAMPLE PROBLEM 12.26

A 7500-kg concrete slab, 5 m × 4 m, rests on five steel I beams, 5 m long, spaced 1-m apart. The beams are built in at each end. A 1000-kg diesel motor generator set rests at the center of the slab. The I beams have a mass of 52.09 kg/m and have an area moment of inertia of 94.48×10^6 (mm)^4. Using numerical methods, state variables, and transfer matrices, determine a value for the natural frequency.

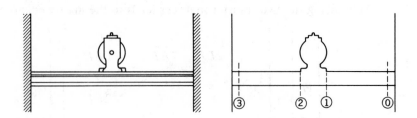

Solution:
This system can be modeled by dividing the beam in two sections, each section being equal to one half of the total beam. The concrete slab is

modeled by including $\frac{3}{8}$ of its total mass at the center of the span. Thus,

$$M = \tfrac{3}{8}[7500 + 5(5)(52.09) + 1000] = 4300.84 \text{ kg}$$

Our solution will only be as good as the validity of this model.

Using dimensionless transfer matrices, it will only be necessary to evaluate $m\omega^2 l^3 / EI$. For $\omega^2 = 10^4$ s^2, $l = 2.5$ m,

$$\frac{m\omega^2 l^2}{EI} = \frac{(4300.84)(10^4)(2.5)^3}{(5)(205 \times 10^9)(94.48 \times 10^{-6})} = 6.9392$$

Starting with station 0 at the right wall and station 3 at the left,

$$\{\bar{z}_3\} = [F_{3-2}][P_{2-1}][F_{1-0}]\{\bar{z}_0\}$$

$[P_{2-1}]$ is the dimensionless transfer matrix for the diesel engine and the effective mass of the slab

$$[P_{2-1}] = \begin{bmatrix} 1 & 0 & 0 & 0 \\ 0 & 1 & 0 & 0 \\ 0 & 0 & 0 & 0 \\ 6.9392 & 0 & 0 & 1 \end{bmatrix}$$

$[F_{3-2}]$ and $[F_{1-0}]$ are identical transfer matrices for one half of the beam

$$[F_{1-1}] = \begin{bmatrix} 1 & 1 & \tfrac{1}{2} & \tfrac{1}{6} \\ 0 & 1 & 1 & \tfrac{1}{2} \\ 0 & 0 & 1 & 1 \\ 0 & 0 & 0 & 1 \end{bmatrix} = [F_{3-2}]$$

Using matrix multiplication,

$$\begin{bmatrix} 1 & 1 & \tfrac{1}{2} & \tfrac{1}{6} \\ 0 & 1 & 1 & \tfrac{1}{2} \\ 0 & 0 & 1 & 1 \\ 0 & 0 & 0 & 1 \end{bmatrix} \begin{Bmatrix} \overline{Y}_0 \\ \overline{\Theta}_0 \\ \overline{M}_0 \\ V_0 \end{Bmatrix} = \{\bar{z}_1\}$$

$$\begin{bmatrix} 1 & 0 & 0 & 0 \\ 0 & 1 & 0 & 0 \\ 0 & 0 & 1 & 0 \\ 6.9392 & 0 & 0 & 1 \end{bmatrix} \begin{bmatrix} 1 & 1 & \tfrac{1}{2} & \tfrac{1}{6} \\ 0 & 1 & 1 & \tfrac{1}{2} \\ 0 & 0 & 1 & 1 \\ 6.9392 & 6.9392 & 3.4696 & 2.1565 \end{bmatrix} \begin{Bmatrix} \overline{Y}_0 \\ \overline{\Theta}_0 \\ \overline{M}_0 \\ \overline{V}_0 \end{Bmatrix} = \{\bar{z}_2\}$$

$$
\begin{bmatrix}
1 & 1 & \frac{1}{2} & \frac{1}{6} \\
0 & 1 & 1 & \frac{1}{2} \\
0 & 0 & 1 & 1 \\
0 & 0 & 0 & 1
\end{bmatrix}
\begin{bmatrix}
2.1565 & 3.1565 & 2.5783 & 1.5261 \\
3.4696 & 4.4696 & 3.7348 & 2.5783 \\
6.9392 & 6.9392 & 4.4696 & 3.1565 \\
6.9392 & 6.9392 & 3.4696 & 2.1565
\end{bmatrix}
\begin{Bmatrix}
\overline{Y}_0 \\
\overline{\Theta}_0 \\
\overline{M}_0 \\
\overline{V}_0
\end{Bmatrix}
= \{z_3\}
$$

Since $\overline{Y}_3 = \overline{Y}_0 = 0$ and $\overline{\Theta}_3 = \overline{\Theta}_0 = 0$ as boundary conditions, these yield the matrix statement that

$$
\begin{Bmatrix} 0 \\ 0 \end{Bmatrix}
=
\begin{bmatrix} u_{13} & u_{14} \\ u_{23} & u_{24} \end{bmatrix}
\begin{Bmatrix} \overline{M}_0 \\ \overline{V}_0 \end{Bmatrix}
$$

For $\omega^2 = 10^4$ to be a natural frequency, $u_{13}u_{24} - u_{23}u_{14} = 0$. A quick check of the final matrix shows

$$
u_{13}u_{24} - u_{23}u_{14} = (2.5783)(2.5783) - (3.7348)(1.5261) = 0.9480
$$

which means that $\omega^2 = 10^4$ is not a natural frequency. Calling $y(\omega^2) = u_{13}u_{24} - u_{23}u_{14}$, and plotting $y(\omega^2)$ as a function of ω^2

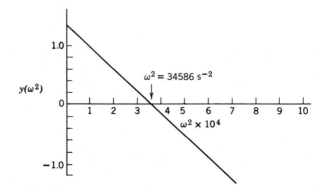

This equation has only one root, since we have modeled the system with only one mass. The natural frequency is $\omega^2 = 34{,}586 \text{ s}^{-2}$ or 1776 rpm, which may cause some difficulty in operating the motor generator set.

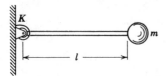

PROBLEM 12.27 An elastic horizontal bar with a flexural rigidity EI has a mass m at one end and pivots about the other in a frictionless bearing. A torsional spring with a modulus K supports the bar and mass in a horizontal position.

(a) Determine the transfer matrix for the torsional spring.

(b) Determine the natural frequency for the system.

Answer (b) $\omega_n^2 = \dfrac{1}{m\left(\dfrac{l^3}{3EI} + \dfrac{l^2}{K}\right)}$

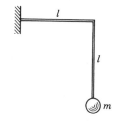

PROBLEM 12.28 An elastic bar is bent at its center in a sharp right angle. One half of the bar is horizontal, the other half is vertical. The vertical half has a mass m at its lower free end. The horizontal half is cantilevered from a wall into which it is rigidly fixed. Determine the natural frequency of motion in the vertical plane.

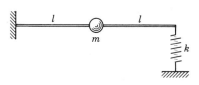

PROBLEM 12.29 Determine the natural frequency of a beam that is built-in at one end and supported on a spring that has a stiffness of k at the other.

Answer: $\omega^2 = \dfrac{EI}{ml^3}\left[\dfrac{3 + kl^3/EI}{1 + 7kl^3/12EI}\right]$

PROBLEM 12.30 The fuselage of a jet aircraft has a mass of 5000 kg. The half-wing span is 6 m and the flexural rigidity of a half-wing is known to be 3×10^6 N·m². If extra fuel is carried in two wing-tip tanks, each having a mass of 300 kg, determine the value of the natural frequency of flexural vibration, using matrix iteration.

Answer: $\omega^2 = 93.12$ s^{-2}

PROBLEM 12.31 A turbine-generator is set on a steel foundation. To a first approximation, the installation can be described as a concentrated mass on a beam supported on flexible columns. Determine the natural frequency of free vibration. See the figure on the next page.

Answer: $\omega^2 = \dfrac{k}{m\left[\dfrac{1}{2} + \dfrac{1}{6}\dfrac{kl^3}{EI}\right]}$

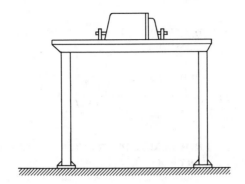

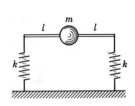

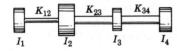

PROBLEM 12.32 For the four-mass torsional system, determine the natural frequencies of torsional vibration using state vectors and transfer matrices. *Hint*: As a point of departure, assume $\omega^2 = 1000$ s^{-2}.

$I_1 = 200$ kg·m^2 $K_{12} = 10^5$ N·m/rad

$I_2 = 400$ kg·m^2 $K_{23} = 3 \times 10^5$ N·m/rad

$I_3 = 100$ kg·m^2 $K_{34} = 4 \times 10^5$ N·m/rad

$I_4 = 300$ kg·m^2

Answer: $\omega_1^2 = 805.5$ s^{-2}

$\omega_2^2 = 1225$ s^{-2}

$\omega_3^2 = 8100$ s^{-2}

CHAPTER THIRTEEN

DISTRIBUTED SYSTEMS

13.1. INTRODUCTION

In a discrete system, elasticity and mass are modeled as discrete properties. Many problems can be solved using discrete systems, but there are disadvantages, the most obvious being that mass and elasticity cannot always be separated in mathematical models of the real systems. An alternate method of modeling would be to distribute elasticity and mass. Systems where elasticity and mass are considered to be distributed parameters are *distributed systems*. Beams, rods, shafts, cables, and strings can be accurately modeled as distributed systems.

Three assumptions are necessary to make a mathematical model of a distributed system. The first is that the material must be homogeneous; the second is that it must be elastic, which means that it follows Hooke's law; and the last is that it must be isotropic. These assumptions are restricting, but they are necessary.

Three very good examples will be given of distributed systems. They have identical mathematical form and are widely printed in texts, although the practical applications are limited.

Longitudinal Vibration of a Uniform Rod. Let us consider the transmission of longitudinal stress along an elastic rod that has a uniform cross-sectional area A and an elastic modulus E. In Figure 13.1a, a cross-sectional element, which is an infinitesimal slice with a thickness dx, has

Fig. 13.1

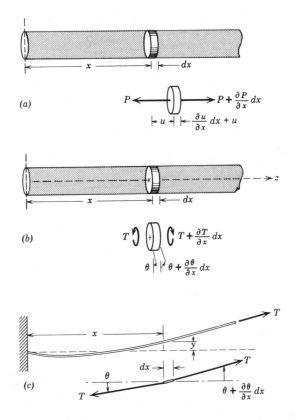

moved a distance u under the elastic stress due to the force P, from its equilibrium position x, as measured from some arbitrary longitudinal coordinate. The element has also been stretched in elastic strain, so that its width has increased by $(\partial u/\partial x)dx = \varepsilon_x dx$, ε_x being the strain in the x direction. The partial derivative is used since the displacement u is a function of both longitudinal displacement and time. Summing the forces in the positive x direction,

$$\sum \mathbf{F} = m\ddot{\mathbf{x}}$$

$$P + \frac{\partial P}{\partial x}dx - P = \frac{\mu}{g}A\,dx\frac{\partial^2 u}{\partial t^2}$$

The acceleration of the element is $\partial^2 u/\partial t^2$, and μ is the weight density of the rod. The force $P = EA\varepsilon_x$. Substituting for P and ε_x, and cancelling dx,

$$\frac{\partial}{\partial x}\left(EA\frac{\partial u}{\partial x}\right) = \frac{A\mu}{g}\frac{\partial^2 u}{\partial t^2}$$

For a uniform rod, the area and the modulus of elasticity are constant. Therefore,

$$\frac{Eg}{\mu}\frac{\partial^2 u}{\partial x^2}=\frac{\partial^2 u}{\partial t^2}$$ (13.1)

Equation 13.1 is the equation of motion for the propagation of longitudinal stress in a longitudinal rod.

Torsional Vibration of a Uniform Rod. In the same way, we can find the equation of motion for the transmission of shear stress in a longitudinal rod. The cross-sectional area is again A and uniform. The modulus of rigidity in torsion is G. In Figure 13.1b, a cross-sectional element has been displaced angularly θ from its equilibrium position. The element is also twisted in elastic strain an amount $\gamma = r(\partial\theta/\partial x)$, r being the radius of the rod. Summing moments applied to the element dx about the geometric axis of the rod,

$$\sum \mathbf{M}_z = I\ddot{\theta}$$

$$T+\frac{\partial T}{\partial x}dx - T = \frac{\mu}{g}J\,dx\frac{\partial^2\theta}{\partial t^2}$$

The mass moment of inertia of the thin element about its geometric axis is the product of the mass per unit length and the polar second moment of area about the same axis. Canceling dx and replacing the torque T by the elastic equation

$$T = GJ\frac{\gamma}{r} = GJ\frac{\partial\theta}{\partial x}$$

$$\frac{\partial}{\partial x}\left(GJ\frac{\partial\theta}{\partial x}\right) = \frac{\mu J}{g}\frac{\partial^2\theta}{\partial t^2}$$

For a uniform rod, neither the polar second moment of area nor the modulus of rigidity vary with x, and

$$\frac{Gg}{\mu}\frac{\partial^2\theta}{\partial x^2}=\frac{\partial^2\theta}{\partial t^2}$$ (13.2)

Equation 13.2 is the equation of motion for the propagation of shear stress in a longitudinal rod.

Transverse Vibration of a String or Cable. The lateral vibrations of a taut string or cable have a similar equation of motion. Consider a string or cable to be stretched under a tension T between two fixed points, as in Figure 13.1c. The cable has a weight per unit length of w. Summing the forces in the y direction, which is transverse to the string,

$$T \sin\left(\theta + \frac{\partial \theta}{\partial x}dx\right) - T \sin\theta = \frac{w}{g}dx\frac{\partial^2 y}{\partial t^2}$$

Using trigonometry,

$$\sin\left(\theta + \frac{\partial \theta}{\partial x}dx\right) = \sin\theta + \cos\theta\frac{\partial \theta}{\partial x}dx$$

For small lateral displacements, the slope of the curvature can be replaced by $\partial y/\partial x$

$$T \cos\theta\frac{\partial^2 y}{\partial x^2} = \frac{w}{g}\frac{\partial^2 y}{\partial t^2} \tag{13.3}$$

$$\frac{gH}{w}\frac{\partial^2 y}{\partial x^2} = \frac{\partial^2 y}{\partial t^2}$$

This is the equation of motion for a vibrating string.

If the distance between supports is so great that the string sags, $\cos\theta$ is not zero, but, $T \cos\theta = H$, the horizontal component of the string tension. It is a constant, which is easily verified by simple statics.

13.2. THE WAVE EQUATION

All three problems are mathematically identical. They all involve a function $\phi(x, t)$, which is expressed as a function of two variables, x and t, according to the partial differential equation of motion,

$$c^2\frac{\partial^2 \phi}{\partial x^2} = \frac{\partial^2 \phi}{\partial t^2} \tag{13.4}$$

It is known as the wave equation.

It is called the wave equation because it can be described as a standing wave, if we use a coordinate system that moves with the disturbance. In Figure 13.2, a traveling wave moves in the positive x direction with a velocity c and a shape that remains unchanged as it travels. In a coordinate system that moves in the same direction with the wave and with the same velocity c, the displacement in the moving coordinate system is

$$z = x - ct$$

The disturbance is stated as a function of the displacement in the moving coordinate system, $\phi(x, t) = f(z) = f(x - ct)$. To show that this is a solution to the wave equation, take the partial derivatives

$$\frac{\partial \phi}{\partial t} = \frac{\partial \phi}{\partial x}\frac{dx}{dt} = c\frac{\partial \phi}{\partial x}$$

Fig. 13.2

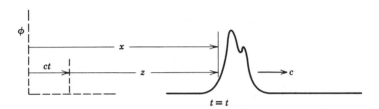

and

$$\frac{\partial^2 \phi}{\partial t^2} = \frac{\partial}{\partial x}\left(\frac{\partial \phi}{\partial t}\right)\frac{dx}{\partial t} = c^2 \frac{\partial^2 \phi}{\partial x^2}$$

This is the wave equation in one dimension. Any and all wave forms that have a constant shape in a coordinate system moving with a velocity c would have this equation of motion.

Going back to our original three problems, we can compute the three velocities of propagation for traveling waves.

For a longitudinal stress waves in a horizontal rod: $c = \sqrt{\dfrac{Eg}{\mu}}$

For shear stress waves in horizontal rod: $c = \sqrt{\dfrac{Gg}{\mu}}$ (13.5)

And, for transverse waves in a string: $c = \sqrt{\dfrac{Hg}{w}}$

One subtle point should be made. If the propagation of elastic stress in a horizontal rod involves both axial and shear stress in the distur-bance, the shear stress waves will fall behind, distorting the traveling wave shape or separate entirely, which is a violation of one of the original assumptions, that the wave form remained unchanged as it traveled.

Solutions to the Wave Equation.   The wave equation can also be solved by the classic method of separation of variables, if we let the function

$\phi(x, t)$ be the product of two separate functions, one of x and one of t.

$$\phi(x, t) = f_1(t)f_2(x)$$

$$\frac{\partial^2 \phi}{\partial t^2} = \frac{\partial^2 f_1(t)}{\partial t^2} f_2(x)$$

$$\frac{\partial^2 \phi}{\partial x^2} = \frac{\partial^2 f_2(x)}{\partial x^2} f_1(t)$$

Substituting in the differential equations of motion,

$$\frac{\partial^2 f_1(t)}{\partial t^2} f_2(x) = c^2 \frac{\partial^2 f_2(x)}{\partial x^2} f_1(t)$$

The functions can be separated by division, which leads to the statement

$$\frac{1}{f_1(t)} \frac{d^2 f_1(t)}{dt^2} = \frac{c^2}{f_2(x)} \frac{d^2 f_2(x)}{dx^2}$$

Since the functions are separated, partial derivatives are no longer needed.

The only way that two functions of independent variables x and t can be equal to each other is for them to be equal to a constant. The constant is arbitrary and for reasons of convenience and foresight, the constant is selected to be $-\omega_n^2$.

$$\frac{1}{f_1(t)} \frac{d^2 f_1(t)}{dt^2} = -\omega_n^2$$

$$\frac{c^2}{f_2(x)} \frac{d^2 f_2(x)}{dx^2} = -\omega_n^2$$

These lead to two separate differential equations of motion,

$$\frac{d^2 f_1(t)}{dt^2} + \omega_n^2 f_1(t) = 0$$

from which,

$$f_1(t) = A \cos \omega_n t + B \sin \omega_n t \tag{13.6}$$

And,

$$\frac{d^2 f_2(x)}{dx^2} + \left(\frac{\omega_n}{c}\right)^2 f_2(x) = 0$$

from which

$$f_2(x) = C \cos \frac{\omega_n}{c} x + D \sin \frac{\omega_n}{c} x \tag{13.7}$$

The complete solution for $\phi(x, t)$ is thus

$$\phi(x, t) = (A \cos \omega_n t + B \sin \omega_n t)\left(C \cos \frac{\omega_n}{c} x + D \sin \frac{\omega_n}{c} x\right) \quad (13.8)$$

There are four arbitrary constants, determined by the initial conditions and boundary conditions.

EXAMPLE PROBLEM 13.1

Determine the displacement of a lateral cross-section of a uniform rod, with one end fixed and the other free, if the rod is vibrating longitudinally.

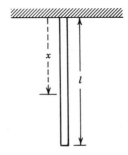

Solution:
The equation of motion for the longitudinal displacement $u(x, t)$ of a lateral cross-section of a uniform rod is

$$c^2 \frac{\partial^2 u}{\partial x^2} = \frac{\partial^2 u}{\partial t^2}$$

From equation 13.8, the displacement u is

$$u(x, t) = (A \cos \omega_n t + B \sin \omega_n t)\left(C \cos \frac{\omega_n x}{c} + D \sin \frac{\omega_n x}{c}\right)$$

For one of the boundary conditions, $u = 0$ at $x = 0$, since the rod is built in at the fixed end

$$0 = (A \cos \omega_n t + B \sin \omega_n t)C$$

and $C = 0$. For the free end of the rod, there can be no applied force. Using this as the second boundary condition, $EA(\partial u/\partial x) = 0$ at $x = l$

$$\frac{\partial u}{\partial x} = 0 = (A \cos \omega_n t + B \sin \omega_n t)D\frac{\omega_n}{c}\cos\frac{\omega_n l}{c}$$

This can only be true if

$$\cos\frac{\omega_n l}{c} = 0$$

This is the frequency equation. Its roots are the characteristic values for this particular problem.

$$\frac{\omega_n l}{c} = \frac{\pi}{2}, \frac{3\pi}{2}, \frac{5\pi}{2}, \cdots \frac{n\pi}{2} \qquad n \text{ being odd}$$

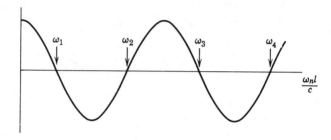

Rearranging terms, the natural frequencies are

$$\omega_n = \frac{n\pi}{2l}\sqrt{\frac{Eg}{\mu}} \qquad n \text{ being odd}$$

and,

$$\omega_1 = \frac{\pi}{2l}\sqrt{\frac{Eg}{\mu}}$$

$$\omega_2 = \frac{3\pi}{2l}\sqrt{\frac{Eg}{\mu}}$$

$$\omega_3 = \frac{5\pi}{2l}\sqrt{\frac{Eg}{\mu}}$$

The displacement u_n represents displacement in a principal mode.

$$u_n = \sin\frac{n\pi x}{2l}(A_n \cos \omega_n t + B_n \sin \omega_n t)$$

The total displacement u will exhibit all modes. Consequently, the total displacement will be the sum

$$u(x, t) = \sum_{n=1,3,5}^{\infty} \sin\frac{n\pi x}{2l}(A_n \cos \omega_n t + B_n \sin \omega_n t)$$

The arbitrary constants A_n and B_n must be determined by the initial conditions.

EXAMPLE PROBLEM 13.2

A uniform rod, with one end fixed and the other free, is stretched under a static load and suddenly released from rest at time $t=0$. From these initial conditions, determine the longitudinal displacement $u(x, t)$.

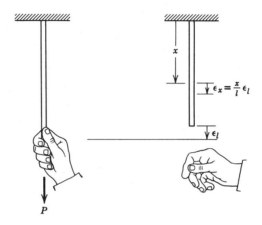

Solution:
At the time of the sudden release of the load, the longitudinal displacement is linearly proportional to the distance x from the fixed end of the rod. At $x=0$, the displacement $u=0$. At $x=l$, the displacement $u=\varepsilon l$, ε being the longitudinal strain. One boundary condition is $u=\varepsilon x$ at $t=0$, and the other is that $\dot{u}=0$ at $t=0$.

The displacement $u(x, t)$ of a uniform rod vibrating longitudinally is

$$u(x, t) = \sum_{n=1,3,5}^{\infty} \sin\frac{n\pi x}{2l}(A_n \cos \omega_n t + B_n \sin \omega_n t)$$

and,

$$\dot{u}(x, t) = \sum_{n=1,3,5}^{\infty} \omega_n \sin\frac{n\pi x}{2l}(B_n \cos \omega_n t - A_n \sin \omega_n t)$$

From the first boundary condition,

$$\varepsilon x = \sum_{n=1,3,5}^{\infty} A_n \sin\frac{n\pi x}{2l}$$

and, from the second,

$$0 = \sum_{n=1,3,5}^{\infty} B_n \omega_n \sin \frac{n\pi x}{2l}$$

Recalling our knowledge of Fourier coefficients,

$$A_n = \frac{2}{l} \int_0^l \varepsilon x \cos \frac{n\pi x}{2l} dx$$

$$= \frac{2\varepsilon}{l} \left(\frac{2l}{n\pi}\right)^2 \int_0^l \left(\frac{n\pi x}{2l}\right) \cos \frac{n\pi x}{2l} d\left(\frac{n\pi x}{2l}\right)$$

$$= \frac{8\varepsilon l}{n^2 \pi^2} \left[\sin \frac{n\pi x}{2l} \right]_0^l = \frac{8\varepsilon l}{n^2 \pi^2} \sin \frac{n\pi}{2}$$

Since n is odd, the $\sin n\pi/2 = \pm 1$, or

$$A_n = \frac{8\varepsilon l}{n^2 \pi^2} (-1)^{(n-1)/2}$$

and,

$$B_n = 0$$

The displacement is then

$$u(x, t) = \frac{8\varepsilon l}{\pi^2} \sum_{n=1,3,5}^{\infty} \frac{(-1)^{(n-1)/2}}{n^2} \sin \frac{n\pi x}{2l} \cos \omega_n t$$

EXAMPLE PROBLEM 13.3

Determine the displacement of a lateral cross-section of a uniform rod that vibrates longitudinally, if one end is fixed and the other end supports a concentrated mass M.

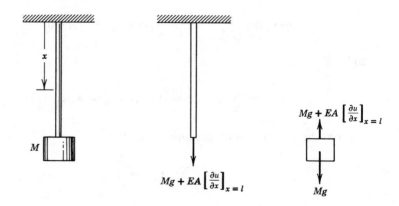

Solution:

The equation of motion for the longitudinal displacement $u(x, t)$ of a lateral cross-section of the uniform rod is again

$$c^2\frac{\partial^2 u}{\partial x^2} = \frac{\partial^2 u}{\partial t^2}$$

and,

$$u(x, t) = (A \cos \omega_n t + B \sin \omega_n t)\left(C \cos\frac{\omega_n x}{c} + D \sin\frac{\omega_n x}{c}\right)$$

The boundary condition at the fixed end of the rod is $u = 0$ at $x = 0$, or

$$0 = (A \cos \omega_n t + B \sin \omega_n t)C$$
$$C = 0$$

At the lower end of the rod, the rod is displaced by the force

$$Mg + EA\left[\frac{\partial u}{\partial x}\right]_{x=l}$$

The second term is an elastic force that accelerates the mass M.

$$-EA\left[\frac{\partial u}{\partial x}\right]_{x=l} = M\frac{\partial^2 u}{\partial t^2}$$

This is the second boundary condition.

$$-EA(A \cos \omega_n t + B \sin \omega_n t)\left(D \cos\frac{\omega_n l}{c}\right)\frac{\omega_n}{c}$$

$$= -M(A \cos \omega_n t + B \sin \omega_n t)\omega_n^2 D \sin\frac{\omega_n l}{c}$$

from which,

$$\tan\frac{\omega_n l}{c} = \frac{EA}{M\omega_n c}$$

Recalling that $c^2 = Eg/\mu$,

$$\frac{\omega_n l}{c}\tan\frac{\omega_n l}{c} = \frac{A\mu l}{Mg}$$

This is the frequency equation. Its roots are the characteristic values for this problem, but they are less easily determined because the equation is transcendental.

Rearranging terms:

$$\tan\frac{\omega_n l}{c} = \frac{A\mu l}{Mg}\frac{c}{\omega_n l}$$

If we let

$$\tan\frac{\omega_n l}{c} = \phi_1 \quad \text{and} \quad \frac{A\mu l}{Mg}\frac{c}{\omega_n l} = \phi_2$$

This transcendental equation is satisfied where $\phi_1 = \phi_2$. Plotting both as a function of $\omega_n l/c$, for $A\mu l = Mg$. The first roots are $\omega_1 l/c = 0.860$, $\omega_2 l/c = 3.460$, and $\omega_3 l/c = 6.437$.

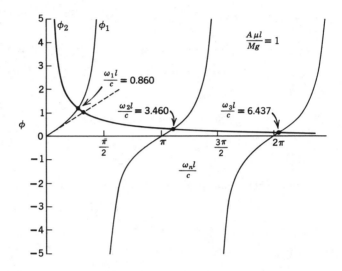

For the first natural frequency, if we had replaced $\tan \omega_n l/c$ by $\omega_n l/c$, which is the straight dashed line,

$$\frac{\omega_n l}{c} = \frac{A\mu l}{Mg}\frac{c}{\omega_n l}$$

or,

$$\omega_n^2 = \frac{c^2}{l^2}\frac{A\mu l}{Mg} = w\frac{Eg}{\mu l^2}\frac{A\mu l}{Mg}$$

$$= \frac{AE}{Ml}$$

This is logical, since the effective modulus $k_i = AE/l$.

The error in making this approximation is 14%, which is not too great, considering our assumption that the mass of the rod was the same as the mass of the weight attached to the rod. For $A\mu l/mg = 0.1$, $\omega_1 l/c = 0.3113$. Making the same approximation, that $\tan \omega_n l/c = \omega_n l/c$ the error is only 1.5%. In each case, the approximate answer results in a higher calculation of the natural frequency than the more correct calculation. This could be expected.

PROBLEM 13.4 Determine the time it takes for a transverse wave to travel along a transmission line, from one tower to another, if the horizontal component of the cable tension is 25,000 N, the towers are 300 m apart, and the cable has a mass of 1.5 kg/m of length.

Answer: 2.32 s

PROBLEM 13.5 Determine the velocity of (a) longitudinal waves and (b) shear (torsional) waves in a solid circular steel bar. (Use $E = 205 \times 10^9$ N/m² and $G = 82.5 \times 10^9$ N/m².)

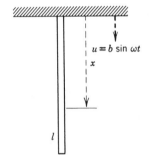

PROBLEM 13.6 The upper end of a uniform rod for a deep well reciprocating pump receives a motion $u = b \sin \omega t$. Describe the steady state vibration of the rod, $u(x)$, in terms of the amplitude b at the upper end and the natural frequencies of the rod. Discount the effect of lifting a column of fluid in the well.

Answer: $u = b\left(\cos\dfrac{\omega x}{c} + \tan\dfrac{\omega l}{c}\sin\dfrac{\omega x}{c}\right)\sin \omega t$

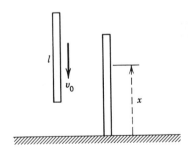

PROBLEM 13.7 A longitudinal bar impacts a rigid surface and this results in longitudinal vibration of the bar. Determine the pressure between the longitudinal bar and the rigid surface while the bar remains in contact with the surface, if the impact occurred with the bar having an initial velocity of v_0. Plot the variation of force with time before rebound occurs.

Answer: $P = \dfrac{4EAv_0}{\pi c}\displaystyle\sum_{1,3,5}^{\infty}\dfrac{1}{n}\sin\dfrac{n\pi c}{2l}t$

PROBLEM 13.8 Two large masses with mass moments of inertia I_1 and I_2 are fixed at either end of a uniform shaft with a length l and polar area moment of inertia J. Find the frequency equation for the torsional vibration of the system.

Answer: $\dfrac{\omega_n l}{c}\left[\tan\dfrac{\omega_n l}{c}\right] = \dfrac{\mu l J}{g}\left(\dfrac{I_1 + I_2}{I_1 I_2}\right)$

13.3. TRANSVERSE VIBRATION OF UNIFORM BEAMS

The transverse vibration of uniform beams is another vibration problem in which elasticity and mass are distributed.

There are several very important applications of this problem. One is the critical speeds of rotating shafts and rotors, and another is the transverse vibration of suspended cables, such as transmission lines.

Fig. 13.3

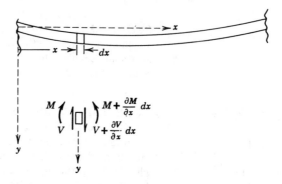

In Figure 13.3, a thin element of a beam has been isolated. As the beam flexes in transverse vibrations, the element will move back and forth in the y direction, and it will also rotate, very slightly. Ignoring the inertial properties of the element in rotation, and summing moments,

$$\sum \mathbf{M} = 0$$

$$V\,dx = \frac{\partial M}{\partial x}\,dx$$

or

$$V = \frac{\partial M}{\partial x}$$

In the y direction, the force summation is

$$\sum \mathbf{F} = dm\frac{\partial^2 \mathbf{y}}{\partial t^2}$$

$$\frac{\partial V}{\partial x}\,dx = \frac{w}{g}\,dx\frac{\partial^2 y}{\partial t^2}$$

or, combining the two equations,

$$\frac{\partial^2 M}{\partial x^2} = \frac{w}{g}\frac{\partial^2 y}{\partial t^2} \tag{13.9}$$

This equation looks like the wave equation, but it is not. The second partial of the moment M with respect to x is equal to the second partial of the displacement y with respect to time. One additional step is needed: the statement of bending moment in terms of the change in slope.

$$M = -EI\frac{\partial^2 y}{\partial x^2}$$

Substituting in equation 13.9

$$-\frac{\partial^2}{\partial x^2}\left(EI\frac{\partial^2 y}{\partial x^2}\right) = \frac{w}{g}\frac{\partial^2 y}{\partial t^2} \tag{13.10}$$

If the beam is uniform, flexural rigidity is constant with displacement, and

$$-EI\frac{\partial^4 y}{\partial x^4} = \frac{w}{g}\frac{\partial^2 y}{\partial t^2} \tag{13.11}$$

This is a fourth-order partial differential equation, but it is solved in the same way as the wave equation, by separating variables. If we let the function $y(x, t)$ be the product of two separate functions, one of x and one of t,

$$y(x, t) = f_1(t)f_2(x)$$

$$\frac{d^2 y}{dt^2} = \frac{d^2 f_1(t)}{dt^2}f_2(x)$$

$$\frac{d^4 y}{dx^4} = \frac{d^4 f_2(x)}{dx^4}f_1(t)$$

Substituting in equation 13.11 and separating the variables,

$$\frac{1}{f_1(t)}\frac{d^2 f_1(t)}{dt^2} = -\frac{EIg}{wf_2(x)}\frac{d^4 f_2(x)}{dx^4} = -\omega_n^2$$

For these two now independent functions to be equal, they must be equal to a constant, conveniently $-\omega_n^2$. As before,

$$\frac{d^2 f_1(t)}{dt^2} + \omega_n^2 f_1(t) = 0$$

or,

$$f_1(t) = A\ \cos\ \omega_n t + B\ \sin\ \omega_n t \tag{13.12}$$

Likewise,

$$\frac{d^4 f_2(x)}{dx^4} - \frac{w\omega_n^2}{EIg}f_2(x) = 0 \tag{13.13}$$

This is different. It is a fourth-order differential equation. Calling $\beta^4 = w\omega_n^2/EIg$, the solution is

$$f_2(x) = C_1 \cosh \beta x + C_2 \sinh \beta x + C_3 \cos \beta x + C_4 \sin \beta x$$

Differentiation will verify $f_2(x)$ as a solution. The complete solution for $y = (x, t)$ is

$$y(x, t) = (A \cos \omega_n t + B \sin \omega_n t)(C_1 \cosh \beta x + C_2 \sinh \beta x$$
$$+ C_3 \cos \beta x + C_4 \sin \beta x) \tag{13.14}$$

In solving this general equation for the transverse vibration of a beam, we will need two initial conditions and four boundary conditions. With the four boundary conditions, we will also find the characteristic values of the frequency equation, which will be in terms of some specific values of βl, which is dimensionless. That is, the natural frequencies will be in terms of βl.

$$\omega_n = \beta^2 l^2 \sqrt{\frac{EIg}{wl^4}} \tag{13.15}$$

13.4. ROTATION AND SHEAR EFFECTS

In the equation of motion for the transverse vibration of a beam, the inertial properties of a section of the beam in rotation were ignored. If we had included them, the equations of motion in the θ direction would be

$$\sum \mathbf{M} = \frac{\mu I}{g} dx \frac{\partial^2 \boldsymbol{\theta}}{\partial t^2}$$

$$\left(V + \frac{\partial V}{\partial x}\frac{dx}{2}\right) + V\frac{dx}{2} - \frac{\partial M}{\partial x}dx = \frac{\mu I}{g} dx \frac{\partial^2 \theta}{\partial t^2}$$

or,

$$V = \frac{\partial M}{\partial x} + \frac{\mu I}{g}\frac{\partial^2 \theta}{\partial t^2}$$

In this instance, I is the second moment of area and not mass moment inertia. The symbol is conventional, but misleading.

In the y direction, the equation of motion would be unchanged.

$$\sum \mathbf{F} = dm \frac{\partial^2 \mathbf{y}}{\partial t^2}$$

$$\frac{\partial V}{\partial x} = \frac{w}{g}\frac{\partial^2 y}{\partial t^2}$$

Combining, and substituting $\theta = \dfrac{\partial y}{\partial x}$

$$-\frac{\partial^2}{\partial x^2}\left(EI\frac{\partial^2 y}{\partial x^2}\right) + \frac{\mu I}{g}\frac{\partial^4 y}{\partial t^2 \partial x^2} = \frac{w}{g}\frac{\partial^2 y}{\partial t^2} \qquad (13.16)$$

This has one additional term more than the simpler equation 13.11, and that term is a cross-derivative. It is a correction to the simpler equation, and it is of some importance in beams with deep sections.

Shear is another effect that has an influence on the equations of motion. In the presence of large shear forces, the beam element is skewed. That is, not all of the change in the slope of the beam is due to bending. Some of the change is due to skewing from shear. As a consequence, $M > EI(\partial^2 y / \partial x^2)$. This leads to a very involved equation of motion, which is a few percent more accurate. In most cases the increased accuracy including shear and rotation is much less than the modeling errors. In general, the correction for shear is larger than the correction for rotation, but both can be ignored if the wavelength is long compared to the depth of the beam section. What it is necessary to realize is that these corrections do exist.

13.5. THE EFFECT OF AXIAL LOADING

Lord Rayleigh considered the case of the vibration of a uniform beam with axial loading, setting up the equation of motion for the beam without solving it. Essentially, vibrating cables and guy wires are examples of this problem. Occasionally, the problem has appeared elsewhere, but the applications are limited. It is, however, a very interesting variation of the problem of transverse vibration of uniform beams. The natural frequency of a vibrating cable can be found by considering the cable to be the equivalent of a string. But this neglects the elastic properties of the cable, and the flexure of cables near points of support is the cause of failures. Consequently, flexural rigidity cannot be ignored.

The regular shedding of vortices from the cable in a light wind is generally accepted as the basic cause of cable vibration. However, there is very little known connection between this cause and its ultimate effect: fatigue damage to the cable. Basically, the cause of wind-induced or aeolian vibration is a simple one. Under certain conditions, as air flows over a cylindrical cable or wire, the flow pattern is not symmetrical, but is regularly disturbed by the formation of vortices at the rear of the cable. This asymmetrical flow arrangement alters the pressure distribution around the cable. As vortices or eddies are areas of reduced pressure, their periodic formation results in a pressure differential that alternately forces the cable upward and downard. If this periodicity corresponds to a resonant frequency, the amplitude of vibration will build up to where the energy imparted to the cable by the wind is matched by the energy dissipated through hysteresis in the cable. It is this alternating flexure of the cable, when severe enough, which results in fatigue failures at clamps, socket eyes, and wherever else a cable or guy wire is highly stressed. Sympathetic vibration of structural members may also bring about their fatigue.

Fig. 13.4

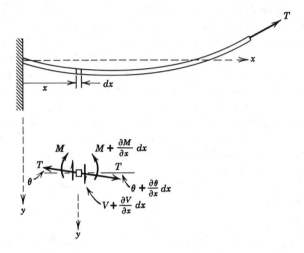

Referring to Figure 13.4, a small element of the cable, dx, is subjected to tensile forces, shear, and bending moments. These unbalanced forces cause acceleration of the cable element in the vertical direction. The equation of motion of the vibrating cable is

$$\sum \mathbf{F} = \frac{w}{g} dx \frac{\partial^2 \mathbf{y}}{\partial t^2}$$

$$\frac{\partial V}{\partial x} dx + H \frac{\partial^2 y}{\partial x^2} dx = \frac{w}{g} dx \frac{\partial^2 y}{\partial t^2}$$

Using the general relations for deflection, shear, and bending moment again, the final form of the equation of motion is a partial differential equation of the fourth order. ($H = T \cos \theta$)

$$+ \frac{\partial^2}{\partial x^2}\left(EI \frac{\partial^2 y}{\partial x^2}\right) - H \frac{\partial^2 y}{\partial x^2} dx + \frac{w}{g} \frac{\partial^2 y}{\partial t^2} = 0 \qquad (13.17)$$

Again, the solution of equation 13.17 may be obtained by the classical method of separating the variables. If the solution can be written as a product of two separate functions that individually depend on x and t, then y can be expressed as

$$y(x, t) = f_1(t)f_2(x)$$

$$\frac{\partial^2 y}{\partial t^2} = \frac{\partial^2 f_1(t)}{\partial t^2}f_2(x) \qquad \frac{\partial^2 y}{\partial x^2} = \frac{\partial^2 f_2(x)}{\partial x^2}f_1(t) \qquad \frac{\partial^4 y}{\partial x^4} = \frac{\partial^4 f_2(x)}{\partial x^4}f_1(t)$$

The component differential equations are

$$\frac{d^2 f_1(t)}{dt^2} + \omega_n^2 f_1(t) = 0 \qquad (13.18)$$

$$EI\frac{d^4 f_2(x)}{dx^4} - H\frac{d^2 f_2(x)}{dx^2} - \frac{w}{g}\omega_n^2 f_2(x) = 0 \qquad (13.19)$$

The time-dependent solution is, of course

$$f_1(t) = A \cos \omega_n t + B \sin \omega_n t$$

Using the usual exponential substitution, $f_2 = Ce^{rx}$ in the time-independent solution results in the characteristic equation,

$$r^4 - \frac{H}{EI}r^2 - \frac{w\omega_n^2}{EIg} = 0$$

Calling κ and λ the quadratic roots of the characteristic equation,

$$\kappa^2 = \left[\frac{H^2}{4(EI)^2} + \frac{w\omega_n^2}{g(EI)^2}\right]^{1/2} + \frac{H}{2EI}$$

$$\lambda^2 = \left[\frac{H^2}{4(EI)^2} + \frac{w\omega_n^2}{g(EI)^2}\right]^{1/2} - \frac{H}{2EI} \qquad (13.20)$$

The solution can be expressed either as an exponential or as circular and hyperbolic functions. Since we have used circular and hyperbolic functions before, it is not too difficult to show that $f_2(x)$ is a solution if,

$$f_2(x) = C_1 \cosh \kappa x + C_2 \sinh \kappa x + C_3 \cos \lambda x + C_4 \sin \lambda x \qquad (13.21)$$

The arbitrary constants, C_1, C_2, C_3, and C_4 are determined by satisfying the proper boundary conditions.

For a simply supported span, the boundary conditions for pinned ends are such that the vertical displacement and the bending moment are both zero for all time. Imposing these boundary conditions will lead to the frequency equation, from which the natural frequencies are

$$\omega_n^2 = \lambda^2\left[\lambda^2\left(\frac{EIg}{w}\right) + \frac{Hg}{w}\right] \qquad (13.22)$$

The term λ must be found from the solution of the frequency equation. Neglecting flexural rigidity, the frequency equation will degenerate to the

Fig. 13.5

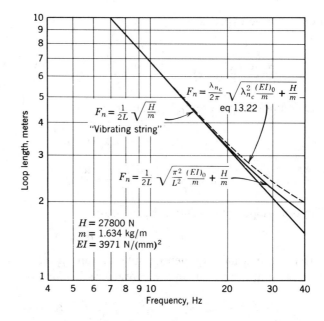

$$F_n = \frac{\lambda_{n_c}}{2\pi} \sqrt{\lambda_{n_c}^2 \frac{(EI)_0}{m} + \frac{H}{m}}$$
eq 13.22

$$F_n = \frac{1}{2L} \sqrt{\frac{H}{m}}$$
"Vibrating string"

$$F_n = \frac{1}{2L} \sqrt{\frac{\pi^2}{L^2} \frac{(EI)_0}{m} + \frac{H}{m}}$$

$H = 27800$ N
$m = 1.634$ kg/m
$EI = 3971$ N/(mm)^2

Loop length, meters

Frequency, Hz

standard case of the vibrating string. Neglecting axial tension, the equation will degenerate to the familiar case of the simply supported uniform beam. In Figure 13.5 a comparison of frequencies and loop lengths has been made between the case of the vibrating string and a typical cable. For end conditions, one is simply supported and the other is clamped. In both cases, the addition of flexural rigidity raises the natural frequency.

Figure 13.6 shows qualitatively the shapes that the deflection and bending moment diagrams would have for the imposed clamped and pinned boundary conditions. Note that, in contrast to the case of the pinned loop (a), a high bending moment is apparent for the clamped end (b). The

Fig. 13.6

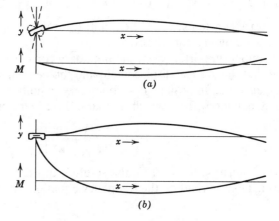

(a)

(b)

bending moment, which is alternating between a negative maximum and a positive maximum, during resonance, is the primary cause of the fatigue failure of cable strands.

EXAMPLE PROBLEM 13.9

Determine the natural frequencies of lateral vibration of a simply supported uniform beam.

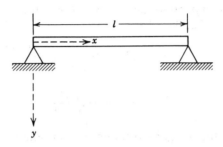

Solution:
The deflection $y(x, t)$ is described by the equation

$$y(x, t) = (A \cos \omega_n t + B \sin \omega_n t)$$
$$\times (C_1 \cosh \beta x + C_2 \sinh \beta x + C_3 \cos \beta x + C_4 \sin \beta x)$$

The boundary conditions are the same at $x=0$ and $x=l$; the deflection and bending moment are both zero. Using these conditions, at $x=0$

$$y(0, t) = 0 = (A \cos \omega_n t + B \sin \omega_n t)$$
$$\times (C_1 \overset{1}{\cancel{\cosh}} \beta x + C_2 \overset{0}{\cancel{\sinh}} \beta x + C_3 \overset{1}{\cancel{\cos}} \beta x + C_4 \overset{0}{\cancel{\sin}} \beta x)$$
$$0 = C_1 + C_3$$

$$M = -EI\frac{\partial^2 y}{\partial x^2} = 0 = -EI\beta^2 (A \cos \omega_n t + \beta \sin \omega_n t)$$
$$\times (C_1 \overset{1}{\cancel{\cosh}} \beta x + C_2 \overset{0}{\cancel{\sinh}} \beta x - C_3 \overset{1}{\cancel{\cos}} \beta x - C_4 \overset{0}{\cancel{\sin}} \beta x)$$
$$0 = C_1 - C_3$$

Hence, $C_1 = C_3 = 0$. At $x=l$

$$y(l, t) = 0 = (A \cos \omega_n t + B \sin \omega_n t)(C_2 \sinh \beta l + C_4 \sin \beta l)$$

and

$$M(l, t) = -EI\frac{\partial^2 y}{\partial x^2} = 0$$
$$= -EI\beta^2 (A \cos \omega_n t + B \sin \omega_n t)(C_2 \sinh \beta l - C_4 \sin \beta l)$$

These two equations can be stated

$$\begin{bmatrix} \sinh \beta l & \sin \beta l \\ \sinh \beta l & -\sin \beta l \end{bmatrix} \begin{Bmatrix} C_2 \\ C_4 \end{Bmatrix} = \begin{Bmatrix} 0 \\ 0 \end{Bmatrix}$$

Barring the trivial solution $C_2 = C_4 = 0$, setting the determinant equal to zero yields the frequency equation

$$2 \sinh \beta l \sin \beta l = 0$$

Since the $\sinh \beta l > 0$ for all values of $\beta l \neq 0$, the only roots to this equation are $\beta l = 0,\ \pi,\ 2\pi,\ 3\pi,\dots,n\pi$.

$$\omega_n^2 = \beta^4 l^4 \frac{EIg}{wl^4} = n^4 \pi^4 \frac{EIg}{wl^4}$$

$$\omega_n = n^2 \pi^2 \sqrt{\frac{EIg}{wl^4}}$$

where n is the integer indicating the harmonic.

EXAMPLE PROBLEM 13.10

Determine the natural frequencies of lateral vibration of a cantilevered uniform beam.

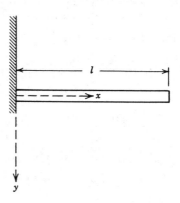

Solution:
The deflection will be the same as the simply supported beam.

$$y(x,\ t) = (A \cos \omega_n t + B \sin \omega_n t)$$
$$\times (C_1 \cosh \beta x + C_2 \sinh \beta x + C_3 \cos \beta x + C_4 \sin \beta x)$$

The arbitrary constants C_1, C_2, C_3, and C_4 will be different since the boundary conditions are different. At the built-in end, $y = 0$ and $dy/dx = 0$. At the free end, the bending moment and shear are zero. At $x = 0$,

$$y(0, t) = 0 = (A \cos \omega_n t + B \sin \omega_n t)$$
$$\times (C_1 \overset{1}{\cosh} \beta x + C_2 \overset{0}{\sinh} \beta x + C_3 \overset{1}{\cos} \beta x + C_4 \overset{0}{\sin} \beta x)$$
$$0 = C_1 + C_3$$

$$\left(\frac{\partial y}{\partial x}\right)(0, t) = 0 = \beta(A \cos \omega_n t + B \sin \omega_n t)$$
$$\times (C_1 \overset{0}{\sinh} \beta x + C_2 \overset{1}{\cosh} \beta x + C_3 \overset{0}{\sin} \beta x + C_4 \overset{1}{\cos} \beta x)$$
$$0 = C_2 + C_4$$

Rewriting the deflection, substituting $C_3 = -C_1$ and $C_4 = -C_2$

$$y = (A \cos \omega_n t + B \sin \omega_n t)$$
$$\times [C_1(\cosh \beta x - \cos \beta x) + C_2(\sinh \beta x - \sin \beta x)]$$

Imposing the boundary conditions at $x = l$,

$$M = -EI\frac{\partial y^2}{\partial x^2} = 0 = -EI\beta^2(A \cos \omega_n t + B \sin \omega_n t)$$
$$\times [C_1(\cosh \beta l + \cos \beta l) + C_2(\sinh \beta l + \sin \beta l)]$$
$$V = -EI\frac{\partial^2 y}{\partial x^3} = 0 = -EI\beta^3(A \cos \omega_n t + B \sin \omega_n t)$$
$$\times [C_1(\sinh \beta l - \sin \beta l) + C_2(\cosh \beta l + \cos \beta l)]$$

These equations can be summarized,

$$\begin{bmatrix} (\cosh \beta l + \cos \beta l) & (\sinh \beta l + \sin \beta l) \\ (\sinh \beta l - \sin \beta l) & (\cosh \beta l + \cos \beta l) \end{bmatrix} \begin{Bmatrix} C_1 \\ C_2 \end{Bmatrix} = \begin{Bmatrix} 0 \\ 0 \end{Bmatrix}$$

which leads to the frequency equation

$$\cosh \beta l \cos \beta l = -1$$

This is also a transcendental equation. If we let $\cos \beta l = \phi_1$ and $-1/\cosh \beta l = \phi_2$, the equation is satisfied at $\phi_1 = \phi_2$ (see graph on page 420.) From the intersections,

$$\beta_1 l = 1.875 \quad \beta_2 l = 4.694 \quad \beta_3 l = 7.855 \quad \beta_4 l = 10.996 \quad \beta_5 l = 14.137$$

For higher modes,

$$\beta_n l = \frac{(2n-1)}{2}\pi$$

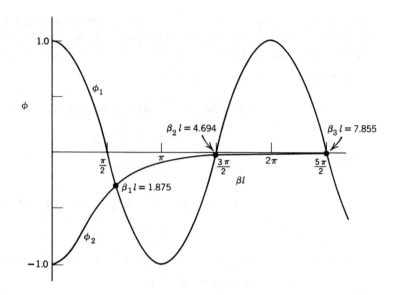

Substituting the proper value for βl will yield the natural frequency

$$\omega_n = \beta^2 l^2 \sqrt{\frac{EIg}{wl^4}}$$

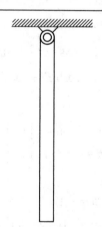

PROBLEM 13.11 Determine an expression for natural frequencies for the lateral vibration of the uniform beam suspended as a pendulum.

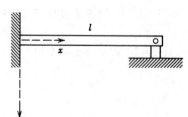

PROBLEM 13.12 Determine an expression for the natural frequencies of a fixed-pinned bar in lateral vibration.

Answer: $\tan \beta l = \tanh \beta l$

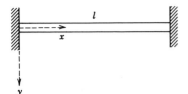

PROBLEM 13.13 Determine the expression for the natural frequencies for the lateral vibration of a uniform beam fixed at both ends.

Answer: $\cos \beta l \cosh \beta l = 1$

PROBLEM 13.14 A platform consists of two I beams, built in at each end. The beams are standard shape with a depth of 114.25 mm, length 1.525 m, area 3852 $(mm)^2$, moment of inertia 25.05×10^6 $(mm)^4$. E is 205×10^9 N/m^2. The mass of these beams is 30.5 kg/m. A uniformly distributed load of 14,235 N is the floor of the platform. What are the first two natural frequencies of the structure?

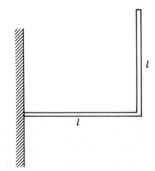

PROBLEM 13.15 A frame consists of a uniform bar fixed at right angles to another bar that is built into the wall. Determine the natural frequencies of the frame.

CHAPTER FOURTEEN

RANDOM
VIBRATION

14.1. INTRODUCTION

Random vibration is the term used for vibration that cannot be described as periodic. There is no pattern to frequency or amplitude, and the usual methods of analyzing periodic motion through harmonics do not apply. The motion can be treated statistically, however, and a considerable body of knowledge of random vibration has been assembled using statistics as a base.

Random vibration has a language all of its own. Measured data are different from conventional vibration data, and response is stated in terms of the probability that some amplitude or acceleration will occur. The terms *narrow band* and *wide band* are both used to describe random vibration. These are qualitative descriptions that are rather loosely used to describe the vibration spectrum.

If motion is truly random, there will be no dominant frequency and no dominant amplitude. Over a given time interval, τ, it can be described by the *mean square* amplitude,

$$\overline{x^2} = \frac{1}{\tau}\int_0^\tau x^2 \, dt \tag{14.1}$$

or the *root mean square* or rms value,

$$\overline{x} = \sqrt{\frac{1}{\tau}\int_0^\tau x^2 \, dt}$$

To simplify analysis, we will assume that this is a stationary process, which means that the root mean square is independent of the time at which the measurement was taken.

14.2. RANDOM VIBRATION IN A SINGLE DEGREE OF FREEDOM SYSTEM

Assuming a stationary process, the response of a single degree of freedom, linear system to random excitation can be predicted quite easily. If an excitation occurs at a frequency ω_1, the response to a sinusoidal force $F_1 \sin \omega_1 t$ is $r_1 = e_1 y_1$, where y_1 is the amplitude ratio at $\omega = \omega_1$, and e_1 is the excitation. In this case $e_1 = F_1/k \sin \omega_1 t$. If the excitation occurs at a frequency ω_2, the response is $r_2 = e_2 y_2$. If these two excitations occur together, it is possible to add them by superpositioning, since we are dealing with a linear system. The root mean square response is

$$\bar{r}_{1+2} = \sqrt{(e_1 y_1)^2 + (e_2 y_2)^2}$$

For n discrete excitations as in Figure 14.1a, all occurring simultaneously, the root mean square response is the square root of the sum of the squares of all the products of excitation and amplitude ratios.

$$\bar{r}_n = \sqrt{\sum_{n=1}^{\infty} (e_n y_n)^2} \tag{14.2}$$

Fig. 14.1

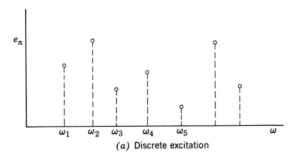

(a) Discrete excitation

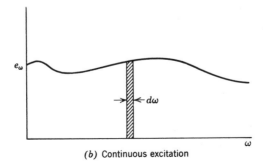

(b) Continuous excitation

In each case, whether due to one, two, or n discrete excitations, if the excitation is in terms of amplitude or the parameter F/k, the response is an amplitude, if the excitation is an acceleration, the response is an acceleration. The amplitude ratio is dimensionless.

For a continuous spectrum of frequencies as in Figure 14.1b, an integral replaces the discrete sum. The excitation is now a continuous function of frequency rather than a discrete expression of one frequency alone.

$$\overline{R} = \sqrt{\int_0^\infty (e_\omega y_\omega)^2 \, d\omega} \tag{14.3}$$

The symbol $\overline{R}$ is used to denote the response of a random spectrum.

To be correct, the excitation e_ω must be in amplitude units or acceleration units per radian per second to the one half power. e_ω^2 is the *mean amplitude spectral density squared*, m^2-s, if the response is an amplitude or the *mean acceleration spectral density squared*, g^2-s if the response is in terms of acceleration. Occasionally, this will also be called the *power spectral density*, but this term is deceptive and should be avoided. It is sometimes shortened to simply *spectral density*.

The mean square spectral density is a measurable quantity. An output, such as the voltage from a seismic instrument, is fed through a band pass filter, squared, and averaged over a long time interval τ. The result is a measured quantity that is the mean square spectral density of the output of the seismic instrument. It is also a function of the bandwidth that the

Fig. 14.2

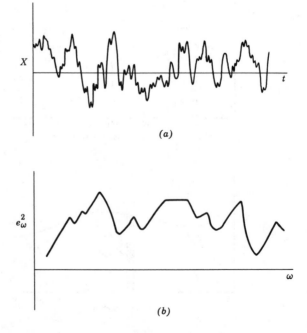

(a)

(b)

band pass filter will admit, its center frequency and the time interval τ, but all these are known. The output can be indicated on a meter or recorded.

Wide band random vibration is a stationary random process in which the mean square spectral density is relatively constant in value over a wide spectrum, usually at least half the total spectrum. Figure 14.2a shows wide band random vibration as it would appear as a function of time, and Figure 14.2b is a typical plot of mean square spectral density. The vibration environment associated with the firing of a rocket is typically a wide band random process.

Narrow band random vibration is a stationary random process in which the mean square spectral density is measurably significant over a narrow range of frequencies. This is typical of resonant systems where a forcing function with one or more closely related frequencies forces motion. The response exhibits *beats* and the amplitude varies widely, but the frequency does not. Figure 14.3a and b show narrow band random vibration, using

Fig. 14.3

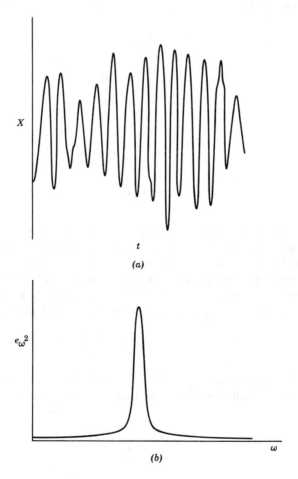

(a)

(b)

actual data from the wind-induced vibration of transmission line cables. Here, the periodic shedding of vortices induces a strong resonant forced vibration which is truly a narrow band random process.

14.3. RESPONSE TO RANDOM EXCITATION

Two kinds of problems are clinically solved using random vibration. The first is to estimate statistically the probable time to failure, given a certain vibration spectrum. The second is to predict the system response to a certain spectral density of excitation, knowing the resonant characteristics of the system.

A random excitation that is constant for all frequencies is called "*white noise*," since it contains all frequencies, all with uniform amplitude. The response of a single degree of freedom system to white noise is one of the most important relations concerning random vibration that a vibration engineer has.

$$\overline{R}^2 = \int_0^\infty (e_\omega y_\omega)^2 \, d\omega$$

$$= e_\omega^2 \int_0^\infty (y_\omega^2) \, d\omega$$

$$= e_\omega^2 \omega_n \int_0^\infty \frac{d\dfrac{\omega}{\omega_n}}{\left(1 - \dfrac{\omega^2}{\omega_n^2}\right)^2 + \left(2\zeta\dfrac{\omega}{\omega_n}\right)^2}$$

$$= e_\omega^2 \omega_n \frac{\pi}{2} Q \qquad (14.4)$$

It is simple and direct, and many random excitations can be approximated as white noise avoiding a more complicated analysis. It requires only the level of excitation, the natural frequency ω_n and the level of damping, expressed by the factor Q. When the mean square response is desired in terms of displacement units and the excitation is known in terms of acceleration units squared per radian per second, dividing by ω_n^4

$$\overline{x}^2 = \frac{\pi e_\omega^2 Q}{2\omega_n^3} \qquad (14.5)$$

If an approximation of white noise cannot be made, the response can be numerically calculated, and equation 14.3 becomes

$$\overline{R} = \sqrt{\sum (e_\omega y_\omega)^2 \, \Delta\omega} \qquad (14.6)$$

In this case, suitable bandwidths $\Delta\omega$ must be chosen. The excitation e_ω and the amplification factor y_ω apply only over each band width. The sample problems show how this is applied.

EXAMPLE PROBLEM 14.1

By actual test, an aircraft instrument package has been damaged by sinusoidal vibration that exceeded the limits of 2.5 g. Determine the necessary static deflection of isolators that could protect the package from damage from a random vibration of 0.1 g^2/Hz from 5 to 2000 Hz. Assume that the isolators are made of synthetic rubber for which $\zeta = 0.05$.

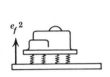
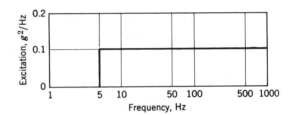

Solution:
This is a simple problem involving "white noise" at constant density of $e_f^2 = 0.1$ g^2/cps. The root mean square response, $\overline{R}$ will be

$$\overline{R} = \sqrt{e^2\omega_n\frac{\pi}{2}Q} = 2.5 \text{ g}$$

Stated in units of g^2-s/rad,

$$e^2 = \frac{0.1}{2\pi}$$

The damping factor is expressed as Q,

$$Q = \frac{1}{2\zeta} = \frac{1}{2(0.05)} = 10$$

Solving for the natural frequency, ω_n

$$(2.5)^2 = \left(\frac{0.1}{2\pi}\right)\omega_n\left(\frac{\pi}{2}\right)(10)$$

$$\omega_n = 25 \text{ s}^{-1}$$

Going back to simple vibration theory, the static deflection will be

$$\omega_n^2 = \frac{g}{\Delta_{st}}$$

or,

$$\Delta_{st} = \frac{9.806}{(25)^2} = 0.0157 \text{ m}$$

EXAMPLE PROBLEM 14.2

An instrument is mounted on a spacecraft structure in such a manner that the response in one direction exhibits the shown modal characteristics. This has been predetermined in a resonance test, where an accelerometer measured the acceleration and displacement over an assigned frequency spectrum. In summary, there are two peaks between 20 to 1000 Hz.

(a) If σ is the standard or root mean square amplitude of vibration, what is the length of time before the amplitude first exceeds 5σ, with a 0.99 reliability?

(b) Determine the mean square response of the instrument to a constant acceleration spectrum density of 0.01 g²/Hz.

(c) Determine the mean square response of the instrument to a variable spectral density.

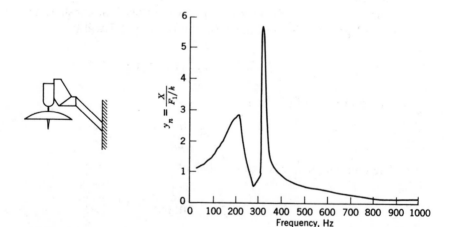

Solution:

(a) Assuming a stationary Gaussian process with zero mean, the frequency that the amplitude x exceeds the root mean square amplitude σ, is

$$v = f e^{-x^2/2\sigma^2}$$

f is the midband frequency which in this case $f = 510$ cps. For the reliability to be 0.99, the probability that the event will occur is $P(t) = 0.01$. The period between successive times when the amplitude exceeds 5 times the mean square value is then

$$\tau = \frac{P(t)}{v} = \frac{0.01e^{(5)^2/2}}{510} = 5.5 \text{ s}$$

If we had desired 0.999 reliability, which is a probability of 0.001, we could have only counted on 0.55 s. In all likelihood, the time would be much longer, but random vibration is studied statistically, and a statement of reliability and probability is necessary. In the design of spacecraft hardware, where the failure of any one of a long series of components would mean the failure of the total system, reliabilities of 0.999 and 0.9999 are not uncommon. Since this chain is cumulative, the general effect is that testing is very expensive, and components are over designed, and redundant components are used if the component is at all critical.

(b) For the mean square response to a constant acceleration spectral density of 0.01 g^2/Hz, the modal response curve is divided into finite bands, with a constant amplitude ratio for the selected bandwidth. The mean square response is then tabulated by squaring the amplitude ratio y_n, squaring the excitation e_n and multiplying the product $(e_n y_n)^2$ by the bandwidth Δf_n. The mean square response is then the square root of the sum. See figure on next page.

$$\overline{R} = \sqrt{\sum_{20}^{1000} (e_n y_n)^2 \cdot \Delta f_n} = 5.146 \text{ g}$$

Random Vibration Response to Constant Spectrum Density

f_n	y_n	e_n	Δf_n	$(e_n v_n)^2 \, \Delta f_n$
60	1.3	0.1	80	1.352
125	1.8	0.1	50	1.620
175	2.6	0.1	50	3.380
210	2.8	0.1	20	1.568
235	1.7	0.1	30	0.869
275	0.5	0.1	50	0.125
325	5.7	0.1	50	16.245
375	1.1	0.1	50	0.605
450	0.6	0.1	100	0.360
550	0.4	0.1	100	0.160
700	0.2	0.1	200	0.160
900	0.1	0.1	200	0.040

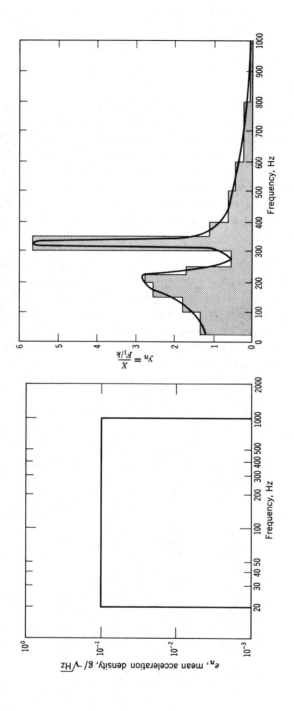

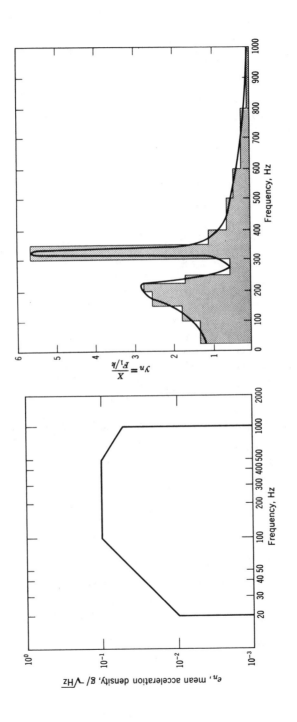

Briefly, this states that the mean square response will exceed 5.146g 32% of the time, 10.29g 4.6% of the time, and 15.44g 0.27% of the time. It is up to the engineer to judge whether his or her design will meet this environment.

(c) For the mean square response to a variable acceleration spectral density, the same procedure is used. Since the acceleration density is less at the highest and lowest frequencies, the mean square response is less. See the figure on previous page.

f_n	y_n	e_n	Δf_n	$(e_n v_n)^2 \, \Delta f_n$
60	1.3	0.037	80	0.185
125	1.8	0.100	50	1.620
175	2.6	0.100	50	3.380
210	2.8	0.100	20	1.568
235	1.7	0.100	30	0.869
275	0.5	0.100	50	0.125
325	5.7	0.100	50	16.245
375	1.1	0.100	50	0.605
450	0.6	0.100	100	0.360
550	0.4	0.080	100	0.102
700	0.2	0.070	200	0.039
900	0.1	0.055	200	0.006

Tabulating, and again taking the square root of the sum $(e_n y_n)^2 \, \Delta f_n$,

$$\overline{R} = \sqrt{\sum_{20}^{1000} (e_n y_n)^2 \, \Delta f_n} = 4.462 \text{ g}$$

PROBLEM 14.3 Determine the root mean square displacement for the space frame of Problem 7.9 in response to white noise with a constant excitation density of 4 g^2/Hz.

PROBLEM 14.4 An electronic instrument package is mounted on four rubber isolators that deflect 5 mm under static load. What is the "white noise" excitation that can be sustained if the response is to be limited to 4 g? Damping is known to be $\zeta = 0.05$.

Answer: $e^2 = 0.144$ g^2/Hz

PROBLEM 14.5 The solar panels for a spacecraft can be modeled as two symmetrically placed cantilevered beams. During resonance testing, the structure shows a marked peak at 82 Hz, but no higher mode was observed below 1000 Hz. Damping is hysteretic at $\zeta = 0.1$.

(a) Determine the root mean square acceleration in response to white noise with a constant excitation density of 0.1 g²/Hz.

(b) Determine the root mean square acceleration in response to a constant excitation density of 0.1 g²/Hz, if the excitation is limited to a spectrum from 20 to 200 Hz only.

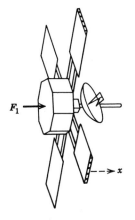

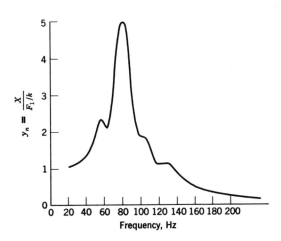

PROBLEM 14.6 What would be the added effect of the spectrum from 200 to 1000 Hz, if no resonance peak is observed between those frequencies?

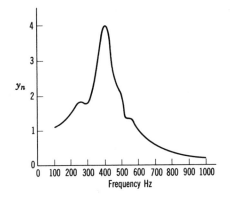

PROBLEM 14.7 Determine the root mean square acceleration response of the high gain antenna shown in the figure of Problem 14.5 to a constant excitation density of 0.1 g²/Hz over the entire spectrum from 20 Hz to 1000 Hz, if the antenna shows a resonance peak at 400 Hz.

Answer: $\overline{R} = 17.7$ g

REFERENCES

There are many good books on the subject of mechanical vibration. Rather than list them all, which I could do, and which would be of doubtful utility, I recommend seven. These I consider to be the best.

1. *Vibration Problems in Engineering*, S. Timoshenko, D. H. Young, and W. Weaver, Jr., Wiley, New York, 1974, 4th Edition.
2. *Mechanical Vibration*, J. P. Den Hartog, McGraw-Hill, New York, 1956, 4th Edition.
3. *Engineering Vibrations*, L. S. Jacobsen and R. S. Ayre, McGraw-Hill, New York, 1958.
4. *The Mechanics of Vibration*, R. F. D. Bishop and D. C. Johnson, Cambridge University Press, New York, 1960.
5. *Shock and Vibration Handbook*, C. M. Harris, and C. E. Crede, McGraw-Hill, New York, 1961.
6. *Elements of Vibration Analysis*, L. Meirovitch, McGraw-Hill, New York, 1986, 2nd Edition.
7. *Theory of Vibrations with Applications*, W. T. Thomson, Prentice-Hall, Englewood Cliffs, NJ, 1988, 3rd Edition.

INDEX

437